Quantitative Analysis

AUTOMATION AND PRODUCTION SYSTEMS
Methodologies and Applications

A series edited by **Hamid R. Parsaei,** University of Louisville, Kentucky, USA

About the series editor

Hamid R. Parsaei, PhD, PE, is a professor of industrial engineering and director of the Manufacturing Research Group at the University of Louisville in Kentucky. He received his MS and PhD in industrial engineering from Western Michigan University and the University of Texas at Arlington, respectively. His teaching and research interests include aspects of engineering economy, robotics, concurrent engineering, computer integrated manufacturing systems and productivity analysis.

This book is part of a series. The publisher will accept continuation orders which may be cancelled at any time and which provide for automatic billing and shipping of each title in the series upon publication. Please write for details.

Quantitative Analysis

An Introduction

Roy M. Chiulli

CRC Press
Taylor & Francis Group
Boca Raton London New York

CRC Press is an imprint of the
Taylor & Francis Group, an **informa** business

First published in 1999 by Gordon and Breach Science Publishers

Published in 2020 by CRC Press
Taylor & Francis Group
6000 Broken Sound Parkway NW, Suite 300
Boca Raton, FL 3487-2742

First issued in paperback 2020

ISBN-13: 978-0-367-57920-3 (pbk)
ISBN-13: 978-90-5699-629-1 (hbk)

Visit the Taylor & Francis Web site at
http://www.taylorandfrancis.com

and the CRC Press Web site at
http://www.crcpress.com

British Library Cataloguing in Publication Data

Chiulli, Roy M.
Quantitative analysis : an introduction. – (Automation &
production systems. Methodologies & applications ; v. 2 –
ISSN 1028-141X)
1. Mathematical optimization 2. Decision-making –
Mathematical models 3. Production management – Mathematical
models
I. Title
519.3

To

my parents *Rosindo and Mary*, who are
with me in my heart forever.

my wife *Joan*, whose understanding
and patience made this book possible.

and

my daughters *Mariana and Irina*,
who make every day a joy for Joan and me.

If a man will begin with certainties he shall end in doubts; but if he will be content to begin with doubts he shall end in certainties.

—Francis Bacon

CONTENTS

SERIES PREFACE

Automation and Production Systems: Methodologies and Applications provides a scientific and practical base for researchers, practitioners and students involved in the design and implementation of advanced manufacturing systems. It features a balance between state-of-the-art research and usefully reported applications. Among the topics that will be presented as part of this series are: strategic management and application of advanced manufacturing systems; cellular manufacturing systems; group technology; design simulation and virtual manufacturing; advanced materials handling systems; quantitative analysis; intelligent feature-base process planning; advanced genetic algorithms; computer-aided process planning; simultaneous engineering; economic evaluation of advanced technologies; and concurrent design of products.

This series encompasses three categories of books: reference books that include material of an archival nature; graduate- and postgraduate-level texts, reference and tutorial books that could also be utilized by professional engineers; and monographs concerning newly developing trends in applications for manufacturing and automation engineering. The volumes are written for graduate students in industrial and manufacturing engineering, as well as researchers and practitioners interested in product and process design, manufacturing systems design integration and control, and engineering management and economics.

Hamid R. Parsaei

PREFACE

The intent of *Quantitative Analysis: An Introduction* is to provide a survey of optimization techniques most often used in decision making. Since a survey cannot treat any individual topic in detail, readers who wish further knowledge in any particular area addressed in this volume are encouraged to examine books that provide a fuller treatment.

Optimization techniques presented here are not complex and may be mastered quickly. Their simplicity does not diminish their power. In fact, simplicity should be an objective in decision making. The two most important factors in solving a problem effectively are: (1) Propose the simplest problem formulation that captures the essential features of the system being examined; and (2) Solve the problem with the simplest technique appropriate. Proposing an elegant formulation of a problem that cannot be solved within the resources available (time, manpower and money) is not helpful to decision making. For example, equations governing moving objects are complex and difficult to solve when all interactions among celestial bodies are considered. And yet the essentials of physical laws of motion can be modeled satisfactorily using simple equations. This allows NASA to plan and execute interplanetary missions based on solutions to simple, linearized equations of motion.

Basic information on simulation is provided in chapter 11. It is recognized that simulation is a powerful optimization tool for decision making. Simulations can capture the complexities and probabilistic nature of a system to a high level of detail. However, the use of simulation has significant drawbacks to a decision maker. Simulations typically require a large investment of resources (time, manpower and money) to formulate, debug and make enough production runs to gain statistical significance. Moreover, simulations often require a great deal of data to faithfully model the specific details of the system. This data, which typically must be gathered by a population survey or through expert opinion, is also expensive and time-consuming to obtain. While some systems can only be satisfactorily modeled using simulation, it is essential to limit its use to these cases. The techniques addressed in this volume provide quick and easily obtainable solutions. It is obvious that an approximate solution provided in a timely manner is much more useful to a decision maker than an exact solution provided after the decision has been made. It is suggested that analysts use these simple techniques, whenever possible, to provide timely responses to decision makers and use simulation, when appropriate, to *refine* the solutions.

An essential aspect of the application of decision making is the realization that the solution arising from quantitative techniques such as those provided in

Quantitative Analysis: An Introduction (or from simulation) cannot be considered "optimal" until the special knowledge of the decision maker is applied. Typically, decisions are affected by political, social, legal and regulatory constraints difficult or impossible to quantify in a satisfactory manner. However, quantitative techniques do provide valuable information to the decision maker to be considered along with all other appropriate factors when deciding upon the true optimum solution.

"Probability is the very guide of life."

Cicero

1 | Probability & Statistics

1.1 Basic Probability

Definition: <u>Probability</u> represents the proportion of the time an outcome will occur *in the long run.*

Example 1: Consider the toss of a fair die. What is the probability of rolling a "6"?

Answer: There are six equally likely outcomes {1,2,3,4,5,6} and the probability of the outcome "6" is 1/6.

For any given sequence of tosses of the die, the number of "6's" will not exactly be one-sixth of the total tosses due to the statistical variation (randomness) of the process. However, for a large number of tosses of a fair die (in the long run), the proportion of "6's" will approach 1/6.

Notes: Two basic axioms of probability are as follows:

• The probability of any single outcome is between 0 and 1, inclusive.

• The sum of the probabilities of all possible outcomes equals 1.

For Example 1, define O_i = The outcome of "i" on a roll of a fair die. Then:

$$P(O_1) = P(O_2) = \cdots = P(O_6) = \frac{1}{6}$$

$$P(O_1) + P(O_2) + \cdots + P(O_6) = 1$$

Typically, the process being examined will be more complex than the process of rolling a die and be capable of many outcomes. However, the set of all possible outcomes for any process (experiment, system) is called the <u>sample space</u>.

Example 2: The arcade at the Big Top Circus has a game where three darts are thrown at a large number of Blue (B) and Green (G) balloons arranged randomly on a target. Each throw will either a Blue or a Green balloon. Provide the sample space of all possible outcomes for the three throws.

Answer: There are eight possible outcomes:

$$\frac{2\ choices}{(Throw\ \#1)} \cdot \frac{2\ choices}{(Throw\ \#2)} \cdot \frac{2\ choices}{(Throw\ \#3)} = 8\ possible\ outcomes\ (permutations)$$

$$\{B, B, B\} = O_1$$
$$\{B, B, G\} = O_2$$
$$\{B, G, B\} = O_3$$
$$\{B, G, G\} = O_4$$
$$\{G, B, B\} = O_5$$
$$\{G, B, G\} = O_6$$
$$\{G, G, B\} = O_7$$
$$\{G, G, G\} = O_8$$

If we assume that Blue and Green are equally likely to be hit (i.e. $P(B) = P(G) = \frac{1}{2}$), then:

$$P(O_1) = P(O_2) = \cdots = P(O_8) = \frac{1}{8}$$

The outcomes may also be enumerated using an outcome tree as shown in Figures 1. The probability tree is shown in Figure 2.

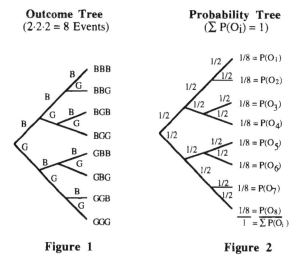

Outcome Tree
($2 \cdot 2 \cdot 2 = 8$ Events)

Probability Tree
($\Sigma P(O_i) = 1$)

Figure 1   **Figure 2**

If we assume that the number of Blue and Green balloons is such that Blue is hit $\frac{2}{3}$ of the time, (i.e. $P(B) = \frac{2}{3}$ and $P(G) = \frac{1}{3}$), then the outcome/probability tree structure is as shown in Figure 3.

2

Outcome/Probability Tree

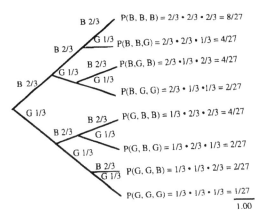

Figure 3

1.2 Events

An <u>Event</u> is defined as a subset of possible outcomes in the sample space.

Example 3: In Example 2 where a group of three balloons is hit, what outcomes represent the Event E = "At least two green in the group of three balloons"? What is P(E)?

Answer: The outcomes for Event E = $\{O_4, O_6, O_7, O_8\}$. The probability of an Event E is the sum of the probabilities of the outcomes that represent the event.

$$P(E) = \Sigma\, P(O_i) \qquad \text{for all } O_i \text{ belonging to Event E}$$

If P(G) = P(B) = 1/2,

$$
\begin{aligned}
P(E) &= P(O_4) + P(O_6) + P(O_7) + P(O_8) \\
&= P(B, G, G) + P(G, B, G) + P(G, G, B) + P(G, G, G) \\
&= (\tfrac{1}{2} \cdot \tfrac{1}{2} \cdot \tfrac{1}{2}) + (\tfrac{1}{2} \cdot \tfrac{1}{2} \cdot \tfrac{1}{2}) + (\tfrac{1}{2} \cdot \tfrac{1}{2} \cdot \tfrac{1}{2}) + (\tfrac{1}{2} \cdot \tfrac{1}{2} \cdot \tfrac{1}{2}) \\
&= \tfrac{1}{8} + \tfrac{1}{8} + \tfrac{1}{8} + \tfrac{1}{8} \\
&= \tfrac{1}{2}
\end{aligned}
$$

If P(B) = 2/3 and P(G) = 1/3,

$$
\begin{aligned}
P(E) &= P(O_4) + P(O_6) + P(O_7) + P(O_8) \\
&= P(B, G, G) + P(G, B, G) + P(G, G, B) + P(G, G, G) \\
&= (\tfrac{2}{3} \cdot \tfrac{1}{3} \cdot \tfrac{1}{3}) + (\tfrac{1}{3} \cdot \tfrac{2}{3} \cdot \tfrac{1}{3}) + (\tfrac{1}{3} \cdot \tfrac{1}{3} \cdot \tfrac{2}{3}) + (\tfrac{1}{3} \cdot \tfrac{1}{3} \cdot \tfrac{1}{3}) \\
&= \tfrac{2}{27} + \tfrac{2}{27} + \tfrac{2}{27} + \tfrac{1}{27}
\end{aligned}
$$

$= \frac{7}{27}$.

Example 4: At the Big Top Circus in Example 2 where a group of three balloons is hit, determine the following:

- P(No Green)
- P(All Same Color)
- P(Exactly Two Green)

Assume an equal number of Blue and Green balloons (i.e. P(B) = P(G) = 1/2).

Answer: Calculate the probability after determination of the outcomes that represent the Event:

$P(No\ Green) = P(B,\ B,\ B) = \frac{1}{2} \cdot \frac{1}{2} \cdot \frac{1}{2} = \frac{1}{8}$

$P(All\ Same\ Color) = P(B,\ B,\ B) + P(G,\ G,\ G) = \frac{1}{8} + \frac{1}{8} = \frac{1}{4}$

$P(Exactly\ Two\ Green) = P(B,\ G,\ G) + P(G,\ B,\ G) + P(G,\ G,\ B)$
$$= \frac{1}{8} + \frac{1}{8} + \frac{1}{8}$$
$$= \frac{3}{8} \ .$$

1.3   Combining Events

Examine the Event "E or F". In set notation, this is written "$E \cup F$" ("E union F").

"$E \cup F$" = The set of outcomes in Event E or Event F or both.

Example 5: At the Big Top Circus in Example 2, let the Event E = "Exactly one Blue in the group of three balloons hit" and let the Event F = "The first balloon hit is Green".

- Find the set of outcomes in the Event "$E \cup F$"
- Find $P(E \cup F)$

Answer: Define the outcomes in Event E and Event F individually:

$E = \{O_4,\ O_6,\ O_7\} = \{(B,\ G,\ G),\ (G,\ B,\ G),\ (G,\ G,\ B)\}$

$F = \{O_5,\ O_6,\ O_7,\ O_8\} = \{(G,\ B,\ B),\ (G,\ B,\ G),\ (G,\ G,\ B),\ (G,\ G,\ G,)\}$

Then the Event "$E \cup F$" and the $P(E \cup F)$ are given as follows:

$E \cup F = \{O_4,\ O_5,\ O_6,\ O_7,\ O_8\}$

$P(E \cup F) = \frac{1}{8} + \frac{1}{8} + \frac{1}{8} + \frac{1}{8} + \frac{1}{8} = \frac{5}{8}$.

Examine the Event "E and F". In set notation, this is written $"E \cap F"$ ("E intersection F").

$"E \cap F"$ = The set of outcomes in both Event E and Event F.

For Example 5, $"E \cap F" = \{O_6, O_7\}$ and the $P(E \cap F) = 2/8$.

Note: The probability of $"E \cup F"$ can also be determined by the *set theory equation*:

$P(E \cup F) = P(E) + P(F) - P(E \cap F)$

which can be verified using a <u>Venn Diagram</u> (see Figure 4).

Venn Diagram

Figure 4

For Example 5, the set theory equation is applied to compute $P(E \cup F)$ as follows:

$E = $ *"Exactly one Blue"* $= \{O_4, O_6, O_7\} \Rightarrow P(E) = 3/8$

$F = $ *" First balloon Green"* $= \{O_5, O_6, O_7, O_8\} \Rightarrow P(F) = 4/8$

$(E \cap F) = \{O_6, O_7\} \Rightarrow P(E \cap F) = 2/8$

$P(E \cup F) = P(E) + P(F) - P(E \cap F)$
$\qquad = 3/8 + 4/8 - 2/8$
$\qquad = 5/8$

Note: To determine $P(E \cup F)$, one can enumerate the set of outcomes for $E \cup F$ or use the set theory equation $P(E \cup F) = P(E) + P(F) - P(E \cap F)$. Both methods give the same probability $5/8$ for Example 5.

If $E \cap F = \phi$, the empty or "null" set, Events E and F are <u>mutually exclusive</u> (no outcomes in common) and $P(E \cap F) = 0$. The set theory equation simplifies as follows:

$P(E \cup F) = P(E) + P(F) - P(E \cap F)$

$\qquad = P(E) + P(F) - 0$

$\qquad = P(E) + P(F)$

5

Example 6: At the Big Top Circus in Example 2, let Event E = "Exactly one Blue balloon among the group of three balloons hit" and Event F = "Exactly one Green balloon among the group of three balloons hit" Determine $P(E \cup F)$.

Answer: Determine the outcomes and probabilities associated with Events E and F:

$$E = \{O_4, O_6, O_7\} \Rightarrow P(E) = 3/8$$

$$F = \{O_2, O_3, O_5\} \Rightarrow P(F) = 3/8$$

$$E \cap F = \phi \Rightarrow P(E \cap F) = 0$$

$$P(E \cup F) = P(E) + P(F) - P(E \cap F) = P(E) + P(F) = 3/8 + 3/8 = 6/8$$

1.4 Complements

Define Event $\hat{E}$ = "All the outcomes in the sample space *not* belonging to Event E ". Then Event $\hat{E}$ ("E hat") is called the complement of Event E .

$$E \cap \hat{E} = \phi, \text{ so that } P(E \cap \hat{E}) = P(\phi) = 0$$

$$E \cup \hat{E} = U \text{ (the Entire Sample Space), so that } P(E \cup \hat{E}) = P(U) = 1$$

Then, the following two equalities for events and complements hold:

(1) $\quad P(E \cup \hat{E}) = P(E) + P(\hat{E}) - P(E \cap \hat{E})$

$$= P(E) + P(\hat{E})$$

(2) $\quad P(E \cup \hat{E}) = P(U)$

$$= 1$$

Equations (1) and (2) imply the following relationship between an Event E and its complement Event $\hat{E}$:

$$P(E) + P(\hat{E}) = 1$$

$$\Rightarrow P(E) = 1 - P(\hat{E})$$

Note: This relationship can be used as an alternate way to compute $P(E)$. Compute the probability of the complement Event $\hat{E}$ and subtract the result from 1.

Example 7: At the Big Top Circus in Example 2, let the Event E = "At least one Blue among the group of three balloons hit". Determine the following:

- Outcomes in E
- Outcomes in $\hat{E}$
- $P(E)$
- $P(\hat{E})$

Answer: The outcomes in Event $E = \{O_1, O_2, O_3, O_4, O_5, O_6, O_7\}$. Then, the outcomes in the complement of Event E are given by:

Event $\hat{E}$ = "No Blue among the three balloons hit" = (G,G,G) = $\{O_8\}$

The probability of Event E may be computed summing the probabilities of the individual outcomes in Event E:

$$P(E) = P(O_1) + P(O_2) + P(O_3) + \cdots + P(O_7) = 7/8 \ .$$

The alternate method of computing the probability of Event E is:

$$P(E) = 1 - P(\hat{E}) \ .$$

The probability of Event $\hat{E}$ is given by:

$$P(\hat{E}) = P(O_8) = 1/8 \ .$$

Hence, the alternate method to compute the $P(E)$ is given by:

$$P(E) = 1 - P(\hat{E}) = 1 - 1/8 = 7/8 \ .$$

Note: $P(At\ least\ one) = 1 - P(None)$ since the events "At least one" and "None" are complements. The outcomes associated with the event "None" are usually much easier to enumerate those associated with the event "At least one". Hence, when computing the probability of the event "At least one", it is usually easier to compute the probability of the event "None" and then to subtract that probability from 1.

1.5 Expected Value and Standard Deviation

Let X be a variable which assigns a *numerical value* to the outcomes of a stochastic (random) process. Then X is called a <u>Random Variable</u> (RV).

Example 8: Let the Random Variable X = "The number of dots on a fair die when rolled". What is the numerical value associated with each outcome?

Answer: The possible outcomes (dots) of a fair die when rolled are $\{1,2,3,4,5,6\}$. The numerical values assigned by the Random Variable X are 1, 2, 3, 4, 5, and 6, respectively.

<u>Note:</u> The variable Y = "The number of dots *squared* on a fair die when rolled" would also be a Random Variable whose numerical values are {1, 4, 9, 16, 25, 36}.

Define the <u>Expected Value</u> of a Random Variable X, denoted by μ or *EX*, as:

$$\mu = EX = \Sigma\, x_i \cdot P(X = x_i) \qquad \text{for all possible numerical values } x_i \text{ of the RV X}$$

Example 9: For the fair die in Example 8, find the Expected Value of the Random Variable X.

Answer: The possible numerical values x_i of the Random Variable X are {1, 2, 3, 4, 5, 6}. The probabilities that the Random Variable will take on each of these numerical values is 1/6, 1/6, 1/6, 1/6, 1/6, and 1/6, respectively.

$$\mu = EX = 1 \cdot (\tfrac{1}{6}) + 2 \cdot (\tfrac{1}{6}) + 3 \cdot (\tfrac{1}{6}) + 4 \cdot (\tfrac{1}{6}) + 5 \cdot (\tfrac{1}{6}) + 6 \cdot (\tfrac{1}{6})$$

$$= {}^{21}\!/_{6}$$

$$= 3.5$$

The Expected Value when a fair die is rolled is 3.5. Of course, 3.5 is *not* a possible value of the Random Variable. It is not proper to interpret the Expected Value of a Random Variable as the most likely value of the Random Variable (though it is possible that the Expected Value and the most likely value are the same). The proper interpretation of Expected Value = 3.5 is as follows:

> "If a fair die is rolled a large number of times and the sum of the rolled numerical values is divided by the number of rolls, that value (which is the average or mean value) will approach the Expected Value 3.5".

How close the set of numerical values of a Random Variable is to its Expected Value defines the <u>standard deviation</u> (σ) of the Random Variable. Normal distributions with large and small standard deviations are shown notionally in Figures 5 and 6, respectively.

Large Standard Deviation σ

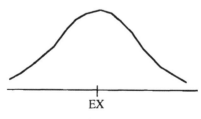

EX

Figure 5

Small Standard Deviation σ

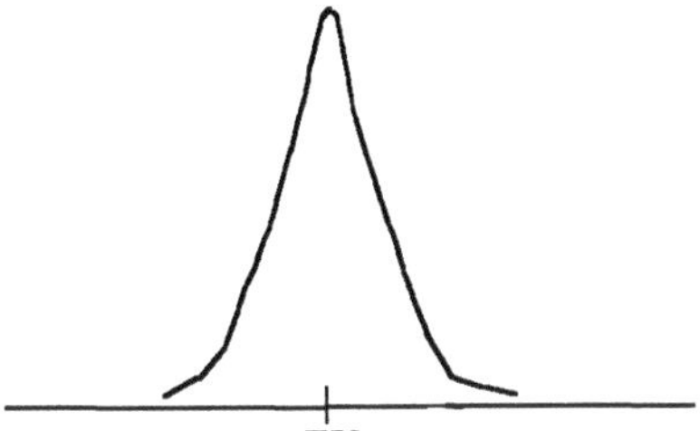

EX

Figure 6

Note: It is typically quite difficult to solve quantitative problems involving stochastic processes. The standard approach is to *simplify* the problem by approximating the Random Variables with a constant (deterministic) value when this is a reasonable assumption. If the standard deviation of a Random Variable is small, it is usually reasonable to approximate the Random Variable by its constant Expected Value.

Define the variance, (σ^2), of a Random Variable X to be:

$Variance\ (\sigma^2) = \Sigma(x_i - EX)^2 \cdot P(X = x_i)$ for all possible numerical values x_i of the RV X

$Standard\ Deviation\ (\sigma) = [Variance]^{1/2}$

Example 10: Determine the Variance and Standard Deviation of the Random Variable X = "The dots on a fair die when rolled".

Answer: The possible numerical outcome values of the RV X = {1, 2, 3, 4, 5, 6} all with probability 1/6. The Expected Value of X equals 3.5 from Example 9. Then,

$$Variance\ X = \sigma^2 = (1 - 3.5)^2 \cdot \tfrac{1}{6} + (2 - 3.5)^2 \cdot \tfrac{1}{6} + \cdots + (6 - 3.5)^2 \cdot \tfrac{1}{6}$$

$$= {}^{6.25}\!/_6 + {}^{2.25}\!/_6 + {}^{0.25}\!/_6 + {}^{0.25}\!/_6 + {}^{2.25}\!/_6 + {}^{6.25}\!/_6$$

$$= {}^{17.5}\!/_6$$

$$= 2.916$$

$$Standard\ Deviation\ X = \sigma = [Variance]^{1/2}$$

$$= 1.71$$

Note: The following is a review of the notation specified in this presentation for a Random Variable X:

$\mu = EX = \Sigma\, x_i \cdot P(x_i) = Expected\ Value\ X$

$\sigma^2 = \Sigma(x_i - EX)^2 \cdot P(x_i) = Variance\ X$

$\sigma = [Variance]^{1/2} = Standard\ Deviation\ X$

1.6 Normal (Gaussian) Distribution

The Normal distribution (or Gaussian distribution) is a continuous $(-\infty, +\infty)$ distribution completely specified by two parameters: its Expected Value μ and its Standard Deviation σ and is denoted by $N(\mu, \sigma^2)$.

The Normal distribution $N(\mu, \sigma^2)$ is symmetric about its Expected Value μ (see Figure 7).

Normal Distribution $N(\mu, \sigma^2)$

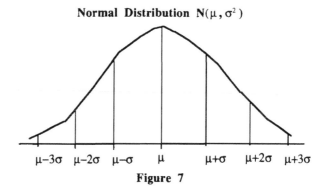

Figure 7

The peak of the Normal curve occurs at the Expected Value μ (see Figure 8).

Normal Distributions With Different Expected Values (μ)

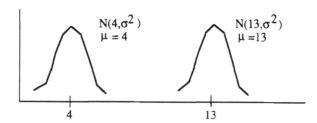

Figure 8

The shape (height) of the Normal curve is determined by its Standard Deviation σ. The smaller the Standard Deviation, the taller the Normal curve and the less dispersion of the outcome values around the Expected Value μ (see Figure 9).

Normal Distributions With Different
Expected Values μ and Standard Deviations σ

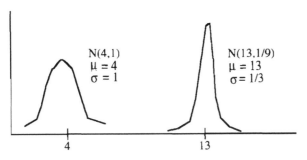

Figure 9

<u>Note:</u> The area (total probability) under a Normal curve is always equal to 1. This property is equivalent to $\Sigma P(x_i) = 1$ for discrete distributions.

A tabular set of Normal curve values are typically found in reference books (see Table 1a and Table 1b. These tables provide the cumulative values of $P(Z \le z)$ for the Random Variable $Z \sim N(0,1)$ called the <u>Standard Normal distribution</u>.

<u>Note:</u> Tables are used because no *closed form* expression exists for computing the cumulative values (e.g. $P(Z \le z)$) for Random Variables with a Normal distribution.

If the Random Variable $X \sim N(\mu, \sigma^2)$, then the Random Variable $Z = [(X - \mu)/\sigma] \sim N(0, 1)$. This property allows probability computations for *any* Normal Random Variable using only the table of Standard Normal Distribution $N(0, 1)$ values.

Example 11: Let X be a Normal Random Variable with Expected Value $\mu = 5$ and Standard Deviation $\sigma = 2$ [i.e. $X \sim N(5,4)$]. Determine $P(X \le 7)$.

Answer: Since $\mu = 5$ and $\sigma = 2$,

$$P(X \le 7) \quad = \quad P[(X - 5)/2 \le (7 - 5)/2]$$

$$= \quad P[Z \le 1] \quad \text{where } Z = (X - 5)/2 \sim N(0,1)$$

$$= \quad 0.84 \quad \quad (Table \ 1b)$$

<u>Note:</u> Properties of any Random Variable $X \sim N(\mu, \sigma^2)$ are as follows (see Figure 7):

- $P(X \le \mu) = P(X \ge \mu) = 0.50$

11

- $P(X < \infty) = P(X > -\infty) = 1$

- $P(\mu - \sigma < X < \mu + \sigma) = 0.68$ $P(X < \mu + \sigma) = 0.84$

- $P(\mu - 2\sigma < X < \mu + 2\sigma) = 0.95$ $P(X < \mu + 2\sigma) = 0.977$

- $P(\mu - 3\sigma < X < \mu + 3\sigma) = 0.997$ $P(X < \mu + 3\sigma) = 0.9986$

Example 12: Let $X \sim N(\mu, \sigma^2)$. Determine $P(\mu - 1.31\,\sigma < X < \mu + 1.31\,\sigma)$.

Answer: The probability is given by the area under the Normal curve from the Expected Value minus 1.31 Standard Deviations to the Expected Value plus 1.31 Standard Deviations. This probability can be computed as follows:

$P(\mu - 1.31\,\sigma < X < \mu + 1.31\,\sigma)$

$= P[(X < \mu + 1.31\sigma] - P[(X < \mu - 1.31\sigma]$

<u>Note</u>: Graphically, this subtraction of probabilities is displayed by the hatched areas in Figure 10.

Determination of Normal Probabilities

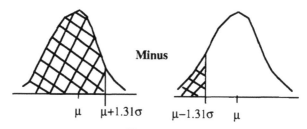

Figure 10

$= P[(X - \mu)/\sigma < 1.31] - P[(X - \mu)/\sigma < -1.31]$

$= P(Z < 1.31) - P(Z < -1.31)$ *where* $Z = (X - \mu)/\sigma \sim N(0, 1)$

$= 0.9049 - 0.0951$ (*Table* 1*b*)

$= 0.81$

<u>Note</u>: Graphically, the result of the subtraction of probabilities is displayed in the hatched area Figure 11.

$$P(\mu - 1.31\sigma < X < \mu + 1.31\sigma)$$

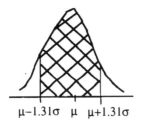

$$\mu - 1.31\sigma \quad \mu \quad \mu + 1.31\sigma$$

Figure 11

1.7 Confidence Intervals

Standard Normal $N(0, 1)$ tables (Table 1a and Table 1b) can be used to compute confidence intervals for all Normal Random Variables.

Example 12: Find a 90% confidence interval for the Random Variable $X \sim N(5, 4)$

Answer: A confidence interval is typically given as a *symmetric* interval around the Expected Value of the Random Variable. In general, one must determine the values a and b such that:

(1) $P(a \le X \le b) = 0.90$ $\qquad\qquad$ *where* $X \sim N(5, 4)$

Since $Z_{.05}$ and $Z_{.95}$ are the 5th and 95th cumulative distribution percentiles, respectively, of the Standard Normal Random Variable Z (see Table 1b), then by definition:

$P(Z_{.05} \le Z \le Z_{.95}) = 0.90$ $\qquad\qquad$ *where* $Z \sim N(0, 1)$

Using the transformation $Z = (X - \mu)/\sigma$,

$P(Z_{.05} \le (X - \mu)/\sigma \le Z_{.95}) = 0.90$ $\qquad\qquad$ *where* $(X - \mu)/\sigma \sim N(0, 1)$

so that:

(2) $P(\mu + Z_{.05}\sigma \le X \le \mu + Z_{.95}\sigma) = 0.90$

Both equations (1) and (2) are representations of a symmetric 95% confidence interval around the Expected Value. This implies that:

$a = \mu + Z_{.05}\sigma$

$b = \mu + Z_{.95}\sigma$

For $X \sim N(\mu, \sigma^2)$, the symmetric interval around the Expected Value μ of the Random Variable given by $(\mu + Z_{.05}\sigma, \mu + Z_{.95}\sigma)$ is a 90% confidence interval.

Since (from Table 1b) $Z_{.05} = -1.645$ and $Z_{.95} = 1.645$ (note the symmetry in that the values are equal but opposite in sign), the interval $(\mu - 1.645\sigma, \mu + 1.645\sigma)$ is a 90% confidence interval for $X \sim N(\mu, \sigma^2)$.

For $X \sim N(5, 4)$ where $\mu = 5$ and $\sigma = 2$, the following are all representations of a symmetric 90% confidence interval around the Expected Value 5:

- $[5 + Z_{.05}(2), 5 + Z_{.95}(2)]$

- $[5 - 1.645(2), 5 + 1.645(2)]$

- $(5 +/- 3.29)$

- $(1.71, 8.29)$.

1.8 Poisson Distribution

The Poisson distribution for a stochastic process is completely specified by one parameter: its Expected Value λ. Unlike the continuous Normal distribution, the Poisson distribution is *discrete* and takes on only integer values zero or larger (see Figure 12).

For a Random Variable X with a Poisson distribution of Expected Value λ:

- *Possible values* $= \{0, 1, 2, 3, 4, \ldots\}$

- $\lambda \geq 0$

- *The Variance of X also equals* λ

The probability distribution for a Poisson Random Variable X is given by:

$$P(X = k) = (\lambda^k \cdot e^{-\lambda})/k! \qquad k = 0, 1, 2, 3, 4, \ldots\ldots$$

where $k! = (k) \cdot (k-1) \cdot (k-2) \cdots (3) \cdot (2) \cdot (1)$ ("*k factorial*")
$0! = 1$ (*0 factorial is defined to be* 1)
$e \approx 2.71828$ (*e is a transcendental constant*)
$EX = \lambda$
Standard Deviation $X = \lambda^{1/2}$

<u>Note</u>: The Poisson distribution is *not symmetric* about its Expected Value λ. The Poisson distribution has a *tail* in the direction of $+\infty$ (see Figure 12). The value of the Expected Value λ is greater than or equal to zero but need not be an integer. Hence, the Expected Value λ need *not* be a possible value for the Poisson distribution.

Poisson Distribution

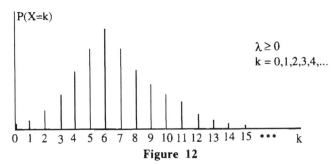

Figure 12

<u>Note</u>: The Poisson distribution is typically used in Queueing (see Chapter 10) to describe the process of customers arriving at a bank or a check-in line where λ is the expected number of customer arrivals per unit time.

Example 13: If customers arrive at a bank according to a Poisson distribution of Expected Value $\lambda = 5$ customers per hour, what is the probability of exactly two arrivals in an hour? What is the probability of less than or equal to two arrivals in an hour?

Answer: For a Poisson process of Expected Value λ,

$$P(X = k) = (\lambda^k \cdot e^{-\lambda})/k! \qquad k = 0, 1, 2, 3, 4, \ldots .$$

Then, for $\lambda = 5$ customers per hour,

$$P(X = 2) = 5^2 e^{-5}/2!$$

$$P(X \le 2) = P(X = 0) + P(X = 1) + P(X = 2)$$

$$= 5^0 e^{-5}/0! + 5^1 e^{-5}/1! + 5^2 e^{-5}/2!$$

Use a calculator or the entries in Table 4 (N = 0, 1, 2 and $\lambda = 5$) to determine the Poisson probabilities:

$$P(X = 0) = 0.007$$
$$P(X = 1) = 0.034$$
$$P(X = 2) = 0.084$$

Then the probability of exactly 2 arrivals in a hour is given by:

$P(X = 2) = 0.084$

and the probability of less than or equal to two arrivals in a hour is given by:

$P(X \leq 2) = 0.007 + 0.034 + 0.084 = 0.125$.

Note: No closed form expression exists for the cumulative probability $P(X \leq k)$ when X has a Poisson distribution. A Poisson table (such as Table 4) or a calculator must be used to compute cumulative Poisson probabilities.

1.9 Exponential Distribution

The Exponential distribution for a stochastic process is a *continuous*, non-negative (≥ 0) distribution completely specified by one parameter: its Expected Value $1/\mu$ (see Figure 13).

The probability distribution f(x) for an Exponential Random Variable X is given by:

$$f(x) = P(X = x) = \mu e^{-\mu x} \qquad x \geq 0$$

> *Expected Value* $= 1/\mu$
> *Variance* $= 1/\mu^2$
> *Standard Deviation* $= 1/\mu$

Exponential Distribution

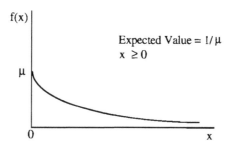

Figure 13

Note: At the origin, the Exponential function takes on the value μ. This can be verified by substituting x = 0 into the probability distribution $f(x)$. That is, $f(0) = \mu e^{-\mu 0} = \mu$.

Unlike the Normal and Poisson distributions, the Exponential cumulative distribution $F(x) = P(X \leq x)$ *can be written as a closed form expression:*

$$F(x) = 1 - e^{-\mu x} \qquad x \geq 0$$

where:

$$F(0) = P(X \leq 0) = 1 - e^{-0} = 0$$

$$F(\infty) = P(X < \infty) = 1 - e^{-\infty} = 1$$

A comparison of Exponential distributions with Expected Values ($1/\mu$) equal to 1/2, 2/3, and 1 is given in Figure 14.

Exponential Distributions With Expected Values of 1/2, 2/3, and 1

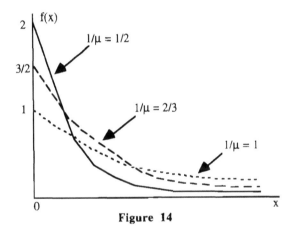

Figure 14

<u>Note:</u> Since the standard deviation ($1/\mu$) of the Exponential distribution equals the Expected Value ($1/\mu$), a small mean implies a small standard deviation (and more of the probability clustered close to the origin).

The Exponential distribution is typically used in Queueing (see Chapter 10) to describe the length of service time for a customer at a bank or check-in line where $1/\mu$ is the expected service time per customer.

Example 14: Service times at a bank are Exponentially distributed with an Expected Value $1/\mu$ of 20 minutes. What is the probability a customer has service time less than 30 minutes? Between 20 and 30 minutes?

Answer: For an Exponential distribution,

$$F(x) = P(X \le x) = 1 - e^{-\mu x} \qquad\qquad x \ge 0$$

For an Exponential distribution with $1/\mu = 20$ minutes, $\mu = 1/20$ minutes so that:

$$P(X \le 30 \; minutes) = 1 - e^{-(1/20)(30)}$$

$$= 1 - e^{-1.5}$$

$$= 1 - 0.2231 \qquad\qquad Table\ 5$$

$$= 0.7769$$

$$P(20 \; min \le X \le 30 \; min) = P(X \le 30 \; min) - P(X \le 20 \; min)$$

$$= [1 - e^{-(1/20)(30)}] - [1 - e^{-(1/20)(20)}]$$

$$= e^{-1} - e^{-1.5}$$

$$= 0.1448 \qquad\qquad Table\ 5$$

Note: If customers arrive randomly with a rate which has a Poisson distribution with Expected Value λ (say 5 customers per hour), then the time between customer arrivals, called the interarrival time, has an Exponential distribution with Expected Value $1/\lambda$ (0.2 hours per customer). This is a useful property in Queueing models (see Chapter 10.

1.10   Statistics Definitions

A statistic is a *function* of a sample of N numerical outcome values of a Random Variable.

Consider N numerical outcome values for the Random Variable $X = \{x_1, x_2, \cdots, x_N\}$. Then the sample mean and sample standard deviation are well-known statistics:

$$\bar{X} = Sample\ Mean = \Sigma\ x_i / N \qquad\qquad for\ \ i = 1, 2, \ldots, N$$

$$s^2 = \Sigma\ (x_i - \bar{X})^2 \ / \ (N - 1) \qquad\qquad for\ \ i = 1, 2, \ldots, N$$

$$s = Sample\ Standard\ Deviation$$

Notes: The following are two often asked questions in statistics:

• Why are there only $N - 1$ degrees of freedom for s^2 ? Since the Mean $\bar{X}$ of the sample is known, Σx_i is a *known* value $(= N\bar{X})$. Thus, if $N - 1$ values of the sample

$X = \{x_1, x_2, \ldots, x_{N-1}\}$ are given, then the *Nth* value of the sample, x_N, is *known* and must equal to ($N\bar{X} - \Sigma x_i$) for $i = 1$ *to* $N - 1$. Thus the Nth value of the sample is fixed rather than *random* and a degree of freedom in the data is lost.

• What is the difference between μ and $\bar{X}$ and between σ and s? The values μ and σ represent the *fixed*, theoretical (perhaps unknown) values for the Expected Value and standard deviation, respectively, of a Random Variable (RV). The sample mean $\bar{X}$ and the sample standard deviation s are *estimates* of the Expected Value μ and the standard deviation σ, respectively, of a RV given a sample of N numerical outcome values. As the sample size N becomes large, $\bar{X}$ and s will approach μ and σ, respectively.

1.11 Goodness of Fit Test (Chi-Square Test)

The Chi-Square (χ^2) Test can be used to determine whether N numerical outcome values $= \{x_1, x_2, \ldots, x_N\}$ for a Random Variable were generated by a hypothesized distribution.

Note: A quick overview of the Chi-Square Test is as follows: The *actual* numerical outcome values a Random Variable are grouped into categories or *cells*. The hypothesized distribution of the Random Variable is used to determine how many of the N numerical outcomes are *expected* in each of the cells. The actual number is compared to the expected number in each cell to determine the goodness of fit to the hypothesized distribution.

Computations for the Chi-Square Test

Let A_i = The actual # numerical outcomes of a Random Variable in cell i

Let E_i = The expected # numerical outcomes of a Random Variable in cell i

Note: The number of *expected* outcomes in each cell must be at least 5 (this is required for statistical significance of the cell). The Chi-Square Test requires that cells be *pooled* (combined) until there are at least 5 expected outcomes for each cell.

The χ^2 statistic is defined as:

$$\chi^2 = \Sigma \left[(A_i - E_i)^2 / E_i\right] \quad for \quad i = 1 \ to \ M$$

where M = # cells (after pooling, if required)

The χ^2 statistic follows the Chi-Square distribution with $M - k - 1$ degrees of freedom.

M = # cells (after pooling, if required)

k = # parameters estimated to determine the expected # outcomes in each cell

Note: Remember that 1 degree of freedom is lost since the mean $\bar{X}$ of the sample is known (so that when the first N-1 outcomes are given, the Nth outcome is derivable and is not random).

The number of *expected* values in each cell must be determined. For N total outcome values, the expected number, E_i, outcomes in cell i (where there are M cells) is given by:

$$E_i = N \cdot p_i \qquad p_i = probability \; an \; outcome \; belongs \; to \; cell \; i \; (i = 1 \; to \; M)$$

where $\quad p_1 + p_2 + \cdots + p_M = 1$

$\qquad$ (*Probability of being in one of M cells is 1*)

$$E_1 + E_2 + \cdots + E_M = N$$

$\qquad$ (# *expected outcomes* = # *actual outcomes* = N)

Notes: There are two cases for determination of the M - k - 1 degrees of freedom (df) for the Chi-Square Test depending on whether parameters must be estimated to compute the E_i.

• Case #1: The parameters of the hypothesized distribution that allow the determination of the expected number of outcomes E_i in each cell i are *given*. For this case, k=0.

$\qquad \Rightarrow \chi^2$ *statistic has M* $- 1$ *degrees of freedom*

• Case #2. The k parameters of the hypothesized distribution that allow the determination of the expected number of outcomes E_i in each cell must be *computed*.

$\qquad$ For example:

$\qquad\qquad \lambda$ *for   the Poisson distribution* $(k = 1)$
$\qquad\qquad 1/\mu$ *for   the Exponential distribution* $(k = 1)$
$\qquad\qquad \mu, \sigma$ *for   the Normal distribution* $(k = 2)$
$\qquad\qquad p$ *for   the Binomial distribution* $(k = 1)$
$\qquad\qquad p$ *for   the Uniform* (*equally likely*) *distribution* $(k = 1)$

$\qquad \Rightarrow \chi^2$ *statistic has M* $- k - 1$ *degrees of freedom*

Example 15 (Case #1): Test the hypothesis that the following set of numerical outcome values (where A_i represents the actual number of "i's" in 30 throws of a die) was generated by a *fair* die.

i	1	2	3	4	5	6
A_i	5	7	3	4	6	5

20

Answer: The total number of outcomes $N = \Sigma A_i$ is 30. The data is grouped into cells corresponding to each face of the die. It is *given* in the Example description that the die is fair so that the probability for each face is $\frac{1}{6}$. The probability p_i for each face did not have to be estimated from the outcome values and thus $k = 0$.

Since $p = \frac{1}{6}$ for all faces is given, the expected number of the 30 outcomes occurring in each cell (see Figure 15) is determined as follows:

$$E_i = N \cdot p_i = 30 \cdot (\tfrac{1}{6}) = 5 \; outcomes \qquad for \; all \; 6 \; cells \; (faces)$$

Actual vs. Expected Outcomes in Each Cell

i	1	2	3	4	5	6
A_i	5	7	3	4	6	5
E_i	5	5	5	5	5	5

Figure 15

Note: Each cell has at least 5 expected outcomes so that no pooling of cells is required. For this Case #1 example, $M = 6$ so that χ^2 has $M - 1 = 6 - 1 = 5$ degrees of freedom (df):

$$\chi^2 \quad = \Sigma \, (A_i - E_i)^2 \, / \, E_i \qquad for \; i = 1 \; to \; 6 \; cells$$

$$= (5 - 5)^2/5 + (7 - 5)^2/5 + (3 - 5)^2/5 + (4 - 5)^2/5 + (6 - 5)^2/5 + (5 - 5)^2/ 5$$

$$= 10/5$$

$$= 2$$

Note: The *smaller* the value of the χ^2 statistic, the better the goodness of fit between the actual and expected data.

From Table 3 with 5 degrees of freedom, the χ^2 95th percentile is as follows:

- $\chi^2_{5,\,.95} = 1.14$

If the χ^2 statistic value computed were *less* than 1.14 , the die could be considered fair (p=1/6 for all faces) with 95% confidence. However, the χ^2 statistic is too large. From Table 3 for 5 df, the χ^2 90th and 75th percentiles are as follows:

- $\chi^2_{5,\,.90} = 1.61$
- $\chi^2_{5,\,.75} = 2.67$

Since the computed statistic $\chi^2 = 2.0 \leq \chi^2_{5, .75} = 2.67$, the die can be considered fair with more than 75% (probably around 85%) confidence.

The hypothesis in Example 15 that the outcome values were generated by a fair die would probably not be rejected (unless high confidence in the die being fair was required).

Example 16 (Case #2): The number of typographical mistakes per page in the first printing of a new book (267 pages) was found to have the following frequency:

Mistakes on a page	0	1	2	3	≥ 4
# Pages	221	34	11	0	1

Test the hypothesis that the number of mistakes on a page is generated by a Poisson distribution.

Answer: In order to determine the expected # outcomes (i. e. pages) in each cell (i.e. number of mistakes on a page), the value of the theoretical Expected Value λ of the Poisson Random Variable X = "Number of mistakes on a page" must be *estimated* using the sample mean $\bar{X}$.

The total number of outcomes (pages) $N = 221 + 34 + 11 + 0 + 1 = 267$. The sample mean $\bar{X}$ (the average number of mistakes on a page) for the sample numerical outcome values is given by:

$$\bar{X} = \Sigma\, x_i\, /\, N$$

$$= [221(0) + 34(1) + 11(2) + 0(3) + 1(4)\ total\ errors]\ /\ [267\ pages]$$

$$= 60\ total\ errors\ /\ 267\ pages$$

$$= 0.225\ errors\ per\ page$$

For a Poisson distribution with $\lambda = 0.225$, a calculator can be used to determine the probability of being in each of the 5 cells (denoting the number of mistakes on a page):

$$P(X = k) = \lambda^k e^{-\lambda}/k! \quad k = 0, 1, 2, 3, 4 \ldots$$

$$P(X = 0\ mistakes) = (0.225)^0 e^{-0.225}/0! = 0.7985$$

$$P(X = 1\ mistake) = (0.225)^1 e^{-0.225}/1! = 0.1796$$

$$P(X = 2\ mistakes) = (0.225)^2 e^{-0.225}/2! = 0.0202$$

$$P(X = 3\ mistakes) = (0.225)^3 e^{-0.225}/3! = 0.00015$$

22

$P(X \geq 4 \text{ mistakes}) = \sim 0.0002$

Note: The complement of $\{X \geq 4\}$ is $\{X=0, X=1, X=2, X=3\}$. Hence, $P(X \geq 4)$ $= 1 - P(X = 0) - P(X = 1) - P(X = 2) - P(X = 3)$.

The *expected* number in each cell (corresponding to 0, 1, 2, 3, ≥ 4 mistakes per page) for the $N = 267$ outcomes (pages) is given by:

$$E_0 = N \cdot p_0 = (267) \cdot (0.7985) = \quad 213.20 \; Pages$$
$$E_1 = N \cdot p_1 = (267) \cdot (0.1796) = \quad 47.95 \;\; Pages$$
$$E_2 = N \cdot p_2 = (267) \cdot (0.0202) = \quad 5.39 \;\;\; Pages$$
$$E_3 = N \cdot p_3 = (267) \cdot (0.00015) = \;\; 0.40 \;\;\; Pages$$
$$E_{\geq 4} = N \cdot p_{\geq 4} = (267) \cdot (0.0002) = \underline{0.06} \;\;\; Pages$$
$$\qquad\qquad\qquad\qquad\qquad\qquad 267 \quad Pages \; (= N \; outcomes)$$

The comparison between the actual and the expected number of pages with each number of mistakes on a page is given in Figure 16.

Actual vs. Expected Outcomes in Each Cell

Mistakes on a Page	0	1	2	3	≥ 4
# Actual Pages	221	34	11	0	1
# Expected Pages	213.2	47.95	5.39	0.40	0.06

Figure 16

The last three cells must be *pooled* to increase the # expected outcomes (pages) to be at least 5 in each cell. Contiguous cells (e.g. 2, 3, and ≥ 4) are typically pooled to maximize # cells remaining $(M = 3)$. Since the theoretical (unknown) Expected Value λ of the Poisson distribution had to be estimated, $k = 1$ for Example 16. The data for the pooled cells are given in Figure 17.

Actual vs. Expected Outcomes For Pooled Cell

Mistakes on a Page	0	1	≥ 2
# Actual Pages	221	34	12
# Expected Pages	213.2	47.95	5.85

Figure 17

The χ^2 statistic is given by:

$$\chi^2 = \Sigma(A_i - E_i)^2 / E_i$$

$$= (221 - 213.2)^2 / 213.2 + (34 - 47.95)^2 / 47.95 + (12 - 5.85)^2 / 5.85$$

$$= 10.81$$

For this Case #2 example, the χ^2 statistic with $M - k - 1 = 3 - 1 - 1 = 1$ degree of freedom has a computed value of 10.81.

From Table 3, the following are χ^2 0.005 and 0.001 percentiles for 1 df:

- $\chi^2_{1,.005} = 7.88$

- $\chi^2_{1,.001} = 10.83$

Since the χ^2 statistic 10.81 is slightly *less* than 10.83 = $\chi^2_{1,.001}$, there is only slightly *greater* than 0.1% confidence that the number of pages of the new book with the actual number of mistakes per page is generated by a Poisson distribution. One would no doubt reject this hypothesis.

1.12 Goodness of Fit Test (F-Statistic)

To forecast customer demand at a business for the next five years, it seems reasonable to observe customer demand for the last 10 years, fit a line to this demand vs. time data, and then extend that line (i.e. trend) over the next 5 years (see Figure 18):

Forecasting Future Demand

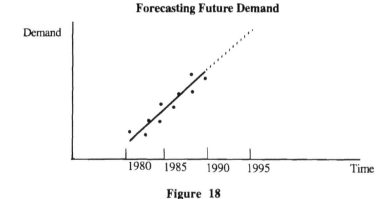

Figure 18

24

Statistically, the future forecasting power (accuracy) of this approach is determined by how well the past ten customer demand data points fit the line. The F-statistic can be used to determine how well a set of data points fits a linear model.

Total deviation: The data is hypothesized to increase linearly over time and will contain deviation from the mean value of the data. The difference between the actual customer demand data points and the *mean* value of the data is called the total deviation in the model (see Figure 19). The total deviation has two components: the deviation *explained* by the linear trend and the deviation *unexplained* (due to randomness) by the linear trend.

Total Deviation in the Data

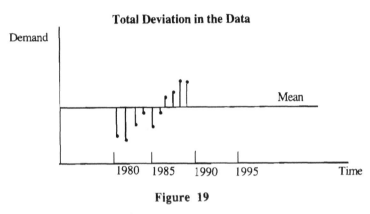

Figure 19

Explained deviation: The difference between the forecasted customer demand data points *(on the line fitted to the data points)* and the mean value of the data (see Figure 20) is the explained deviation (due to linear trend over time) in the data.

Explained Deviation in the Data

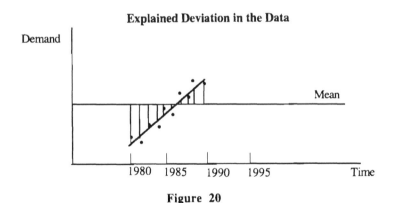

Figure 20

Unexplained deviation: The difference between the actual customer demand data points and the forecasted customer demand data points (on the line fitted to the data points) is the unexplained deviation (due to randomness) in the data (see Figure 21). If all the actual customer demand data points were to lie exactly on the forecasted line, the unexplained deviation would be zero (and the line would explain the data perfectly). To the extent that

25

the customer demand data points do not lie on the forecasted line, the amount of the total deviation in the data that is unexplained increases and the remaining amount of total deviation in the data that is explained decreases.

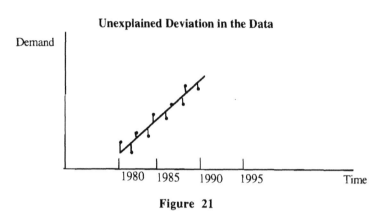

Unexplained Deviation in the Data

Figure 21

Note: The following definitions and relationships between the total deviation, the explained deviation, and the unexplained deviation have been given:

$$Total\ Deviation\ = (Actual\ Data - Mean\ Value)$$
$$Explained\ Deviation\ = (Forecasted\ Data - Mean\ Value)$$
$$Unexplained\ Deviation\ = (Actual\ Data - Forecasted\ Data)$$

$$\Rightarrow Total\ Deviation\ = Explained\ Deviation + Unexplained\ Deviation$$

The F-Statistic is defined as the ratio of the explained *variance* to the unexplained *variance*:

F= [Explained Deviation / 1] / [Unexplained Deviation / (N-2)]

= Explained Variance / Unexplained Variance

where the F-Statistic follows the F Distribution with $v_1 = 1$ degrees of freedom associated with the explained deviation and $v_2 = N - 2$ degrees of freedom associated with the unexplained deviation (where N is the # customer demand data points).

Note: The explained and unexplained variances are determined by dividing the explained and unexplained deviations, respectively, by their applicable number of degrees of freedom (see Section 1.10). Overall, there are N-1 degrees of freedom for the N customer demand data points (the mean being calculated and known accounts for a loss of 1 degree of freedom in the data). The explained deviation has 1 degree of freedom since there is 1 independent variable (time) in this linear regression (demand vs. time). The unexplained deviation is left with the remaining N-2 degree of freedom in the data.

There are detailed formulas (see Section 6.3) to compute the F-statistic using the N customer demand data point values. Computer statistical routines typically provide the value of the F-statistic. The treatment of the F-statistic in this chapter is meant to convey its significance rather than the mechanics of its computation.

Assume for the N = 10 customer demand data points in Figure 18 that the value of the F-statistic (with $v_1 = 1$ and $v_2 = N - 2 = 8$ degrees of freedom) has been provided by a statistical routine to be 9.25.

Note: A *larger* F-statistic denotes a better the fit of the data to the linear model. More of the total variance in the model is explained variance (the numerator of the F-statistic) and less of the total variance is unexplained variance (the denominator of the F-statistic).

To find the 95th and 99th percentiles of the F distribution with 1 and 8 degrees of freedom, use Table 2a and Table 2b, respectively:

- $F_{1, 8, .95} = 5.32$
- $F_{1, 8, .99} = 11.26$

Since the computed value of the F-statistic = 9.25 is greater than $F_{1, 8, .95} = 5.32$ but less than $F_{1, 8, .99} = 11.26$, the data fits a line with better than 95% confidence (but not quite 99% confidence). The hypothesis that the data fits a linear model would no doubt be accepted.

The line fitted to the last 10 years of customer demand can now be extended *with statistical confidence* to forecast demand for the next 5 years.

Note: Overextending the forecast into the future is a very common error in forecasting. It is best to use the past data to forecast the *near* future since it is likely that the assumptions that generated the past demand will not continue far into the future.

1.13 Student t Distribution

When the Expected Value μ and standard deviation σ of a Random Variable (RV) are unknown, the sample mean $\bar{X}$ and sample standard deviation s are used to compute confidence intervals based on N numerical outcomes $\{x_1, x_2, \ldots, x_N\}$.

Consider the estimates for the sample mean and sample standard deviation of an RV (see Section 1.10) which have been derived from N numerical outcomes $\{x_1, x_2, \ldots, x_N\}$:

$\bar{X} = \Sigma x_i / N$

$s = [\Sigma (x_i - \bar{X})^2 / (N - 1)]^{1/2}$

If N>30, the Standard Normal distribution Z~N(0,1) is used to estimate confidence interval (e.g. 95%) around the sample mean $\bar{X}$ for a Random Variable X (See Section 1.7):

$(\bar{X} + Z_{.025} \cdot s, \bar{X} + Z_{.975} \cdot s)$

If N≤30, the <u>Student t distribution</u> is used to estimate confidence intervals (e.g. 95%) around the sample mean $\bar{X}$ of a Random Variable X:

$$(\bar{X} + t_{df,\ 025} \cdot s,\ \bar{X} + t_{df,\ 975} \cdot s)$$

where df is the # degrees of freedom = N - 1

<u>Note:</u> The properties of the Standard Normal distribution and the Student t distribution compare as follows (see Figures 22, 23, and 24):

• Both the Standard Normal distribution and the Student t distribution are continuous distributions which range from -∞ to +∞.

• Both the Standard Normal distribution and Student t distribution have a 50th percentile at the value zero.

• Both the Standard Normal distribution and Student t distribution have a peak at their mean value zero.

• Both the Standard Normal distribution and the Student t distribution are symmetric about the value zero.

• There is only one Standard Normal N(0,1) distribution curve. However, for each value for the degrees of freedom in the data, there is a *different* Student t distribution. All Student t distributions (like the Standard Normal distribution) have an area under the curve of 1.

• When the # degrees of freedom is low (see Figure 22), the Student t distribution is much more dispersed than the Standard Normal distribution and its peak at zero is significantly lower than the peak of the Standard Normal distribution at zero.

Student t (10 df) vs. Standard Normal Distribution

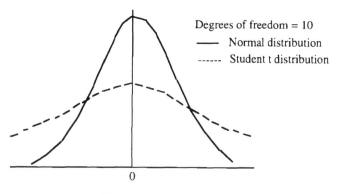

Degrees of freedom = 10
——— Normal distribution
------ Student t distribution

0

Figure 22

• As the number of # degrees of freedom increases (see Figure 23), the Student t distribution becomes less dispersed about its mean of zero and the peak value at the mean of zero increases. As the number of degrees of freedom approaches 30, the curves (and hence the percentiles) associated with the Student t and Standard Normal distributions become almost identical (see Figure 24).

Student t (20 df) vs. Standard Normal Distribution

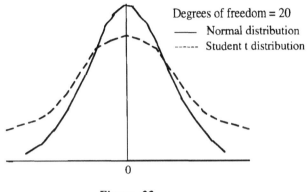

Figure 23

Student t (30 df) vs. Standard Normal Distribution

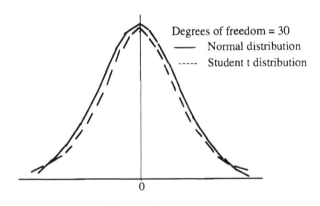

Figure 24

Determination of confidence intervals for small samples ($N \leq 30$) requires that percentiles of the Student t distribution be used. Let the sample size be $N=7$ (and thus $N \leq 30$). Then Table 6 provides the 97.5th percentile (2.45) for the Student t distribution with 6 degrees of freedom. The symmetric property of the Student t distribution is used to determine the 2.5th percentile (-2.45) for 6 degrees of freedom. The 2.5th and 97.5th percentiles are equal but opposite in sign.

- $t_{6,.975} = +2.45$
- $t_{6,.025} = -2.45$

Notes:

• A 95% confidence interval around the sample mean $\bar{X}$ of a Random Variable X for a sample size N=7 is given by:

$$(\bar{X} + t_{df,.025} \cdot s, \ \bar{X} + t_{df,.975} \cdot s)$$

$$(\bar{X} + t_{6,.025} \cdot s, \ \bar{X} + t_{6,.975} \cdot s)$$

$$(\bar{X} - 2.45 \cdot s, \ \bar{X} + 2.45 \cdot s)$$

• If the sample size used to compute $\bar{X}$ and s were greater than 30, the 95% confidence interval around the sample mean $\bar{X}$ of a Random Variable X would be given by:

$$(\bar{X} + Z_{.025} \cdot s, \ \bar{X} + Z_{.975} \cdot s)$$

$$(\bar{X} - 1.96 \cdot s, \ \bar{X} + 1.96 \cdot s)$$

• Confidence intervals of any given percentile will be larger when the Student t distribution must be used (N≤30) than when the Standard Normal distribution can be used (N>30). This is expected due the greater uncertainty associated with the computation of the sample mean and sample standard deviation when the sample size is small.

1.14 Conditional Probability

Suppose that E and F are two Events and that $P(E) > 0$. Then the "Law of Conditional Probability" may be used to compute the probability of Event F given Event E, written as $P(F|E)$, as follows:

$$P(F|E) = \frac{P(E \cap F)}{P(E)}$$

Note: If $P(E) = 0$, the conditional probability of Event F given Event E has *no meaning*.

Example 17: Two fair coins are each flipped once in sequence. Determine the conditional probability that both coins will show a "Heads" after the first coin shows a "Heads".

Answer: Define Event E = "Heads on the flip of the first coin" and Event F = "Heads on the flip of both coins". Example 17 asks for the determination of $P(F|E)$.

For the flips of two fair coins, the four possible outcomes in the sample space are given by:

$O_1 = \{H, H\}$
$O_2 = \{H, T\}$
$O_3 = \{T, H\}$
$O_4 = \{T, T\}$

Each outcome is equally likely and has a probability of 1/4. The Event $E = \{O_1, O_2\}$ and thus the $P(E) = 1/2$. The Event $F = \{O_1\}$ and thus $P(F) = 1/4$. Hence, the Event $E \cap F = \{O_1\}$ and thus $P(E \cap F) = 1/4$.

Then the conditional probability that both coins will show a "Heads" after the first coin shows a "Heads" is given by:

$$P(F|E) = \frac{P(E \cap F)}{P(E)}$$

$$= \frac{\frac{1}{4}}{\frac{1}{2}}$$

$$= \frac{1}{2}$$

Note: Another way to look at this problem of conditional probability is to use the property that probability is proportion in the long run. Suppose this process of flipping two coins is performed 100 times (see Figure 25). On the first flip, one should expect 50 Heads (H_1) and 50 Tails (T_1). For the 50 Heads on the first flip (50 H_1), one should expect 25 Heads on the second flip (25 $H_1 \cap H_2$) and 25 Tails on the second flip (25 $H_1 \cap T_2$). For the 50 Tails on the first flip (50 T_1), one should expect 25 Heads on the second flip (25 $T_1 \cap H_2$) and 25 Tails on the second flip (25 $T_1 \cap T_2$). There are 50 trials with Heads on the first flip (H_1). The proportion that have Heads on the second flip ($H_1 \cap H_2$) is 25/50 or 0.50.

Conditional Probability Using Probability as a Proportion

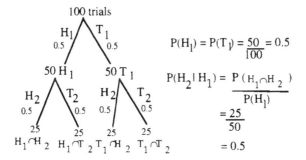

Figure 25: tree showing:

100 trials

H_1 (0.5) / T_1 (0.5)

50 H_1 : H_2 (0.5) / T_2 (0.5)
50 T_1 : H_2 (0.5) / T_2 (0.5)

25 $H_1 \cap H_2$ 25 $H_1 \cap T_2$ 25 $T_1 \cap H_2$ 25 $T_1 \cap T_2$

$P(H_1) = P(T_1) = \dfrac{50}{100} = 0.5$

$P(H_2|H_1) = \dfrac{P(H_1 \cap H_2)}{P(H_1)}$

$= \dfrac{25}{50}$

$= 0.5$

Figure 25

Example 18: Consider a group which is 40% male (M) and 60% female (F). Suppose that 50% of the males and 30% of the females smoke (S). Find the conditional probability that a person in the group is male given that the person smokes.

Answer: The following probabilities are given in the Example 18 description:

$$P(M) = 0.40$$
$$P(F) = 0.60$$
$$P(S|M) = 0.50$$
$$P(S|F) = 0.30$$

Example 18 asks for the conditional probability that the person is male given the person smokes, $P(M|S)$, which is given by:

$$P(M|S) = \frac{P(M \cap S)}{P(S)}$$

Neither the probability $P(M \cap S)$ or $P(S)$ is given in the problem description. However, applying the definition of conditional probability and using the probabilities provided in the description for Example 18 gives the following:

$$P(S|M) = \frac{P(M \cap S)}{P(M)}$$

$$P(M \cap S) = P(S|M) \cdot P(M)$$

$$= (0.50) \cdot (0.40)$$

$$= 0.20$$

The set of all smokers is the set of all smokers who are male or who are female. Mathematically, this set and its probability can be written as follows:

$$S = (M \cup F) \cap S$$

$$S = (M \cap S) \cup (F \cap S)$$

$$P(S) = P(M \cap S) + P(F \cap S)$$

The $P(M \cap S)$ has already been computed. The $P(F \cap S)$ can also be computed by applying the definition of conditional probability and the probabilities provided in the description for Example 18:

$$P(S|F) = \frac{P(F \cap S)}{P(F)}$$

$$P(F \cap S) = P(S|F) \cdot P(F)$$

)

$$= (0.30) \cdot (0.60)$$

$$= 0.18$$

Hence, the probability of being a smoker, $P(S)$, is given by:

$$P(S) = P(M \cap S) + P(F \cap S)$$

$$= 0.20 + 0.18$$

$$= 0.38$$

The probability of the person being male given the person smokes, P(M|S), is then given by:

$$P(M|S) = \frac{P(M \cap S)}{P(S)}$$

$$= \frac{0.20}{0.38}$$

$$= 0.526$$

Note: Another way to look at Example 18 is to consider a group of 100 people (see Figure 26) of which 40 are male (M) and 60 are female (F). Of the 40 males, 50% or 20 smoke (M∩S) and 50% or 20 don't smoke (M∩NS). Of the 60 females, 30% or 18 smoke (F∩S) and 70% or 42 don't smoke (F∩NS). There are a total of 38 smokers. The proportion (probability) of the 38 smokers (S) that are male (M∩S) is 20/38 or 0.526.

Conditional Probability Using Probability as a Proportion

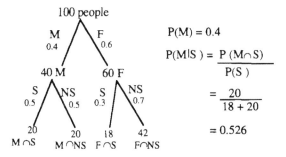

Figure 26

1.15 Bayes' Theorem

Consider the case where a decision maker has identified a set of outcome states E_1, E_2, $\cdots$, E_N with the *a priori* (initial) probabilities $P(E_1)$, $P(E_2)$, $\cdots$, $P(E_N)$ for the outcome states.

When an Event F occurs, the decision maker is able to use the *information* (the occurrence of the event) to update the a priori probabilities to a set of *a posteriori* (updated) probabilities $P(E_1|F)$, $P(E_2|F)$, $\cdots$, $P(E_N|F)$ for the outcome states.

Notes:

• The states E_1, E_2, $\cdots$, E_N must be such that all states are mutually exclusive. That is, $E_i \cap E_j = \phi$ so that $P(E_i \cap E_j) = P(\phi) = 0$ for all states i and j.

• The states E_1, E_2, $\cdots$, E_N must be such that the union of all states equals the sample space U. That is, $E_1 \cup E_2 \cup \cdots \cup E_N = U$ so that $P(E_1 \cup E_2 \cup \cdots \cup E_N) = P(U) = 1$.

Bayes' Theorem uses the information provided in the occurrence of an Event F to update the N a priori probabilities $P(E_i)$ into N a posteriori probabilities $P(E_i|F)$ as follows:

$$P(E_i \mid F) = \frac{P(E_i \cap F)}{P(F)}$$

where using the definition of condition probability:

$$P(E_i \cap F) = P(F \mid E_i) \cdot P(E_i)$$

and using set theory:

$$F = F \cap (E_1 \cup E_2 \cup \cdots \cup E_N)$$
$$F = (F \cap E_1) \cup (F \cap E_2) \cup \cdots \cup (F \cap E_N)$$
$$P(F) = P(F \cap E_1) + P(F \cap E_2) + \cdots + P(F \cap E_N)$$

The probability P(F) can be rewritten using the definition of condition probability:

$$P(F) = P(F|E_1)P(E_1) + P(F|E_2)P(E_2) + \cdots + P(F|E_N)P(E_N)$$

Then, the full form of Bayes' Theorem for updating the a prior probabilities is given by:

$$P(E_i|F) = \frac{P(F|E_i)P(E_i)}{P(F|E_1)P(E_1) + P(F|E_2)P(E_2) + \cdots + P(F|E_N)P(E_N)}$$

for all states E_i, $i = 1, 2, \cdots, N$

Example 19: A given disease occurs in 1 out of 1000 people in the general population. A test for the disease produces a positive result for a person with the disease with probability 0.99. The test also produces a positive result for a healthy person with probability 0.05 (false positive probability 0.05). Given the information in the event that a person produces a positive test result, update the probability this person actually has the disease.

Answer: Without the information inherent in the test result, the *a priori* probabilities of a random person having the disease (D) and of being healthy (H) are given in the example:

$$P(D) = 0.001$$
$$P(H) = 0.999$$

Example 19 asks for the updated probability of having the disease $P(D|TP)$ once a person tests positive for the disease.

When a person has the disease or is healthy, the test result is positive (TP) with the probabilities given in the description for Example 19:

$$P(TP|D) = 0.99$$
$$P(TP|H) = 0.05$$

Given the event that the person's test for the disease is positive (TP), the a priori probability the person has the disease P(D)=0.001 is updated to P(D|TP) using Bayes' Theorem as follows:

$$P(D \mid TP) = \frac{P(D \cap TP)}{P(TP)}$$

$$= \frac{P(D \cap TP)}{P(D \cap TP) + P(H \cap TP)}$$

$$= \frac{P(TP|D)P(D)}{P(TP|D)P(D) + P(TP|H)P(H)}$$

$$= \frac{(0.99)(0.001)}{(0.99)(0.001) + (0.05)(0.999)}$$

$$= \frac{0.00099}{0.05094}$$

$$= 0.0194$$

and hence,

$P(H \mid TP) = 0.9806$

Given the event that the test result is positive, the *a priori* probabilities for having the disease and being healthy ($P(D)=0.001$ and $P(H)=0.999$, respectively) are updated to the *a posteriori* probabilities of $P(D|TP) = 0.0194$ and $P(H|TP) = 0.9806$, respectively.

Note: Another way to look at Example 19 is to consider a group of 100,000 people who take the test for the disease (see Figure 27). In reality, 100 people (1 out of each 1000) will have the disease (D) and 99,900 (999 out of 1000) will be healthy (H). Of the 100 people with the disease, 99 people (99%) will test positive (D∩TP) and only 1 will test negative (D∩TN). Of the 99,900 people who are healthy, 4,995 people (5%) will test positive (H∩TP) and 94,905 will test negative (D∩TN). There are a total of 4,095 + 99 = 5,094 people who test positive (P). Of the 5,094 people who test positive (P), only a small proportion (probability) 99/5094 = 0.0194 actually have the disease (P∩D).

The decision maker should be skeptical whether a person with a positive test result actually has the disease and should perform further tests before accepting the result. Even though the test seems quite reliable (99% for those with the disease), the overwhelming fraction of the population does not have this disease. The preponderance of healthy people leads to many false positives. In fact, over 98% of the positive test results occur for healthy people.

Conditional Probability Using Probability as a Proportion

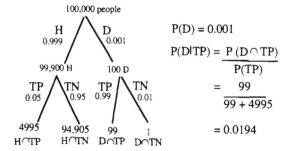

Figure 27

Exercise 1

Probability & Statistics

Problem #1 (Simple Probability)

A fair die is rolled twice.

a) What are the possible outcomes in the sample space?

b) What is the probability of getting "doubles" (the same number on both rolls)?

c) What is the probability of throwing a "7" as the sum of the two throws?

d) What is the probability of getting "doubles" or a "7" as the sum of the two throws?

e) What is the probability of getting "doubles" or "less than 4" as the sum?

f) What is the probability of getting "more than 3" as the sum?

Problem #2 (Expected Value)

Consider the following game. You win $30 dollars is the sum of the two throws of a fair die is less than or equal to 3. You win nothing if the sum of the two throws is greater than 3.

a) What is the Expected Value of this game?

b) Would you pay $2 to play this game? Would you pay $3?

c) What is the Standard Deviation of this game?

Problem #3 (Normal Distribution)

Consider two Normally distributed Random Variables $X_1 \sim N(5, 4)$ and $X_2 \sim N(10, 9)$.

a) Qualitatively, compare the locations and shapes of these Normal Random Variables based on their Means and Standard Deviations.

b) For the Random Variable X_1 , find $P(X_1 < 10)$.

c) Find a 95% confidence interval around the mean for X_2 .

Problem #4 (Poisson Distribution)

Arrivals at a bank are Poisson distributed with a mean rate (λ) of 10 customers per hour.

a) Find the probability of 8 customer arrivals in an hour.

b) Find the probability of 9 to 11 customer arrivals, inclusive, in an hour.

Problem #5 (Exponential Distribution)

The service time in an airport ticket line is exponentially distributed with a mean ($1/\mu$) of 2 hours per customer.

a) Find the probability that the customer service time is less than 90 minutes.

b) Find the probability that the customer service time is between 30 minutes and 90 minutes.

Problem #6 (Statistics)

Consider the outcome values for the Random Variable $X = \{2, 6, 3, 7, 4, 2, 3, 5\}$.

a) Estimate the Mean of the Random Variable X.

b) Estimate the Standard Deviation of the Random Variable X.

Problem #7 (Chi-Square Test)

Consider an actual outcome stream (A_i) grouped into the following 7 cells. The expected outcomes (E_i) in each cell generated by an Exponential distribution are provided (no parameter needs to be estimated to determine the expected outcomes in each cell).

A_i	2	10	2	4	2	6	9
E_i	4	3	8	2	4	8	6

a) Determine the Chi-Square χ^2 statistic for this actual outcome stream.

b) What is the confidence that the actual outcome stream was generated by the Exponential distribution?

Problem #8 (F-Distribution)

Consider the F distribution with v_1 and v_2 degrees of freedom. Using Tables 2a and 2b, find:

a) $F_{1, 10, 99}$ and $F_{1, 20, 99}$.

b) $F_{1, 10, 95}$ and $F_{1, 20, 95}$

Problem #9 (Conditional Probability)

Two fair coins are flipped. Determine the conditional probability that both flips show "Heads" given that at least one of them shows a "Heads".

Problem #10 (Bayes' Theorem)

Cars at an automobile plant are equally likely to be manufactured on Monday, Tuesday, Wednesday, Thursday, or Friday. Cars made on Monday have a 4% chance of being a lemon; cars made on Tuesday, Wednesday, and Thursday have a 1% chance of being a lemon; and cars made on Friday have a 2% chance of being a lemon. Given that a car is purchased and it turns out to be a lemon, determine the probability that the car was manufactured on a Monday.

Chapter 1

Probability & Statistics Additional Solved Problems

Problem #1 (Simple Probability)

Benny Six-Pack decides to buy a $1 lottery ticket to this week's Pick 6 lottery. Each week the state lottery office chooses six numbers randomly from the numbers 1 to 54 (none of the six numbers chosen may be repeated). A winning lottery ticket must have the exact six numbers chosen randomly by the lottery office.

a) What is the number of different possible lottery tickets Benny may choose?

Answer: The first of the six numbers can be any of the 54 possible numbers. The second of the six numbers is limited to the remaining 53 numbers (in order not to repeat the first number chosen), and so on. The number of different possible Pick 6 lottery tickets is given by:

$$\underset{1^{st}\ \#}{\frac{54}{}} \bullet \underset{2^{nd}\ \#}{\frac{53}{}} \bullet \underset{3^{rd}\ \#}{\frac{52}{}} \bullet \underset{4^{th}\ \#}{\frac{51}{}} \bullet \underset{5^{th}\ \#}{\frac{50}{}} \bullet \underset{6^{th}\ \#}{\frac{49}{}}$$

The total number of different possible tickets for the Pick 6 lottery is $54 \bullet 53 \bullet 52 \bullet 51 \bullet 50 \bullet 49 = 18.6$ Billion!

b) Benny chooses six different numbers based on his and his wife's birthdays. What is the probability that Benny has chosen a winning ticket.

Answer: Each of the 18.6 Billion different possible lottery tickets has an equal chance of being the winning ticket. Hence, the exact choice of the six numbers does not affect Benny's chance of winning the lottery since each of the 18.6 Billion choices of six different numbers is equally likely to win. The probability that Benny will buy a winning lottery ticket is then:

$$P(Win) = \frac{1}{18,600,000,000} = 0.000000000053763$$

<u>Note:</u> While the probability of Benny winning the lottery is the same regardless of the six numbers chosen, the amount Benny wins may be affected by his choice of numbers. Many lottery players do as Benny did and choose numbers based on birthdays, anniversaries, etc. Since numbers associated with days and months tend to be 31 or less, lottery tickets chosen by the state lottery office with lower numbers (below 31) tend to have multiple winners more often than those lottery tickets with some higher numbers (above 31). Thus, Benny would be less likely to have to share the lottery money if he chooses lottery tickets with higher numbers included.

c) Benny has his heart set on being a millionaire before he dies and decides to use his life savings to buy 10,000 different lottery tickets. What is the probability that Benny will buy a winning lottery ticket?

Answer: If Benny buys 10,000 different lottery tickets, then 10,000 out of the 18.6 Billion different possible lottery tickets that can be chosen will result in Benny winning the lottery. The probability of Benny winning the lottery is then:

$$P(Win) = \frac{10,000}{18,600,000,000} = 0.00000053763$$

Benny is 10,000 times as likely to win the lottery now than when he bought only 1 ticket. However, the probability of winning the lottery is still small (i.e. 10,000 times a very small number is still small). This strategy may not be one of Benny's brighter ideas.

Problem #2 (Normal Distribution)

At the college Tic Toc Tech, grades on the Midterm for the course Clock Watching are distributed according to a Normal distribution with an Expected Value of 67 points with a standard deviation of 12 points.

a) What percentage of the class will score below 60 points?

Answer: Let the Random Variable X denote the scores on the Clock Watching Midterm. Then X is Normally distributed with Expected Value $\mu=67$ and variance $\sigma^2=144$ (standard deviation $\sigma=12$):

$$X \sim N(67,144)$$

A cumulative distribution lookup table does not exist for a Normal Random Variable with expected value 67 and standard deviation 12. The only lookup table which does exist is for a Standarnd Normal Random Variable $Z\sim N(0,1)$ with Expaected Value 0 and standard deviation 1. Hence, one must transform the Random Variable X into the Standard Normal Variable Z. This is done by subtracting 67 and dividing by 12:

$$P(X < 60) = P(\frac{X - 67}{12} < \frac{60 - 67}{12})$$

$$= P(Z < \frac{60 - 67}{12})$$

$$= P(Z < -0.5833)$$

$$= 0.28 \hspace{3cm} (Table\ 1b)$$

b) What is the 90th percentile for the grades (i. e. the grade such that 90% of the class grades are at or below this grade)?

Answer: The 90th percentile is determined by finding the grade a such that:

$$P(X \leq a) = 0.90$$

Again, there is no cumulative distribution table for a Random Variable with a Normal distribution with Expected Value 60 and standard deviation 12. The Random Variable X is transformed into the Standard Normal Random Variable Z in order to solve for the value a:

$$P(\frac{X-67}{12} \le \frac{a-67}{12}) = 0.90$$

$$P(Z \le \frac{a-67}{12}) = 0.90$$

$$P(Z \le 1.28) = 0.90 \qquad\qquad (Table\ 1b)$$

$$\Rightarrow \frac{a-67}{12} = 1.28$$

$$\Rightarrow a \cong 82$$

c) Find a 90% confidence interval for the scores on the Midterm in Clock Watching at Tic Toc Tech.

Answer: The 90% confidence interval (around the Expected Value 67) is determined by finding the grades a and b such that:

$$P(a \le X \le b) = 0.90$$

Again, no cumulative distribution table exists for a Random Variable with a Normal distribution with Expected Value 60 and standard deviation 12. The Random Variable X is transformed into the Standard Normal Random Variable Z in order to solve for the values a and b:

$$P(\frac{a-67}{12} \le \frac{X-67}{12} \le \frac{b-67}{12}) = 0.90$$

$$(1)\ P(\frac{a-67}{12} \le Z \le \frac{b-67}{12}) = 0.90$$

By definition, the Standard Normal Random Variable $Z \sim N(0,1)$:

$$(2)\ P(Z_{.05} \le Z \le Z_{.95}) = 0.90$$

Equations (1) and (2) imply that:

$$\frac{a-67}{12} = Z_{.05} = -1.64 \qquad\qquad \Rightarrow a \cong 47$$

$$\frac{b-67}{12} = Z_{.95} = 1.64 \qquad\qquad \Rightarrow b \cong 87$$

Since for a Random Variable X with Expected Value 67 and standard deviation 12,

$$P(47 \le X \le 87) \cong 0.90,$$

a 90% confidence interval for the grades on the Midterm is approximately 47 points to 87 points (note the symmetry around the Expected Value of 67 points).

Problem #3 (Chi-Square Test)

The back of a statistics workbook contains a Random Number table with a total of 250 digits consisting of 0's , 1's , ... and 9's. It is expected that any Random Number table will contain an equal number of each of the 10 possible digits. Consider the following comparison of the actual vs. expected number of each of the 10 digits in the 250 digit Random Number table:

	0	1	2	3	4	5	6	7	8	9
Actual Number	36	20	30	35	20	17	31	29	18	14
Expected Number	25	25	25	25	25	25	25	25	25	25

Test the hypothesis that the table at the back of the statistic workbook provides Random Numbers.

Answer: No pooling of cell needs be done since the expected number of occurrences in each cell in greater than or equal to 5. Also, no parameter was estimated in order to determine the expected number in each cell. The hypothesis that the digits are random and hence equally likely (probability = 1/10) was given in the problem statement. The Chi-Square statistic is computed as follows:

$$\chi^2 \quad = \Sigma \, (A_i - E_i)^2 \, / \, E_i \qquad for \;\; i = 1 \; to \; 10 \; cells$$

$$= (36 - 25)^2/25 \; + \; (20 - 25)^2/25 \; + \; \cdots \; + \; (14 - 25)^2/25$$

$$= 23.3$$

For the above comparison table, the number of cells M=10 so that χ^2 has 10 - 1 = 9 degrees of freedom. From Table 3 for 9 degrees of freedom:

- $\chi^2_{9, .005} = 23.60$
- $\chi^2_{5, .010} = 21.67$

Since the χ^2 statistic 23.3 is less than 23.60 = $\chi^2_{9, .005}$, there is only 0.5% confidence that this table provides Random Numbers. One would no doubt reject the hypothesis.

"It is remarkable that a science which began with the consideration of games of chance should have become the most important object of human knowledge."

Laplace, *Theorie Analytique des Probabilites*

2 | Decision Analysis

2.1 Expected Value Decision

Decision analysis selects the *best* action (based on a given criterion) among a set of alternative actions. The components of a decision analysis include:

- *Actions (a set of alternatives available to the decison maker)*
- *Stochastic outcomes of each action (outside the control of the decision maker)*
- *Probabilities of each action – outcome pair*
- *Payoffs (or costs) of each action – outcome pair*

The Expected Value Decision criterion is often used by decision makers to choose the best action in a stochastic (probabilistic) environment. The Expected Value Decision chooses the action which provides the best expected return for the decision maker. When actions by the decision maker results in payoffs, the Expected Value Decision criterion chooses the action with the *maximum expected payoff*. When actions by the decision maker results in costs, the Expected Value Decision criterion chooses the action with the *minimum expected cost*.

Note: The *best* action need not be the *optimal* action for the decision. Decision analysis chooses the optimal action only when the optimal action is among the set of alternative actions defined by the decision maker. Decision analysis does not search for the optimal decision but only evaluates the set of alternative actions which it is given.

Example 1: Formulate the common activity of preparing to go to work on a day when rain is a significant possibility as a decision analysis. Identify a set of alternative actions, stochastic outcomes, probabilities, and payoffs for this activity. Determine the best action using the Expected Value Decision (maximum expected payoff) criterion.

Answer: Each decision maker would formulate this decision analysis at least slightly different based on his own experiences and values. One possible set of *actions* for the decision maker in preparing to go to work when rain is a significant possibility is:

- *Stay Home*
- *Go To Work Without Umbrella*
- *Go To Work With Umbrella*

On a given day, the stochastic *outcomes* (states of weather outside of the control of the decision maker) for these three actions and the *probabilities* for these outcomes might be:

- *Rain*: $P(Rain) = 0.4$
- *No Rain*: $P(No\ Rain) = 0.6$

Note: The same stochastic outcomes (Rain and No Rain) apply to each of the three possible actions identified. In a general decision analysis, each action may have a *different* set of outcomes associated with it. For example, consider the decision whether or not to buy a lottery ticket. If a ticket is purchased, the possible outcomes are Win or Lose. Whereas if no ticket is purchased, the only possible outcome is No Outcome (neither win nor lose).

Assume that each action-outcome pair has the result and *payoff* given in Figure 1.

Action/Outcome Pair Payoffs

Action/Outcome Pair	Result	Payoff
Stay Home / Rain	Get Fired	-100
Stay Home / No Rain	Get Fired	-100
Go Without Umbrella / Rain	Get Wet	-30
Go Without Umbrella / No Rain	Look Smart	+50
Go With Umbrella / Rain	Stay Dry	+30
Go With Umbrella / No Rain	Look Foolish	-10

Figure 1

Notes:

• The implementation of the Expected Value Decision requires that a *numerical value* be given to the result of each action-outcome pair. The assignment of numerical values allows the decision analysis to determine the "best" action *quantitatively*.

• Sometimes there is a *subjective* (based on one's personal values) assignment of numerical values (see Section 2.5) as in Figure 1 rather than *objective* (e.g. based on the cash payments for a game of chance) assignment of numerical values.

• Expected Value Decisions can be formulated either to maximize expected payoff or to minimize expected cost. Decision analyses sometimes have action-outcome pairs which result in a mixture of payoffs *and* costs. When the maximum expected payoff formulation is chosen, any costs are handled as negative payoffs. When the minimum expected cost formulation is chosen, any payoffs are handled as negative costs.

A <u>decision matrix</u> (see Figure 2) is used to present *visually* the actions, outcomes, probabilities, and action-outcome payoffs (or costs) for a decision analysis.

Decision Matrix For Preparing To Go To Work

Outcome Action	Rain (0.4)	No Rain (0.6)
Stay Home	Get Fired (-100)	Get Fired (-100)
Go To Work Without Umbrella	Get Wet (-30)	Look Smart (+50)
Go To Work With Umbrella	Stay Dry (+30)	Look Foolish (-10)

Figure 2

Under the Expected Value Decision criterion, the "best" decision is the action which provides the maximum expected payoff. An action's expected payoff is the summation of the possible payoffs for the action times the probability of each payoff. The expected payoff for the actions Stay Home, Go To Work Without Umbrella, and Go To Work With Umbrella are given by:

$$E(Stay\ Home) = (-100)\cdot(0.4) + (-100)\cdot(0.6) = -100$$

$$E(Go\ To\ Work\ Without\ Umbrella) = (-30)\cdot(0.4) + (+50)\cdot(0.6) = 18$$

$$E(Go\ To\ Work\ With\ Umbrella) = (+30)\cdot(0.4) + (-10)\cdot(0.6) = 6$$

The action Go To Work Without Umbrella is the Expected Value Decision since it provides the maximum expected payoff of 18.

<u>Notes</u>:

• The best action may be quite different if the payoffs associated with action-outcome pairs are changed or the probabilities associated with the stochastic outcomes are changed.

• Since the numerical payoffs are subjective (based on personal values) and the outcome probabilities are forecasts (based on a weather prediction), an analysis should be performed to determine the *sensitivity* of the chosen action to the small changes in the values of the payoffs or outcomes probabilities. If a small change in any payoff or probability changes the best action, this payoff or probability should be examined closely to ensure its accuracy.

• It may happen that the action with the maximum expected payoff *backfires*. For example, one could Go To Work Without Umbrella (the best action in Example 1) and it rains. There is a 40% chance of this outcome. The decision analysis solution only guarantees that given the set of payoffs (one's value system) and the set of outcome probabilities (forecasted weather), the best decision *in the long run* (if this decision were faced many times) is Go To Work Without Umbrella. The *average* payoff of this action is 18.

• Suppose the only possible actions identified in Example 1 were Stay Home and Go To Work With Umbrella. The best action would be Go To Work With Umbrella. In this case, the action To Go To Work Without Umbrella (with a higher expected payoff) was not even among the alternatives. The following limitations hold for the Expected Value Decision:

- An action must be among the alternatives contemplated to be the Expected Value Decision.
- The fact that an action is the Expected Value Decision does not mean that an action with a higher expected payoff (or lower expected cost) does not exist.

2.2 Maximin Decision

It is sometimes the case that the stochastic outcome probabilities are uncertain and can not be assessed to the satisfaction of the decision maker. And yet the decision maker may have to make a decision despite the inability to assess the outcome probabilities satisfactorily.

The Maximin Decision criterion is often used for decisions under uncertainty where the stochastic outcome probabilities can not be satisfactorily assessed. The Maximin Decision criterion chooses the action (among the set of alternatives) where the minimum payoff (for *any* possible outcome for that action) is maximized. That is, the action providing the best worst case payoff is chosen.

Notes:

• Some treatments of decision analysis present the Minimax Decision instead of the Maximin Decision. The Maximin Decision for *payoffs* is equivalent to the Minimax Decision for *costs*. For both these criteria, the criterion is to choose the action which provides the decision maker the best worst case result.

• People buy insurance based on the Minimax Decision criterion. Insurance minimizes the maximum cost to the policy holder due to possible accidents or disasters. If the Expected Value Decision criterion were used, the best action would often be *not* to buy insurance due to the low probability of a major accident or disaster. Insurance companies (with large cash reserves) can suffer occasional large losses. They sell insurance (and make money in the long run) based on the Expected Value Decision criterion.

Example 2: Determine the Maximin Decision (for payoffs) when preparing for work on a day when rain is a significant possibility as given in Example 1.

Answer: Examine the decision matrix for preparing to go to work given in Figure 2 without the probabilities for the stochastic outcomes Rain and No Rain (see Figure 3).

Preparing to Go to Work Decision Matrix

Outcome Action	Rain	No Rain
Stay Home	Get Fired (-100)	Get Fired (-100)
Go To Work **Without Umbrella**	Get Wet (-30)	Look Smart (+50)
Go To Work **With Umbrella**	Stay Dry (+30)	Look Foolish (-10)

Figure 3

48

The minimum payoff associated with any outcome for the three alternative actions are:

- *Stay Home* $\Rightarrow$ -100 (*Rain or No Rain*)
- *Go To Work Without Umbrella* $\Rightarrow$ -30 (*Rain*)
- *Go To Work With Umbrella* $\Rightarrow$ -10 (*No Rain*)

The Maximin Decision is Go To Work With Umbrella since this is the action with the maximum (best) minimum (worst case) payoff. This decision avoids the large negative payoffs (where negative payoffs are costs) associated with the actions Stay Home (Get Fired, Payoff = -100) or Go To Work Without Umbrella (Get Wet, Payoff = -30).

Note: The drawback to the Maximin (or Minimax) Decision is that the probabilities with which outcomes occur are ignored. An action with an outcome providing an undesirable payoff is avoided no matter how small the probability of that outcome. The Maximin Decision typically leads to a quite *conservative* decision (see Example 3).

Example 3: Determine the Maximin Decision for the following game of chance:

Outcome Action	Win (0.99)	Lose (0.01)
Bet $10	Win +$1000	Lose -$10
Don't Bet	$0	$0

Answer: The minimum (worst case) payoff associated with any outcome for the two alternative actions are:

- *Bet* $10 $\Rightarrow$ $-$10 (*Lose*)
- *Bet* $0 $\Rightarrow$ $0

The Maximin Decision is the action Bet $0. This action has the best worst case payoff of $0 and avoids the negative payoff (cost) of -$10. Examine the Expected Value Decision:

$$E(Bet\ \$10) = (+\$1000)\cdot(0.99) + (-\$10)\cdot(0.01) = +\$989.90$$

$$E(Don't\ Bet) = (\$0)\cdot(0.99) + (\$0)\cdot(0.01) = \$0$$

The Expected Value Decision criterion (and logic) strongly point to the action to Bet $10 as the best action. The decision maker needs to be sensitive to the issue of over-conservatism when considering the application of the Maximin Decision criterion. Using the Maximin Decision criterion, an action with a likelihood of a big win was not selected in order to avoid the small probability of losing $10.

Note: Conservatism in decision-making is sometimes appropriate (see Example 4).

Example 4: Determine the best action for the game of chance described by the following decision matrix. Use *either* the Expected Value Decision or the Maximin Decision criterion as appropriate to determine the best action.

Outcome Action	Win (0.9999)	Lose (0.0001)
Bet $10	Win +$200	Lose -$1,000,000
Don't Bet	$0	$0

Answer: The minimum (worst case) payoff for any outcome of the two alternative actions are:

- *Bet* $10 $\Rightarrow$ – $1,000,000 *(Lose)*
- *Don't Bet* $\Rightarrow$ $0

The Maximin Decision is the action Don't Bet. This action has the best worst case payoff of $0 and avoids the large negative payoff (cost) of -$1,000,000. Examine the Expected Value Decision:

$$E(Bet \ \$10) = (+\$200)\cdot(0.9999) + (-\$1,000,000)\cdot(0.0001) = +\$99.98$$
$$E(Don't \ Bet) = (\$0)\cdot(0.9999) + (\$0)\cdot(0.0001) = \$0$$

The action Bet $10 is the Expected Value Decision (maximum expected payoff). However, most decision makers will not risk the loss of $1,000,000 (even though the probability of this happening is small). For most decision makers, the action Don't Bet (the Maximin Decision) is a better decision than the action Bet $10 (the Expected Value Decision).

2.3   Decision Trees

Decision trees are an alternate means of visualizing the decision analysis process. Like the decision matrix, decision trees also include:

- *Actions*
- *Stochastic Outcomes*
- *Probabilities*
- *Payoffs (or Costs)*

Decision matrices and decision trees are both useful for single decisions where each action has the *same* set of possible outcomes and probabilities. Decision trees, however, are a *generalization* of the decision matrix where there need not be the same stochastic outcomes and probabilities for each action. Decision trees have the additional capability of solving *multiple* stage decisions where there may be *different* actions, outcomes, probabilities, and payoffs to consider at each stage of the decision.

Note: Decision trees use the Expected Value Decision criterion (the maximum expected payoff or the minimum expected cost) to determine the best action among the available alternatives. Decision trees are not typically used to implement the Minimax or Maximin decision. For the Maximin or Minimax criteria, the decision maker should determine the action which provides the best worst case payoff or cost, respectively.

Define the following notation for decision trees:

- The diamond symbol "♦" signifies an *action node* and denotes a choice of alternative actions available to the decision maker.

- The circle symbol "o" signifies an *outcome node* and denotes a set of stochastic outcomes (governed by a probability distribution outside the control of the decision maker).

Example 5: (Make vs. Buy Decision Tree). The Pricebusters supermarket chain must decide whether it is less costly (higher profit) to make its own hamburger rolls in-house or to buy the rolls from an outside vendor. The cost to make the hamburger rolls in-house requires a fixed monthly investment of $11,000 and a variable cost of $10 per gross (12 dozen rolls) produced. The cost to order hamburger buns from a vendor is $22 per gross. The demand for hamburger rolls at the supermarket chain has been surveyed monthly over the past two years and been found to conform to the following distribution:

Demand (Gross/Month)	Probability
800	0.30
900	0.50
1000	0.20

Formulate this Make vs. Buy decision as a decision tree and solve for the Expected Value Decision (minimum expected monthly cost).

Answer: The formulation of the decision tree is presented first. Subsequently, the solving of the decision tree (using the Expected Value Decision criterion) will be presented.

Formulation of the Decision Tree

There are two initial actions, Make (M) and Buy (B), which can be chosen by Pricebusters supermarket. These actions are denoted in the decision tree by a "♦" prior to the branches Make and Buy (see Figure 4).

Figure 4

The possible demand outcomes for the Make or Buy actions are identical. The demand can be 800, 900, or 1000 gross/month (stochastic outcomes outside of the control of the decision maker). The outcomes for both the Make and Buy actions are specified in the decision tree by an outcome node ("o") prior to the three possible demands (see Figure 5).

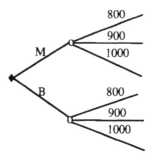

Figure 5

The probability of each action-outcome pair is then added beside each of the three stochastic outcomes for both the Make and Buy decisions (see Figure 6).

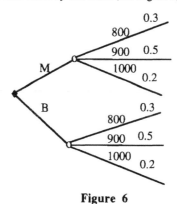

Figure 6

Finally, the cost of each action-outcome pair is added to complete the decision tree. For example, assume Pricebusters supermarket decides to Make the rolls and the demand is 800 gross/month (top outcome branch for the Make action node). Then the cost per month to Pricebusters supermarket is given by the addition of the $11,000 per month fixed cost plus $10 for each of the 800 gross/month required to meet demand:

$$Cost\ (Make,\ Demand = 800) = \$11,000 + \$10 \cdot (800) = \$19,000 / month$$

Likewise, assume Pricebusters supermarket decides to Buy rolls and the demand is 800 gross/month (top outcome branch for the Buy action node). Then the cost per month to Pricebusters supermarket is given by the $22 charged by the outside vendor for each of the 800 gross/month required to meet demand:

$$Cost\ (Buy,\ Demand = 800) = \$22 \cdot (800) = \$17,600 / month$$

The cost for each action-outcome pair is computed and placed next to each of the six branches at the right end of the decision tree (see Figure 7) in order to complete the formulation of the Make vs. Buy decision tree.

Make vs. Buy Decision Tree

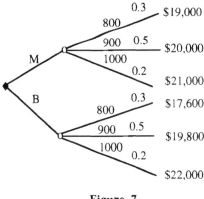

Figure 7

Notes:

• The complete decision tree in Figure 7 displays visually all the elements of the Make vs. Buy decision analysis: actions, outcomes, probabilities, and costs.

• The decision tree is read *forwards* (from left to right). The leftmost "♦" symbol indicates that there is an initial decision to Make or Buy. The "o" symbol after the Make action indicates possible stochastic outcomes (in this case three possible demands with the given probabilities). The "o" symbol after the Buy action indicates possible stochastic outcomes (in this case the same three possible demands with the same probabilities). The costs for the six action-outcome pairs are given at the end of the six decision tree branches.

Solving the Decision Tree

The decision tree is solved by working *backwards* toward the initial action node. At each iteration of the solution process, the rightmost set of stochastic outcome nodes is assigned the *single* value equal to its expected value.

The expected value of the cost per month for the outcome node associated with the Make action is defined as the summation of the possible costs for the Make action times the probability of each cost:

$$E(Make) = (\$19,000) \cdot (0.3) + (\$20,000) \cdot (0.5) + (\$21,000) \cdot (0.2)$$

$$= \$19,900 \,/\, month$$

The expected value of the cost per month for the outcome node associated with the Buy action is defined likewise:

$$E(Buy) = (\$17,600) \cdot (0.3) + (\$19,800) \cdot (0.5) + (\$22,000) \cdot (0.2)$$

$$= \$19,580 \, / \, month$$

The decision tree in Figure 7 is simplified by placing the computed expected values next to the outcome nodes associated with the Make and Buy actions (see Figure 8).

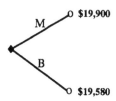

Figure 8

The decision tree in Figure 8 indicates a choice of whether to Make (expected monthly cost $19,900) or to Buy (expected monthly cost $19,580) the required rolls to meet customer demand. The Buy action is the Expected Value Decision since it provides the minimum expected cost per month to Pricebusters supermarket.

Note: The expected cost per month of the Make and Buy decisions are quite close. If there are factors other than cost (e.g. future plans for growth of the Pricebusters supermarket chain's bakery) to be considered, then the action Make which is not the Expected Value Decision may be taken. The minimum expected cost is typically only one factor out of several quantitative and qualitative factors weighed by the decision maker before making the decision whether to Make or Buy.

Assuming the decision is to proceed with the Buy action, the decision tree is *pruned* by placing slashes on the Make action branch (see Figure 9). In general, all actions other than the best action (minimum expected cost) are pruned at action nodes.

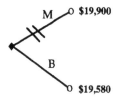

Figure 9

The solution to the decision tree is completed by placing the expected value of the cost of the minimum expected cost Buy action next to the initial action node (see Figure 10).

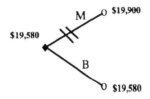

Figure 10

Instead of proceeding step by step through the decision tree (computing expected values for the rightmost sets of outcome nodes, placing expected values next to the outcome nodes, and pruning actions not providing the minimum expected cost), the above steps can be performed on the original complete Make vs. Buy decision tree (see Figure 7). This decision tree computation for the Make vs. Buy decision is illustrated in Figure 11.

Decision Tree Computation for Make vs. Buy Decision

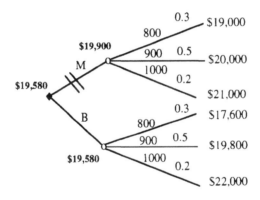

Figure 11

2.4 Sequential Decision Trees

Often a decision process involves <u>sequential decisions</u> in a stochastic environment. Common examples of sequential decisions include:

– *Blackjack ("21")*

– *Daily Stock Market or Commodities Market Transactions*

– *Buy / Don't Buy / Pay For More Information Before Deciding*

Example 6: (Sequential Decision Tree for Blackjack "21"). A card player is in a game of Blackjack ("21") where the object of the game is to come closer than the dealer to a point total of 21 points without going over. The player currently holds two cards: a face card (worth ten points) and a nine. The point total of the current hand is 19 points. Provide the actions and outcomes for the decision tree to decide the best future action(s) for the card player given this hand.

Answer: The actions and outcomes of the decision tree for a sequential Blackjack decision process are given in Figure 12. Example 7 will add probabilities and payoffs for each action-outcome pair and determine the Expected Value Decision.

Note: There are many versions of Blackjack played under different "house" rules. Figure 12 presents one version of a Blackjack game as an example of a sequential decision tree.

Sequential Blackjack ("21") Game Decision Tree
(No Probabilities or Payoffs)

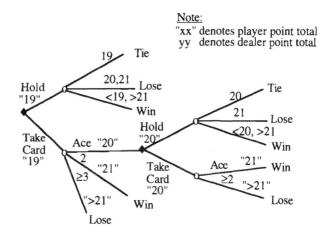

Figure 12

Sequential decision trees are also read from *left to right*. The initial decision for the card player (the decision maker) is a choice among the following actions:

- *Hold* "19"
- *Take Card* "19"

Note that the player's point totals are in quotations (e.g. "19") while the dealer's point totals do not have quotations marks surrounding them (e.g. 19). If the card player's action is to Hold at "19", there are three outcomes possible for the player. These three outcomes depend on the final point total accumulated by the *dealer* (his opponent) who has the option of taking additional cards as long as his/her point total is less than 21:

- *Tie (Dealer ends with* 19)
- *Lose (Dealer ends with* 20 *or* 21)
- *Win (Dealer ends with* <19 *or* >21)

If the card player's action is to Take Card at "19", there are three outcomes possible for the player:

- *Game Continues (Additional card is an Ace = 1; Player now has "20")*
- *Win (Additional card is a 2; Player ends with "21")*
- *Lose (Additional card is ≥ 3; Player ends with ">21")*

If the additional card taken is an Ace and the player's point total is "20", the game continues and a *sequential decision* must be made by the player. The card player has a choice between one of two actions:

- *Hold "20"*
- *Take Card "20"*

If the card player's sequential decision is the action Hold at "20", there are three outcomes possible for the player. These three player outcomes depend on the final point total accumulated by the *dealer* (his opponent) who has the option of taking additional cards as long as his/her point total is less than 21:

- *Tie (Dealer ends with 20)*
- *Lose (Dealer ends with 21 points)*
- *Win (Dealer ends with < 20 or > 21)*

If the card player's action is to Take Card at "20", there are two outcomes possible for the player:

- *Win (Additional card is an Ace = 1; Player ends with "21")*
- *Lose (Additional card is ≥ 2; Player ends with ">21")*

The solution process for solving decision trees with sequential decisions as illustrated in Figure 12 is analogous to the solution process for solving decision trees with single decisions (see the Make vs. Buy Decision Tree given in Figure 7):

- Work *backwards* from right to left.

- For the rightmost outcome nodes in the decision tree, compute the expected value of each outcome node. Designate the expected value of the outcome node to the action which generated that set of outcomes.

- At the rightmost action nodes, choose the action with the maximum expected payoff (or the minimum expected cost). Prune all other actions at each action node. Designate the maximum expected payoff (or minimum expected cost) among the possible actions as the value of the action node.

- The solution process moves to the left alternating between the outcome nodes and action nodes until the *initial* action node is reached. The Expected Value Decision is given by the set of actions not pruned. The value of the initial (leftmost) action node is the value of the Expected Value Decision.

Example 7: (Solving the Sequential Blackjack "21" Decision Tree). Assume for the Blackjack ("21") game in Example 6 that a $5 bet is made by the card player. If the player loses, the $5 bet is lost. If the player wins, the return to the player is a net of $5. If the player and dealer tie, the $5 bet is returned for a net of $0. Payoffs and sample (though not actual) probabilities are added to the Blackjack ("21") decision tree (see Figure 13). Solve this sequential decision tree for the Expected Value Decision.

Sequential Blackjack ("21") Game Decision Tree
(With Probabilities and Payoffs)

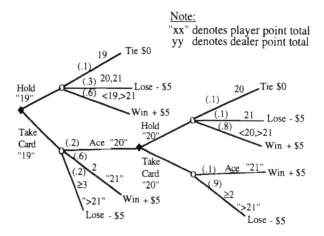

Figure 13

Answer: The solution process proceeds backwards from right to left. First compute the expected value for the rightmost outcome nodes (namely, the outcome nodes associated with the Hold "20" and Take Card "20" actions). The expected value for the outcome node associated with the Hold "20" action is given as follows:

$$E(Hold\ "20") = (0.1) \cdot (\$0) + (0.1) \cdot (-\$5) + (0.8) \cdot (+5) = \$3.50$$

The expected value for the outcome node associated with the rightmost Take Card "20" action is given as follows:

$$E(Take\ Card\ "20") = (0.1) \cdot (\$5) + (0.9) \cdot (-\$5) = -\$4.00$$

The action Hold "20" has the maximum expected payoff and the value $3.50 is designated to the rightmost action node (denoting the choice between the Hold "20" or Take Card "20" actions). The action Take Card "20" is pruned.

The partially solved Blackjack "21" decision tree is given in Figure 14.

Partial Solution to Sequential Blackjack ("21") Decision Tree

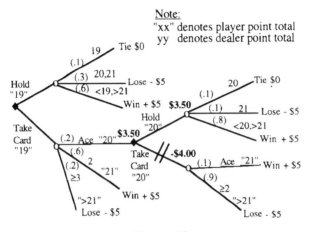

Figure 14

Next examine the expected value associated with the rightmost outcome nodes (namely, those associated with the initial actions Hold "19" and Take Card "19"). The expected value for the outcome node associated with the Hold "19" action is given as follows:

$$E(Hold \text{ "19"}) = (0.1) \cdot (\$0) + (0.3) \cdot (-\$5) + (0.6) \cdot (\$5) = \$1.50$$

The expected value for the outcome node associated with the initial Take Card "19" action is given as follows:

$$E(Take \ Card \text{ "19"}) = (0.2) \cdot (\$3.50) + (0.6) \cdot (\$5) + (0.2) \cdot (-\$5) = \$2.70$$

Note that the payoff of the outcome Ace ($3.50) is the optimal value of the decision tree from that point onwards in the sequential decision process. The action Take Card "19" has the maximum expected payoff and the value $2.70 is designated to the initial action node (with the choice between Hold "19" and Take Card "19" actions). The value of the initial action node is the Expected Value Decision (maximum expected payoff). The action Hold "19" is pruned.

The expected value solution to the Blackjack "21" decision tree is given in Figure 15. This solved decision tree signifies that for this set of payoffs and this set of sample probabilities, the Expected Value Decision for the player is to take a card when holding a hand worth 19 points. If the card taken is an Ace (worth one point), the player should hold at 20 points.

On any individual hand, this strategy could backfire and result in a loss for the player. However, if this situation (a hand worth 19 points) were faced by the card player many times, this sequential decision (the Expected Value Decision) would result in the highest total payoff to the player (with an average net payoff of $2.70 per game).

Expected Value Solution For Sequential Blackjack ("21") Decision Tree

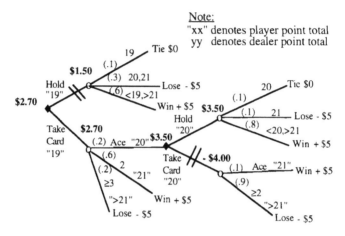

Figure 15

A common decision analysis amenable to sequential decision trees is whether to buy something, not to buy something, or to pay for more information on the stochastic environment before making the decision.

Example 8: (Buy/Don't Buy/Pay for More Information) An investor is given the opportunity to invest $20,000 in any of several oil drilling projects offered by Y. L. Katr, Inc. If a drilling finds oil, there is a payment to the investor of $150,000 (net $130,000). If no oil is found, the investor receives no payment (net -$20,000). Historically, the probability of a randomly chosen oil drill being successful is about 10%.

For a cost of $2,000, the investor can have access to geological survey data for a *particular* oil drilling project. The geological surveys for oil drilling projects offered by Y. L. Katr are either "Excellent" (20% of the projects) or "Good" (80% of the projects). An Excellent geological survey upgrades the probability of finding oil to about 30%. Only a Good geological survey implies that there is about a 5% chance of finding oil. After receiving the geological survey, the investor can then decide to invest an additional $20,000 in that oil drilling project or not to invest (forfeiting the $2,000 cost of the geological survey data).

Determine the action(s) which provides the maximum expected payoff: Invest (choose an oil drilling project without the benefit of geological survey data); Don't Invest (pass on the opportunity offered); or Pay For Geological Survey (for a particular oil drilling project and then decide whether to Invest or Don't Invest).

Answer: The sequential decision tree for Y. L. Katr, Inc. investment which includes actions, outcomes, probabilities, and payoffs is given in Figure 16. As usual, the sequential decision tree is read from left to right (forwards) but is solved from right to left (backwards).

Y. L. Katr, Inc. Sequential Decision Tree

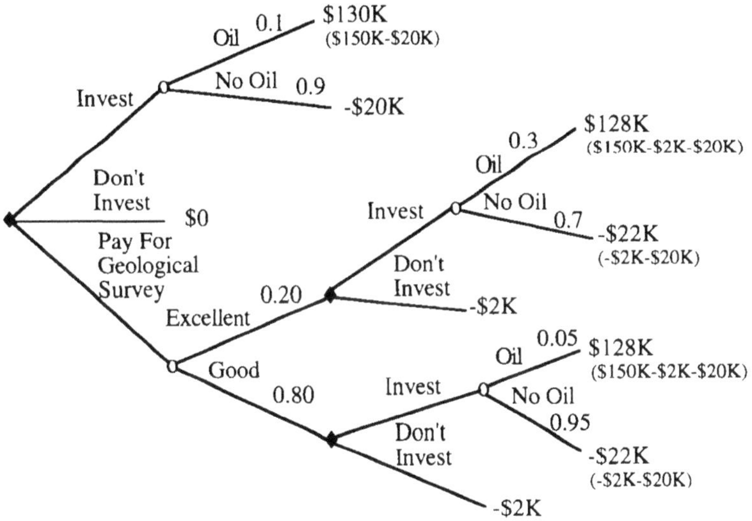

Figure 16

There are three initial alternative actions for the investor (see Figure 16):

- *Invest*
- *Don't Invest*
- *Pay For Geological Survey*

If the initial action taken by the investor is Don't Invest, there is no investment cost and no payment to the investor. Hence, there are no stochastic outcomes and payoffs. The expected value of the Don't Invest action is $0.

If the initial action taken is Invest, there are two possible stochastic outcomes:

- *Oil*
- *No Oil*

For the initial action Invest, the probabilities and payoffs for the stochastic outcomes Oil and No Oil are given in the description of Example 8:

- $P(Oil) = 0.1$, *Payoff* $= \$150K - \$20K = \$130K$
- $P(No\ Oil) = .9$, *Payoff* $= -\$20K$

If the action is Pay For Geological Survey (at additional cost \$2K), the stochastic outcomes for the geological survey and their probabilities are given in the description of Example 8:

61

- *Excellent; P(Excellent) = 0.20*
- *Good; P(Good) = 0.80*

If the outcome of the geological survey is Excellent, the investor has a sequential decision consisting of two alternative actions:

- *Invest*
- *Don't Invest*

If the action taken is Invest, the stochastic outcomes are again Oil and No Oil. The conditional probabilities and payoffs given the survey is Excellent are as follows:

- *P(Oil|Excellent) = 0.30, Payoff = $150K – $2K – $20K = $128K*
- *P(No Oil|Excellent) = 0.70, Payoff = –$2K – $20K = –$22K*

If the action taken is Don't Invest, the investor loses the $2K paid for the geological survey.

If the outcome of the geological survey is Good, the investor has a sequential decision consisting of two alternative actions:

- *Invest*
- *Don't Invest*

If the action taken is Invest, the stochastic outcomes are again Oil and No Oil. The probabilities and payoffs given the survey is Good are as follows:

- *P(Oil|Good) = 0.05, Payoff = $150K – $2K – $20K = $128K*
- *P(No Oil|Good) = 0.95, Payoff = –$2K – $20K = –$22K*

If the action taken is Don't Invest, the investor loses the $2K paid for the geological survey.

The sequential decision tree for Y. L. Katr, Inc. investment which includes actions, outcomes, probabilities, and payoffs as given in Figure 16 can now be solved from right to left (backwards).

The rightmost outcome nodes are the outcome node associated with the decision to Invest after an Excellent geological survey and the outcome node associated with the decision to Invest after a Good geological survey. The expected value of these two rightmost outcome nodes are given by:

- *E(Invest w/ Excellent Geological Survey) = (0.3)·($128K) + (0.7)·(– $22K)*
$$= \$23K$$

- *E(Invest w/ Good Geological Survey) = (0.05)·($128K) + (0.95)·(–$22K)*
$$= -\$14.5K$$

The expected value of $23K is designated as the value of the action Invest after an Excellent geological survey and the expected value -$14.5K is designated as the value of the action Invest after a Good geological survey (see Figure 17). For the action node following an Excellent geological survey, the action Invest has a higher expected value than the action Don't Invest ($23K vs. -$2K). The action Don't Invest is pruned. For the action node following an Good geological survey, the action Don't Invest has a higher expected value than the action Invest (-$2K vs. -$14.5K). The action Invest is pruned.

After evaluating the rightmost outcome nodes, designating the values of the associated action nodes and pruning the lower expected value actions, the partially solved sequential decision tree for Y. L. Katr is given by Figure 17.

Partial Solution to Y. L. Katr, Inc. Sequential Decision Tree

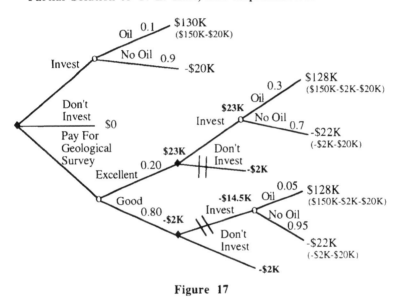

Figure 17

The rightmost outcome nodes not evaluated to this point in the solution process are the node associated with the initial decision whether to Invest and the node associated with the decision whether to Pay For Geological Survey. The expected value associated with these two rightmost unevaluated outcome nodes are as follows:

- $E(Invest) = (0.1) \cdot (\$130K) + (0.90) \cdot (-\$20K) = -\$5K$
- $E(Pay\ For\ Geological\ Survey) = (0.2) \cdot (\$23K) + (0.8) \cdot (-\$2K) = \$3.0K$

Note that the payoff of the outcome Excellent for the geological survey is $23K which is the expected value of the sequential action node. Also the payoff of the outcome Good for the geological survey is -$2K which is the expected value of the sequential action node.

The expected value of the outcome nodes associated with the initial actions Invest and Pay For Geological Survey are designated to the actions Invest and Pay For Geological Survey, respectively. The expected value of the action Don't Invest is simply $0 since there is no investment and no payoff.

For the initial action node whether to Invest, Don't Invest, or to Pay For Geological Survey (before deciding whether to invest), the action Pay For Geological Survey has a higher expected value than Invest ($3K vs. -$5K) and Don't Invest ($3K vs. $0K). The value $3K is assigned to the initial action node. This is the value of the Expected Value Decision. The actions Invest and Don't Invest are pruned. The solved decision tree for the Y. L. Katr sequential oil drilling decision is given in Figure 18.

Solution to Y. L. Katr, Inc. Sequential Decision Tree

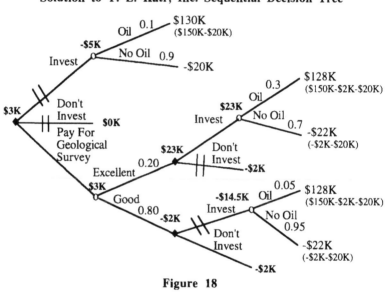

Figure 18

Notes:

• The Expected Value Decision is Pay For Geological Survey before deciding whether to invest. If the result of the geological survey is Excellent, then Invest. If the result of the geological survey is only Good, then Don't Invest. On any particular investment, this sequential decision strategy can backfire and result in a loss. However, over the long run (many occurrences of this investment opportunity), the Expected Value Decision will result in the highest total payoff (with an average net payoff of $3K per investment).

• Examine the Maximin Decision (used for payoffs) for the Y. L. Katr. Inc. sequential decision. For the three alternative actions, the worst case payoffs are given by:

   • *Invest* $\Rightarrow$ – $20K (No Oil)
   • *Don't Invest* $\Rightarrow$ $0
   • *Pay For Geological Survey* $\Rightarrow$ – $22K (No Oil)

Since the actions Invest and Pay For Geological Survey have the potential for loss, the Maximin Decision is Don't Invest with a payoff of $0 (best worst case payoff to the investor). Oil drilling investments may not be appropriate for *conservative* decision makers who apply the Minimax criterion to their decisions.

• In Example 8, the probabilities of finding Oil and No Oil were updated based on an Excellent or Good geological survey. The decision was made to pay for the information in the geological survey on the basis that the information (despite its $2K cost) led to a decision with a *higher* expected value ($3K) than not using a geological survey ($0K). An investor should be willing to pay up to $5K for this geological information.

2.5 Utility & Subjective Probability

Utility

The payoff or cost of an outcome is often difficult to measure or forecast. Instead of using *absolute* terms like dollars for the payoff or cost of an outcome, decision makers must sometimes assess the payoff or cost in terms of the utility of an outcome.

Utility attaches *numerical values* (required to solve decision analyses) to outcomes based on their worth to the individual decision maker (according to his or her own value system). The numerical values are chosen to preserve *relative* preferences. For example, an owner of a business may not be able to determine the exact cost in dollars to his business of two or three customers per day turning away due to slow service. But it is clear that the owner *prefers* that only two customers per day turn away rather than three. The outcome of only two customers per day turning away has a *higher* utility to the owner.

If outcome X_1 is preferred to outcome X_2 due to its higher worth (as perceived by the decision maker), the decision maker will attach to X_1 a higher utility. If a utility function U is defined based on the preferences of the decision maker, then:

$$U(X_1) > U(X_2)$$

If the decision maker is *indifferent* between outcomes X_1 and X_2 (both have the same worth), the decision maker will attach the same utility to each outcome. That is:

$$U(X_1) = U(X_2)$$

Notes:

• The numerical values assigned as the utility of outcomes are *subjective* (based on the value system of the decision maker). The relative size of the numerical values is important but the order of magnitude (e.g. units, tens, etc.) of the numerical values is unimportant. The values must *preserve preferences* (i.e. $U(X_1) > U(X_2)$ if X_1 is preferred to X_2).

• When assigning utilities to outcomes, the set of utilities must be *self-consistent* (that is, preserve preferences over the *entire* set of possible outcomes). For example, assume outcome X_1 is preferred to outcome X_2 which is preferred to outcome X_3 which is preferred to outcome X_4. Since X_1 is preferred to X_4, one can assign utilities of 10 to X_1 and 5 to X_4. Since X_2 is preferred to X_3, one can assign utilities of 6 to X_2 and 4 to X_3.

However, these two assignments are not self-consistent in that although X_3 is preferred to X_4, X_3 has been given a utility of 4 and X_4 has been given a utility of 5.

Example 9: (Consistency of Utility Functions). Consider outcomes denoted by X_1, X_2, X_3 and X_4 where X_1 is preferred to X_2, X_2 is preferred to X_3, and X_3 is preferred to X_4. It is implied by the transitive property that if X_1 is preferred to X_2 and X_2 is preferred to X_3, then X1 is preferred to X_3, etc.). Determine if the following sets of utilities (from three different individuals) are self-consistent:

Outcomes	Set #1 Utilities U(X)	Set #2 Utilities U(X)	Set #3 Utilities U(X)
X_1	22000	80	230
X_2	16000	64	170
X_3	8000	25	70
X_4	5000	15	120

Answer: Since X_1 is preferred to X_2 which is preferred to X_3 which is preferred to X_4, the values of the assigned utilities must be such that:

$$U(X_1) > U(X_2) > U(X_3) > U(X_4)$$

Utility sets #1 and #2 satisfy this criterion and are self-consistent. Utility set #3 does not preserve preferences over the entire set of outcomes. Outcome X_3 is preferred to outcome X_4 and yet:

$$U(X_3) < U(X_4) .$$

Notes:

• The fact that the utilities assigned to X_1, X_2, X_3, and X_4 in Set #1 and Set #2 of Example 9 are generally a factor of a thousand apart in magnitude is acceptable as long as each set of utilities reflect the individual's value system and are self-consistent [i.e. $U(X_1) > U(X_2) > U(X_3) > U(X_4)$].

• When the decision maker is satisfied that the numerical utilities assigned to outcomes are indicative of his or her value system and are self-consistent, these values can be entered into a decision matrix of decision tree. A decision matrix can then be solved for the Maximin/Minimax Decision or the Expected Value Decision. Likewise, a decision tree can be solved for the Expected Value Decision.

• The decision analysis for the activity of preparing to go to work on a day when rain is a significant possibility used utilities as action-outcome payoffs rather than dollars. The Expected Value Decision (see Example 1 in Section 2.1) and Maximin Decision (see Example 2 in Section 2.2) to this decision analysis were solved on the basis of the decision maker's own utility function.

Subjective Probability

A probability distribution (required to solve for the Expected Value Decision) for the set of outcomes associated with an action may also be difficult to measure or forecast reliably.

For example, consider an experimental drug undergoing tests for the probability of effectiveness prior to its release to the general population. There are many factors that must be controlled in selecting a test group to assure that the test group is statistically similar to the general population. These factors may include age, gender, race, ethnicity, smoking habits, stage of the disease, etc. Some of the factors affecting the effectiveness drug may be unknown at the time the test group is chosen.

Rather than spending large sums of money collecting lots of data in order to assure statistical significance for all the appropriate known and unknown factors (if in fact this can be done with confidence), an alternate method of assessing probability would be to use *expert opinion* to assess the probability of effectiveness of the drug. Doctors and other health professionals administering the drug might be *asked* their assessment the effectiveness of the drug. This estimation technique of surveying expert opinion to assess probabilities leads to a subjective probability distribution.

The opinions of the experts are also subjective. For example, without rigorous controls on the population for which a doctor has administered the drug, the experience of the doctor may be anecdotal and not representative of the effectiveness of the drug with the general population. However, it is typically assumed that for a large number of experts with wide experience, the assessment of the experts based on their personal experience will apply to the overall population.

Notes:

• Individuals make subjective probability estimates every day. For example, individuals often decide whether to take an umbrella (based on whether it "looks" like rain that day) and whether to take an alternate route home (based on whether traffic is "likely" to be bad at a given time). Individuals may use their own judgment (based on personal experience) or the expert opinions of TV weather reports and car radio traffic reports.

• When the decision maker is satisfied with the assessment of the subjective probability distribution, these probabilities can be assigned to the outcomes. A decision matrix can then be solved for the Maximin/Minimax Decision or the Expected Value Decision. Likewise, a decision tree can then be solved for the Expected Value Decision.

2.6 Expected Value Of Information

Decision makers seek to increase their confidence in which outcome state will occur so that a decision can be made with greater certainty. There is sometimes the option of *paying for information* which can be used to update the initial state probabilities and increase the confidence in which outcome state will occur. When the information increases the certainty under which a decision is made, the expected value of the decision will increase (see Example 8 where the results of the geological survey increased the certainty of Oil or No Oil and increased the expected value of the oil drilling investment).

To determine the *value* of the additional information, the expected value of the decision using the additional information is compared with the Expected Value Decision (using no additional information). The value of this additional information to the decision maker is the amount by which the expected value of the decision using the additional information *improves* the expected value of the decision using no additional information.

Note: In Example 8, if Pay for Geological Survey were not a possible action, the Expected Value Decision would be Don't Invest with a value $0 (see Figure 18). The opportunity to Pay For Geological Survey increased the value of the Expected Value Decision by $5K (with a net of $3K to the investor since $2K is used to pay for the geological data). The value of the geological survey to the investor is $5K. As long as the investor pays less than $5K for the geological data, the value of the expected value of the decision with the geological survey information will be greater than the value of the Expected Value Decision (without the geological survey information).

There are two concepts often used in decision analysis in conjunction with the expected value of information:

1) Expected Value of Perfect Information (EVPI)

In this scenario, the decision maker pays for information which indicates *with certainty* when an outcome state will occur. With certainty in which outcome will occur at the time a decision must be made, the decision maker always chooses the action with the highest payoff for that outcome.

Computation of the Expected Value of Perfect Information: Define the Expected Value Under Certainty to be the sum of the highest payoff possible for each outcome times the probability that outcome will occur. The Expected Value of Perfect Information (EVPI) is then the amount by which the Expected Value Under Certainty *improves* the Expected Value Decision (EVD) which uses no additional information. For payoffs, the EVPI is given by:

$$EVPI = Expected\ Value\ Under\ Certainty - EVD$$

For costs, the EVPI is given by:

$$EVPI = EVD - Expected\ Value\ Under\ Certainty$$

Example 10: (Expected Value of Perfect Information). Consider the activity of preparing for work on a day when rain is a significant possibility given in Example 2. The following decision matrix was derived for this activity (see Figure 2).

Decision Matrix For Preparing To Go To Work

Outcome Action	Rain (0.4)	No Rain (0.6)
Stay Home	Get Fired (-100)	Get Fired (-100)
Go To Work Without Umbrella	Get Wet (-30)	Look Smart (+50)
Go To Work With Umbrella	Stay Dry (+30)	Look Foolish (-10)

What is the Expected Value of Perfect Information (EVPI)?

Answer: The expected value of the three alternative actions was computed in Example 2 as follows:

$E(Stay\ Home) = (-100) \cdot (0.4) + (-100) \cdot (0.6) = -100$

$E(Go\ To\ Work\ Without\ Umbrella) = (-30) \cdot (0.4) + (+50) \cdot (0.6) = 18$

$E(Go\ To\ Work\ With\ Umbrella) = (+30) \cdot (0.4) + (-10) \cdot (0.6) = 6$

The action Go To Work Without Umbrella is the Expected Value Decision since it provides the maximum expected payoff of 18.

For the two outcomes Rain and No Rain, the best action and the maximum payoff possible is given as follows:

$Rain \Rightarrow Go\ To\ Work\ With\ Umbrella \Rightarrow\ +30$

$No\ Rain \Rightarrow Go\ To\ Work\ Without\ Umbrella \Rightarrow\ +50$

Then the Expected Value Under Certainty (where the outcome is known at the time at the time a decision must be made) is the sum of the maximum payoff for each outcome times the probability the outcome will occur:

$Expected\ Value\ Under\ Certainty = (+30) \cdot (0.4)\ +\ (+50) \cdot (0.6) = +42$

The Expected Value of Perfect Information (EVPI) is the amount the Expected Value Under Certainty improves on the value of the Expected Value Decision:

$EVPI = Expected\ Value\ Under\ Certainty\ - EVD\ =\ 42\ -\ 18\ =\ 24$

A person with this value system should be willing to pay up to 24 for perfect information on whether it will Rain or Not Rain as he or she prepares to go to work.

Note: Perfect information takes the uncertainty out of the decision and always allows the action with the highest payoff to be chosen. Hence, the EVPI is always non-negative and is an *upper bound* on the value of information to the decision maker. That is, no information can have a value to the decision maker greater than the EVPI.

2) Expected Value of Sample Information (EVSI)

In this scenario, the decision maker pays for sample (imperfect) information which provides more confidence in which outcome state will occur (though not certainty in which outcome state will occur as with perfect information). The value of the sample information is the extent to which the information changes the initial *(a priori)* probabilities of the possible outcome states into updated *(a posteriori)* state probabilities resulting in an *improved* expected value to the decision maker. Bayes' Theorem (see Section 1.15) is used to update the initial state probabilities based on the sample information.

Computation of the Expected Value of Sample Information: For each of the alternative actions, compute the expected value of the action using the *updated* outcome probabilities (i.e. the sum of the outcome payoffs times the updated outcome probabilities). Define the

Expected Value Using Updated Probabilities to be the maximum expected payoff using the updated probabilities among the alternative actions.

When the result of outcomes involves *payoffs*, the Expected Value of Sample Information (EVSI) is the amount the Expected Value Using Updated Probabilities is greater than (improves on) the Expected Value Decision (EVD) using no additional information and thus the initial outcome probabilities:

$$EVSI = Expected\ Value\ Using\ Updated\ Probabilities - EVD$$

When the result of outcomes involves *costs*, the Expected Value of Sample Information (EVSI) is defined as the amount the Expected Value Using Updated Probabilities is less than (improves on) the Expected Value Decision:

$$EVSI = EVD - Expected\ Value\ Using\ Updated\ Probabilities$$

The EVSI is always non-negative. The sample information provides more certainty in the outcomes and hence allows a better decision to be made. Since the EVPI is an upper bound on the value of information, it follows that:

$$EVSI \le EVPI.$$

Example 11: (Expected Value of Sample Information). A studio executive must decide on a marketing strategy for a just-completed horror movie. The alternative actions being considered are: Shelf (S) the movie and absorb the $10 Million in already spent production costs; a Targeted (T) Release to just those theaters with the right demographics; or a Wide (W) Release across the nation and worldwide. The decision will naturally be based on the anticipated level of acceptance by the movie-going public and the potential payoffs (or costs) to the studio.

Three possible outcomes for the horror movie have been identified: a Flop (F); a Limited (L) Success; and a Hit (H). Historically, horror movies are a Flop, a Limited Success, and a Hit with probabilities 0.3, 0.5, and 0.2, respectively. The payoffs (or costs) associated with the alternative actions and possible outcomes are given in the following decision matrix:

Horror Movie Payoffs and Costs

	Flop (F) (0.3)	Limited Success (L) (0.5)	Hit (H) (0.2)
Shelf (S)	-$10 M	-$10 M	-$10 M
Targeted Release (T)	-$15 M	$10 M	$20 M
Wide Release (W)	-$30 M	$5 M	$50 M

The decision tree for the horror movie decision is given in Figure 19.

Horror Movie Decision Tree
(Without Sample Information)

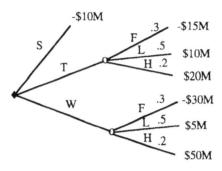

Figure 19

a) What is the Expected Value Decision (using no additional information)?

Answer: The expected value of the payoff for the three alternative actions Shelf (S), Targeted (T) Release, and Wide (W) Release using the initial (a priori) probabilities (see Figure 19) is as follows:

$E(Shelf) = -\$10\ Million$

$E(Targeted) = (-\$15M)\cdot(0.3) + (\$10M)\cdot(0.5) + (\$20M)\cdot(0.2)$
$= \$4.5\ Million$

$E(Wide) = (-\$30M)\cdot(0.3) + (\$5M)\cdot(0.5) + (\$50M)\cdot(0.2)$
$= \$3.5\ Million$

The Expected Value Decision (EVD) is a Targeted (T) Release since this action has the maximum expected value (\$4.5 Million) among the three alternative actions.

b) The studio executive also has the option to test the horror movie with audiences at screenings prior to making the marketing decision. What is the most the executive should be willing to pay for the movie screening information regardless of its quality (i.e. What is the EVPI)?

Answer: Perfect information identifies with certainty the outcome state of the movie and allows the studio executive to choose the action with the highest payoff for that outcome.

When the movie is known to be a Flop (probability 0.3), the best action is to Shelf the movie with payoff (loss) of -$10 Million. When the movie is known to be a Limited Success (probability 0.5), the best action is for a Targeted Release with payoff of $10 Million. When the movie is known to be a Hit (probability 0.2), the best action is for Wide Release with payoff of $50 Million. Hence, the Expected Value Under Certainty is given by:

$$Expected\ Value\ Under\ Certainty = (0.3) \cdot (-\$10M) + (0.5) \cdot (\$10M) + (0.2) \cdot (\$50M)$$
$$= \$12\ Million$$

$$EVPI = Expected\ Value\ Under\ Certainty - EVD$$
$$= \$12M - \$4.5M$$
$$= \$7.5\ Million$$

The studio executive should not be willing to pay over $7.5 Million dollars for results of a movie screening test no matter how accurate the screening test is in predicting whether the horror movie will be a Flop, a Limited Success, or a Hit.

c) The results of the movie screening will be either a Positive (P) or a Negative (N) report (sometimes called an *indication*). Historically, when an horror movie turns out to a Flop, the screening gave a Positive report prior to its release only 30% of the time (i.e. P(P|F) = 0.3) and a Negative report 70% of the time (i.e. P(N|F) = 0.7). In general, the conditional probabilities associated with the movie screening test are as follows:

Conditional Probabilities For Horror Movie Screening Test

	Flop (F)	Limited Success (L)	Hit (T)
Positive (P)	0.3	0.6	0.9
Negative (N)	0.7	0.4	0.1

What is the value of this sample (movie screening) information (EVSI) to the studio executive?

Answer: To determine the value of the sample information, one must determine the change of the initial (a priori) probabilities P(F), P(L), and P(T) into updated (a posteriori) probabilities P(F|P), P(L|P), and P(H|P) given a positive (P) movie screening report. Likewise, one must determine the updated (a posteriori) probabilities P(F|N), P(L|N), and P(H|N) given a negative (N) movie screening report. Bayes' Theorem (an application of the Law of Conditional Probability) is used to determine the updated probabilities. The decision tree for the horror movie decision tree using the sample information is given in Figure 20.

Horror Movie Decision Tree
(Using Sample Information)

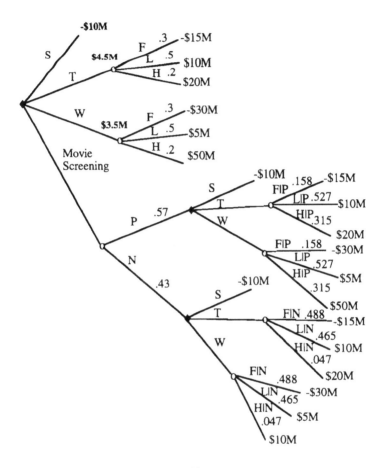

Figure 20

Note: The Law of Conditional Probability (see Section 1.14) provides two alternate ways that probabilities such as $P(F \cap P)$ can be stated:

$$(1)\ P(F|P) = \frac{P(F \cap P)}{P(P)} \quad \Rightarrow \quad P(F \cap P) = P(F|P) \cdot P(P)$$

$$(2)\ P(P|F)\ =\frac{P(F\cap P)}{P(F)}\ \Rightarrow\ P(F\cap P)=P(P|F)\cdot P(P)$$

The second definition of $P(F\cap P)$ is often used to rewrite conditional probabilities to take advantage of probabilities which are known.

When the report is Positive (P) or Negative (N), the probability of the horror move being a Flop (F) is updated using the Bayes' Theorem to P(F|P) and P(F|N) as follows:

$$P(F|P)\ =\frac{P(F\cap P)}{P(P)}=\frac{P(P|F)\cdot P(F)}{P(P)}$$

$$P(F|N)\ =\frac{P(F\cap N)}{P(N)}=\frac{P(F|N)\cdot P(F)}{P(N)}$$

All the above probabilities required to update the probability of the horror movie being a Flop given a positive or a negative screening report have been provided except for the probability of a Positive report P(P) and a Negative report P(N).

Applying set theory, the Events P and N can be written as follows:

$$P = U\cap P = (F\cup L\cup H)\cap P$$
$$N = U\cap N = (F\cup L\cup H)\cap N$$

$$P = (F\cap P)\cup (L\cap P)\cup (H\cap P)$$
$$N = (F\cap N)\cup (L\cap N)\cup (H\cap N)$$

$$P(P) = P(F\cap P)\cup P(L\cap P)\cup P(H\cap P)$$
$$P(N) = P(F\cap N)\cup P(L\cap N)\cup P(H\cap N)$$

The Law of Conditional is used to replace the $P(F\cap P)$, $P(L\cap P)$, $\cdots$, and $P(H\cap N)$ with $P(P|F)\cdot P(F)$, $P(P|L)\cdot P(L)$, $\cdots$, and $P(N|H)\cdot P(H)$, respectively:

$$\begin{aligned}P(P)\ &= P(P|F)\cdot P(F)\ + P(P|L)\cdot P(L)\ + P(P|H)\cdot P(H)\\ &= (0.3)\cdot(0.3)\ + (0.6)\cdot(0.5)\ + (0.9)\cdot(0.2)\\ &= 0.57\end{aligned}$$

$$\begin{aligned}P(N)\ &= P(N|F)\cdot P(F)\ + P(N|L)\cdot P(L)\ + P(N|H)\cdot P(H)\\ &= (0.7)\cdot(0.3)\ + (0.4)\cdot(0.5)\ + (0.1)\cdot(0.2)\\ &= 0.43\end{aligned}$$

The initial probabilities P(F), P(L), and P(H) are updated using Bayes' Theorem from 0.3, 0.5, and 0.2 based on a Positive (P) screening report to a set of probabilities *more certain* of the horror movie being a Hit:

$$P(F|P) = \frac{P(F \cap P)}{P(P)} = \frac{P(P|F) \cdot P(F)}{P(P)} = \frac{(0.3) \cdot (0.3)}{0.57} = 0.158$$

$$P(L|P) = \frac{P(L \cap P)}{P(P)} = \frac{P(P|L) \cdot P(L)}{P(P)} = \frac{(0.6) \cdot (0.5)}{0.57} = 0.527$$

$$P(H|P) = \frac{P(H \cap P)}{P(P)} = \frac{P(P|H) \cdot P(H)}{P(P)} = \frac{(0.9) \cdot (0.2)}{0.57} = 0.315$$

The expected value of the decision based on a *Positive* screening report (positive indication) is computed as follows:

$$E(Shelf) = -\$10 \, Million$$

$$E(Targeted) = (-\$15M) \cdot (0.158) + (\$10M) \cdot (0.527) + (\$5M) \cdot (0.315)$$
$$= \$ 4.475 \, Million$$

$$E(Wide) = (-30M) \cdot (0.158) + (\$5M) \cdot (0.527) + (\$50M) \cdot (0.315)$$
$$= \$13.86 \, Million$$

The Expected Value Decision given a Positive (P) screening report is the action Wide (W) Release with a value of $13.86 Million.

The initial probabilities P(F), P(L), and P(H) are updated using Bayes' Theorem from 0.3, 0.5, and 0.2 based on a Negative (N) screening report to a set of probabilities (0.488, 0.465, and 0.047) *more certain* of the horror movie not being a Hit:

$$P(F|N) = \frac{P(F \cap N)}{P(N)} = \frac{P(N|F) \cdot P(F)}{P(N)} = \frac{(0.7) \cdot (0.3)}{0.43} = 0.488$$

$$P(L|N) = \frac{P(L \cap N)}{P(N)} = \frac{P(N|L) \cdot P(L)}{P(N)} = \frac{(0.4) \cdot (0.5)}{0.43} = 0.465$$

$$P(H|N) = \frac{P(H \cap N)}{P(N)} = \frac{P(N|H) \cdot P(H)}{P(N)} = \frac{(0.1) \cdot (0.2)}{0.43} = 0.047$$

The Expected Value Decision given a *Negative* screening report (negative indication) is computed as follows:

$E(Shelf) = -\$10\ Million$

$E(Targeted) = (-\$15M)\cdot(0.488) + (\$10M)\cdot(0.465) + (\$5M)\cdot(0.047)$
$= -\$2.435\ Million$

$E(Wide) = (-\$30M)\cdot(0.488) + (\$5M)\cdot(0.465) + (\$50M)\cdot(0.047)$
$= -\$9.965\ Million$

The Expected Value Decision given a Negative (N) screening report is the action Targeted (T) Release with a value of -\$2.435 Million.

We have seen that the screening test arrives at a Positive report 57% of the time and at a Negative report 43% of the time for action movies. The expected value of a Positive report and a Negative report is \$13.86 M and -\$2.435 M, respectively. Hence, the Expected Value Using Updated Probabilities is computed as follows:

Expected Value Using Updated Probabilities

$=(0.57)\cdot(\$13.86M) + (0.43)\cdot(-\$2.435M)$

$= \$6.85\ Million$

The Expected Value of the Screening Test (EVSI) is the improvement in the Expected Value Using Updated Probabilities over the Expected Value Decision (using no additional information and thus the initial probabilities):

$EVSI = Expected\ Value\ Using\ Updated\ Probabilities - EVD$

$= \$6.85M - \$4.5M$

$= \$2.35\ Million$

The studio executive should be willing to pay up to \$2.35 Million for the sample screening test information. As expected the Expected Value of Sample Information (\$2.35 Million) is less than the Expected Value of Perfect Information (\$7.5 Million). The solved decision tree for the horror movie decision is given in Figure 21.

Solved Horror Movie Decision Tree

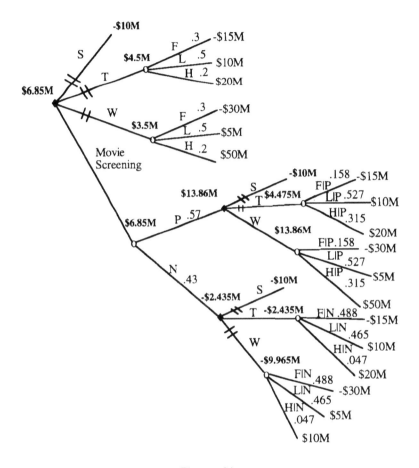

Figure 21

Note: The updated probabilities in Figure 21 can be computed using the definition of probability as proportion in the long run. For a random sample of 100 horror movies (see Figure 22), one would expect 30 to be a Flop (F), 50 to be a Limited Success (L), and 20 to be a Hit (H). For the 30 Flops (F), a movie screening would historically give 30% or 9 a Positive report (P∩F) and 70% or 21 a Negative report (N∩F). For the 50 Limited Successes (L), a movie screening would historically give 60% or 30 a Positive report (P∩L) and 40% or 20 a Negative report (N∩L). For the 20 Hits, a movie screening would historically give 90% or 18 a Positive report (P∩H) and 10% or 2 a Negative report (N∩H). Overall, 9 + 30 + 18 = 57 horror movies (0.57) would be given a Positive report and 21 + 20 + 2 = 43 horror movies (0.43) would be given a positive report. The updated probability P(F|P) is the P(F∩P) divided by the P(P) (i.e. the proportion of those movies given a Positive report which are a Flop). There are 57 Positive reports of which 9 are a

Flop. The probability of a Flop given a positive report P(F|P) is then 9/57 = 0.158. All other conditional probabilities may be computed in a similar fashion.

Conditional Probability Using Probability as a Proportion

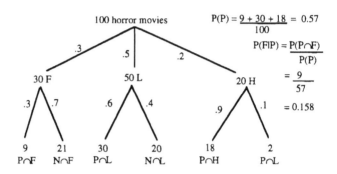

Figure 22

Exercise 2

Decision Analysis

Problem #1 (Decision Matrix)

An airline faces the problem of passenger "no shows" despite reservations. For any given full flight, there will be zero, one, two, or three "no shows" with probability 0.50, 0.25, 0.15, and 0.10, respectively. The airline must decide whether to overbook each full flight by zero, one, two, or three passengers. For each empty seat (caused by "no shows" exceeding overbooking), the airline loses the air fare (average revenue loss $600). For each extra passenger than seats available (caused by overbooking exceeding "no shows"), the airline must book the passenger on a different flight and must suffer the loss of goodwill (estimated loss $1000).

a) Determine the Expected Value (minimum expected cost) Decision among the actions of overbooking the full flight by zero, one, two, or three passengers. Use a decision matrix.

b) Determine the Minimax Decision (the action with the best worst case cost).

c) Formulate the decision tree that also could be used to solve for the Expected Value (minimum expected cost) Decision.

Problem #2 (Decision Tree)

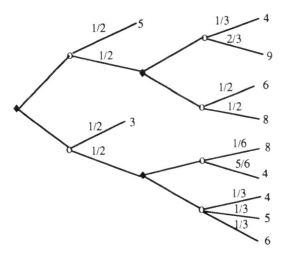

Consider the above sequential decision tree with payoffs and probabilities as given. Determine the Expected Value Decision. That is, solve the decision tree for the maximum expected payoff. Identify the best actions by pruning all others.

Problem #3 (Decision Tree)

A student needs to buy a new high-end computer to complete her college assignments. The college bookstore is offering students the following choices:

> • Buy this model new high-end computer *now* for $6000.

> • Lease a computer for two months ($600 total charge) at which time the college bookstore may have negotiated a 30% discount with Apple 🍎 (there is an 0.8 probability of successful negotiation). If negotiations are not successful, the student is guaranteed to be able to buy this new high-end computer for $6000.

a) Formulate a decision tree with actions, outcomes, payoffs, and probabilities.

b) Solve the decision tree to determine the Expected Value Decision (whether to Buy Now or Lease for two months).

c) How large must the total charge to Lease a computer for two months become in order for the Expected Value Decision to become Buy Now?

Problem #4 (Play/Don't Play/Pay For Information)

Consider a deck of 52 cards with 40 number cards (A-2-3-4-5-6-7-8-9-10 from four suits) worth their number value and 12 face cards (J-Q-K from four suits) worth 10 points each. A card player must decide whether or not to play the following game:

A card is chosen at random from the full deck of 52 cards. A card player has two options:

> • Bet $15 with a return of $60 (net of $45) if the card chosen is a face card and no return (net of -$15) if the card chosen is a number card.

> • Pay an additional $5 for the information whether the card is greater in value than 7. If the card is revealed to be >7, a card player may: 1) Bet $15 with the same return of $60 if the card chosen is a face card and no return if a number card; or 2) Quit forfeiting the $5. If the card is revealed to be ≤7, the player would obviously Quit forfeiting the $5.

a) Formulate a sequential decision tree with actions, outcomes, payoffs, and probabilities.

b) Determine the Expected Value (maximum expected payoff) Decision.

Chapter 2

Decision Analysis Additional Solved Problems

Problem #1 (Decision Analysis)

The Fizz Soft Drink Company has been very successful and demand for its products have increased steadily over the last few years. Fizz has decided to expand and is considering three levels of expansion: Small (S), Medium (M), and Large (L). Due to uncertainty in the marketplace, demand in the next few years for Fizz products may Remain Constant (RC), Increase Moderately (IM) or Increase Significantly (IS). Market research has indicated that the probability that future demand will Remain Constant, Increase Moderately, and Increase Significantly is 0.2, 0.3, and 0.5, respectively. The net revenue per year if Fizz expands Small and demand Remains Constant, Increases Moderately, and Increases Significantly is $-2 Million, $4 Million, and $8 Million, respectively. The net revenue per year if Fizz expands Medium and demand Remains Constant, Increases Moderately, and Increases Significantly is $-4 Million, $3 Million, and $11 Million, respectively. The net revenue per year if Fizz expands Large and demand Remains Constant, Increases Moderately, and Increases Significantly is $-7 Million, $1 Million, and $13 Million, respectively.

a) Formulate a decision matrix for this decision.

Answer: To formulate the decision matrix, first define its elements:

Actions = Expand Small (S), Medium (M), Large (L)

Outcomes = Remain Constant (RC), Increase Moderately (IM), Increase Significantly (IS)

Probabilities = RC (0.2), IM (0.3), IS (0.5)

Payoffs: Net Revenues

       Expand Small/Remain Constant = - $2M
       Expand Small/Increase Moderately = $4 M
       Expand Small/Increase Significantly = $8M

       Expand Medium/Remain Constant = - $4M
       Expand Medium/Increase Moderately = $3 M
       Expand Medium/Increase Significantly = $11 M

       Expand Large/Remain Constant = - $7 M
       Expand Large/Increase Moderately = $1 M
       Expand Large/Increase Significantly = $13 M

The Fizz Soft Drink Company expansion decision matrix is as follows:

Outcome / Action	Remain Constant RC (0.2)	Increase Moderately IM (0.3)	Increase Significantly IS (0.5)
Small (S)	- $2 M	$4 M	$8 M
Medium (M)	- $4 M	$3 M	$11 M
Large (L)	- $7 M	$1M	$13 M

b) Determine the Expected Value Decision (maximum expected payoff).

Answer: The expected value decision is computed as follows:

$$E(Small) = (0.2) \cdot (- \$ 2 \ M) + (0.3) \cdot (\$4 \ M) + (0.5) \cdot (\$8 \ M) = \$4.8 \ M$$
$$E(Medium) = (0.2) \cdot (- \$ 4 \ M) + (0.3) \cdot (\$3 \ M) + (0.5) \cdot (\$11 \ M) = \$5.6 \ M$$
$$E(Large) = (0.2) \cdot (- \$ 7 \ M) + (0.3) \cdot (\$1 \ M) + (0.5) \cdot (\$13 \ M) = \$5.4 \ M$$

The action with the highest expected payoff ($5.6 M) is to expand Medium.

c) Determine the Maximin Decision.

Answer: The Maximin decision is the action which maximizes the minimum (worst case) net revenue:

Worst Case payoff (Small) = - $2M
Worst Case payoff (Medium) = - $4 M
Worst Case payoff (Large) = - $7 M

The Maximin decision (the action with the best worst case payoff) is to expand Small.

d) Determine the Expected Value of Perfect Information (EVPI). Interpret the meaning of the EVPI for this decision.

Answer: The Expected Value of Perfect Information is computed as follows:

For each outcome, compute the best action and maximum payoff:

Remain Constant (0.2) $\Rightarrow$ Expand Small $\Rightarrow$ - $2 M
Increase Moderately (0.3) $\Rightarrow$ Expand Small $\Rightarrow$ $4 M
Increase Significantly (0.5) $\Rightarrow$ Expand Large $\Rightarrow$ $13 M

The Expected Value Under Certainty is given as:

$$\text{Expected Value Under Certainty} = (0.2) \cdot (- \$ 2 \ M) + (0.3) \cdot (\$4M) + (0.5) \cdot (\$13 \ M)$$
$$= \$7.3 \ M$$

The Expected Value of Perfect Information (EVPI) is given by:

EVPI = Expected Value Under Certainty - Expected Value Decision

= $7.3 M - $5.6 M
= $1.7 M

The EVPI is an upper bound on the value of any information to the decision maker.

e) Formulate this decision as a decision tree. Do not solve the decision tree.

Answer: The decision tree is given as follows:

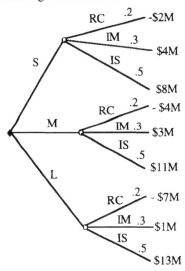

Problem #2 (Decision Trees)

The Schubert Theater Group has been offered the opportunity to buy a new play for production on Broadway. Plays on Broadway typically are either a success enduring for many years or a failure closing quite quickly. Without detailed knowledge of the play, the Schubert must rely on the historical probability that a play is a Success (S) and a Failure (F) which is given by 0.20 and 0.80, respectively. If a play is a success, the historical net revenue to Schubert is $600K. If the play is a failure, the historical net revenue is -300K.

a) Solve the following decision tree to decide whether the best action for the Schubert Theater Group is to Invest (I) or Don't Invest (D) in the new play:

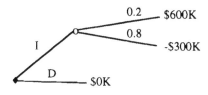

Answer: The solution to the initial decision tree is as follows:

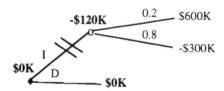

The best action (maximum payoff) is to Don't Invest with an expected value $0K.

b) A third possible action becomes available to the Schubert Theater Group. The noted critic Orson Bean is willing to read the play and provide the Schubert a Positive (P) or a Negative (N) review. In the past, when Orson has given a play script a Positive review, the play has gone on to be an enduring Success with probability 0.90 and a Failure with probability 0.10. When Orson has given a play script a Negative review, the play has gone on to be an enduring success with probability 0.20 and a Failure with probability 0.80.

The decision tree for this sample information scenario is given below. All the conditional probabilities have been computed. Solve the decision tree to find the optimum action(s) and expected value:

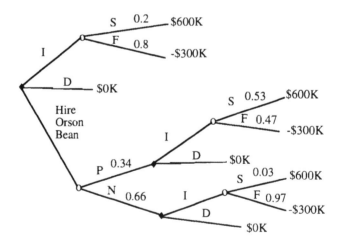

Answer: The solution to the decision tree using the updated probabilities for the sample information is as follows:

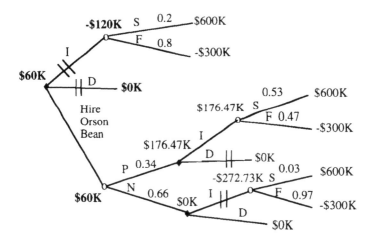

The optimum action is to Hire Orson Bean with a expected value of $60K. If there is a positive review, Invest in the play. If there is a negative review, Don't Invest in the play.

c) Note that the above decision tree assumed that Orson provided the review of the new play free of charge. What is the most the Schubert should be willing to pay Orson for his review of the play. That is, what is the value of his sample information?

Answer: The expected value using the updated probabilities is $60K. The expected value of the decision using no sample information was computed in a) to be $0K (Don't Invest). Hence, the most the Schubert Theater Group should be willing to pay Orson Bean is the increase over the expected value using no information ($60K - $0K = $60K). Mathematically,

EVSI = Expected Value Using Updated Probabilities - Expected Value Decision

= $60K - $0

= $60K

Problem #3 (Decision Analysis)

Consider the following decision matrix consisting of actions, outcomes, probabilities, and payoffs.

Outcome Action	Outcome 1 (0.20)	Outcome 2 (0.30)	Outcome 3 (0.25)	Outcome 4 (0.25)
Action 1	-2	3	8	6
Action 2	4	2	-3	-4
Action 3	6	4	3	2

a) What is the Maximin Decision and its value?

Answer: Examine the worst case payoff for each action:

Action 1 $\Rightarrow$ -2
Action 2 $\Rightarrow$ -4
Action 3 $\Rightarrow$ 2

The Maximin Decision (the action with the best worst case payoff) is Action 3 with a value of 2.

b) What is the Expected Value Decision (EVD) and its value?

Answer: The expected value of each action is computed as follows:

$$Expected\ Value\ (Action\ 1) = (-2)\cdot(0.20) + (3)\cdot(0.30) + (8)\cdot(0.25) + (6)\cdot(0.25)$$
$$= 4.0$$

$$Expected\ Value\ (Action\ 2) = (4)\cdot(0.20) + (2)\cdot(0.30) + (-3)\cdot(0.25) + (-4)\cdot(0.25)$$
$$= -3.5$$

$$Expected\ Value\ (Action\ 3) = (6)\cdot(0.20) + (4)\cdot(0.30) + (3)\cdot(0.25) + (2)\cdot(0.25)$$
$$= 3.65$$

The Expected Value Decision (EVD), the action with the highest expected payoff, is Action 1 which has a value 4.0.

c) What is the Expected Value of Perfect Information (EVPI)?

Answer: The maximum payoff for each outcome is as follows:

Outcome 1 $\Rightarrow$ Action 3 $\Rightarrow$ 6
Outcome 2 $\Rightarrow$ Action 3 $\Rightarrow$ 4
Outcome 3 $\Rightarrow$ Action 1 $\Rightarrow$ 8
Outcome 4 $\Rightarrow$ Action 1 $\Rightarrow$ 6

$$Epected\ Value\ Under\ Certainty = (6)\cdot(0.20) + (4)\cdot(0.3) + (8)\cdot(0.25) + (6)\cdot(0.25)$$
$$= 5.9$$

$$EVPI = Expected\ Value\ Under\ Certainty - EVD$$
$$= 5.9 - 4.0$$
$$= 1.9$$

d) In addition to the above three actions, the decision maker is also given the option to Buy Information. There is a probability that this information is positive (P) or negative (N). Based on whether the information is positive or negative, the probabilities of outcomes 1 through 4 are updated (using conditional probability).

The following decision tree provides the initial three actions denoted by A1 (Action 1), A2 (Action 2), and A3 (Action 3). Also, the additional action to Buy Information with its updated probabilities is provided. Solve the decision tree and determine the Expected Value of Sample Information (EVSI).

Decision Tree to Determine Expected Value of Sample Information (EVSI)

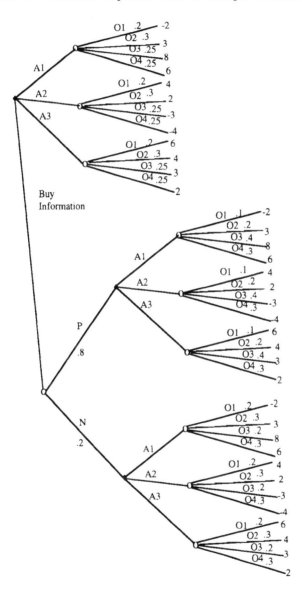

Answer: The Expected Value of Sample Information is computed as follows:

$$Expected\ Value\ Using\ Updated\ Probabilities\ =\ (0.8){\cdot}(5.4) + (0.2){\cdot}(3.9)$$
$$=\ 5.1$$

$$EVSI = Expected\ Value\ Using\ Updated\ Probabilities\ -\ EVD$$
$$=\ 5.1\ -\ 4.0$$
$$=\ 1.1$$

See decision tree on following page.

Solved Expected Value of Sample Information (EVSI) Decision Tree

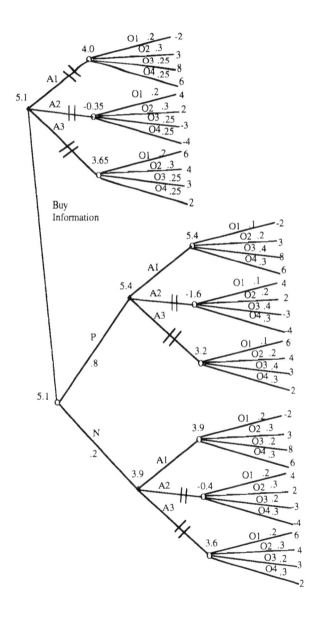

"There is safety in numbers."

Author Unknown

3 | Break-Even Analysis

3.1 Introduction

Definitions

Deterministic decisions involve the choice of actions where (it is assumed) that the parameters of the analysis are *fixed*.

- Deterministic decision: An order of calculators is 600 units. If the monthly demand for calculators at an electronics store is exactly 200 units per month, how often must the electronics store place orders for calculators.

Stochastic decisions involve the choice of actions where there is a *probabilistic* environment (a more realistic case).

- Stochastic decision: An order for 600 calculators just arrived. If the monthly demand for calculators is normally distributed with expected value 200 and standard deviation 20, when does the next order of calculators need to arrive in order to meet all demand with 90% confidence.

Deterministic Decisions

Simple deterministic decisions are ones in which there are only a few parameters and/or few available alternative actions to be evaluated. The parameters of the analysis are assumed to be fixed.

- Simple deterministic decision: The Make vs. Buy Decision (see Section 3.3) where there are only *two* alternative actions (Make and Buy) and a small number of parameters. The parameters of the analysis are assumed to be fixed.

Complex deterministic decisions are ones in which there are a large number of parameters and/or alternative actions to be evaluated. The parameters of the analysis are assumed to be fixed.

91

- Complex deterministic decision: Linear Programming analysis (see Chapter 9) which can sort efficiently through an *infinite* number of possible actions to determine the best action. The parameters of the analysis are assumed to be fixed.

Stochastic Decisions

Simple stochastic decisions are ones in which there are only a few parameters and/or a few available alternative actions to be evaluated. It is assumed there is uncertainty in a small number of the parameters of the analysis.

- Simple stochastic decision: Inventory analysis with stochastic demand (see Chapter 4) where the decision can be formulated with a small number of parameters and only the demand for inventory is stochastic.

Complex stochastic decisions are ones in which there are a large number of parameters and/or alternative actions to be evaluated. There may be uncertainty in numerous parameters of the analysis.

- Complex stochastic decision: Queueing analysis (see Chapter 10) which seeks the minimum number of servers required to meet the demand of customers. A large number of simplifying assumptions must be made to formulate a decision in a Queueing decision analysis framework.

Notes:

• There are only a few techniques (e.g. Linear Programming and Queueing) which can solve a complex decision *analytically* (where the solution can be expressed by an equation or an algorithm involving the input parameters). While some complex decisions can be modeled successfully by a Linear Programming or Queueing formulation, most complex decisions must be analyzed using Simulation (see Appendix A).

• Decisions faced by a decision maker on an everyday basis are often complex (many parameters and alternative actions) and stochastic (probabilistic) in nature. However, there are numerous reasons to attempt to analyze a complex stochastic decision as if the decision were simple (only a few alternatives) and deterministic (fixed parameters):

- *Data Availability:* The detailed data required to formulate complex problems is often difficult to obtain or non-existent.
- *Cost:* The data collection to estimate probability distributions and payoffs is expensive.
- *Time:* Model formulation, initial simulation runs, debugging, production runs, data analyses, etc. is a time consuming.

• Simulation is a powerful technique for solving complex (especially stochastic) decisions but its utility is sometimes limited due to its being data intensive, costly, and time consuming. When a quick, relatively inexpensive solution is required, analytical techniques can be used to give an approximate solution. If resources allow (manpower, money, schedule), Simulation can be applied later in order to refine the solution.

• When simplifying a complex and/or stochastic decision into a simple and/or deterministic decision, the fidelity of the decision process must be examined before applying the results:

- Have the essential characteristics of the problem been simplified or altered (in order to apply a specific solution technique) to the extent that the results are no longer valid?

• In general, the decision maker should attempt to analyze a decision using the *simplest* formulation which captures the essence of the problem and where the solution yields useful results (see Figure 1). That is, simplify as much as possible but no more!

Classification of Decisions

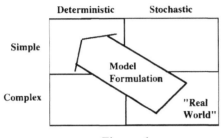

Figure 1

3.2 Break-Even Volume

Definitions

A fixed cost is the portion of the overall production cost that is *independent* of the amount of output produced. The fixed cost occurs even if *no* output is actually produced.

A variable cost is the portion of the overall production cost that *varies directly* with the amount of output produced.

Break-Even Volume (Simple Deterministic Decision)

A business has decided it is more economical to Make (see Section 3.3) a new product (either a manufactured item or a service) in-house. The cost for the business to Make a product typically consists of two components: a *fixed cost* (floor space, personnel, utilities, etc.) to set up and administer the capability to manufacture an item or deliver a service; and a *variable cost* (the marginal cost of manufacturing or delivering one more unit).

The initial interest for a business to Make a product to its customers is based on the expectation that the product can be sold at a price higher than the variable cost to manufacture or deliver the product. However, the final decision whether or not to offer the product is based on whether the *demand* is large enough so that the revenue on the volume of items sold will offset *both* the fixed and variable costs of manufacture or delivery.

The volume of demand where the revenue for the items sold equals the sum of the fixed and variable costs of manufacture or delivery is called the Break-Even Volume. Below this volume of demand, it does not make sense from an economic standpoint to Make the product. Above this volume of demand, it does makes sense from an economic standpoint to Make this product. As usual, other factors (e.g. the product may be a loss leader to increase the volume of demand for other more profitable products) may influence the simple deterministic decision whether or not to offer the product to its customers.

Example 1: (Break Even Volume). Mercy Hospital is contemplating purchasing a Digitalized Orthogonal Gastrointestinal (DOG) scanner. The fixed cost associated with purchasing the DOG, remodeling present facilities at Mercy Hospital to accommodate the DOG, hiring and training of personnel, etc. totals approximately $1M per year over the next ten years (the expected life of the DOG). The variable cost to Mercy Hospital per DOG scan averages $150 while a survey of other hospitals indicates that the going price of a DOG scan averages $400.

a) Determine the Break-Even Volume of yearly demand for DOG scans analytically and graphically.

Answer: For each DOG scan administered by Mercy Hospital, there is a cost of $150 and a revenue of $400 (for a profit of $250). To determine the Break-Even Volume of x scans/year, total revenue is set equal to total cost for the DOG scanner:

$$Total\ Revenue\ =\ Total\ Cost\ (Fixed\ Cost\ +\ Variable\ Cost)$$

$$\$400 \cdot (x\ scans/yr)\ =\ \$1,000,000/yr\ +\ \$150 \cdot (x\ scans/yr)$$

$$\$250 \cdot (x\ scans/yr)\ =\ \$1,000,000/yr$$

$$Break\text{–}Even\ Volume\ =\ 4,000\ scans/year$$

Mercy should pursue the DOG if the volume of demand is forecasted to be above 4,000 scans/year. Graphically, the Break-Even Volume analysis is illustrated in Figure 2.

Mercy Hospital Break-Even Volume Analysis

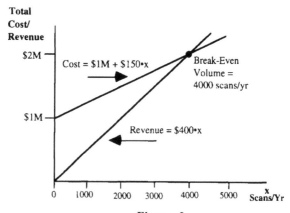

Figure 2

94

b) Assume the expected demand for DOG scans has been forecasted to be 5000 scans/year and Mercy Hospital pursues the DOG. If the yearly demand for scans is actually a Normal Random Variable with expected value 5000 and standard deviation 600, what is the probability Mercy Hospital will lose money in a given year?

Answer: The yearly demand for DOG scans is actually a Normal Random Variable $X \sim N(\mu,\sigma^2)$ with $\mu = 5000$ and $\sigma = 600$. Mercy Hospital will lose money if the yearly demand is less than 4000 scans/year. This probability is computed as follows:

$$P(X < 4000) = P(\frac{X - 5000}{600} < \frac{4000 - 5000}{600})$$

$$= P(Z < \frac{-1000}{600}) \qquad (\frac{X - 5000}{600}) = Z \sim N(0,1)$$

$$= P(Z < -1.66)$$

$$= 0.0485 \qquad (See\ Table\ 1b)$$

3.3 Make vs. Buy Decision

A major factor in the decision for a business whether to <u>Make</u> vs. <u>Buy</u> a product (a simple deterministic decision) is the specification of the fixed cost vs. variable cost (see Figure 3). Again, a product can be a service as well as a manufactured item. For example, a business often must decide between having in-house services (Make) such as payroll, graphics, maintenance, etc. vs. out-sourcing (Buy) these services to vendors.

Fixed Cost vs. Variable Cost

	Make Decision	Buy Decision
Fixed Cost (Set Up Production)	**Yes** (Independent of Output)	**No** (May be a nominal cost)
Variable Cost (Cost per Unit)	**Lower** (Produced In-house)	**Higher** (Purchased from Vendor)

Figure 3

The Make decision entails a fixed cost (e.g. salaries, floor space, utilities, etc.) to set up an output production capability. There is a lower variable cost per unit of output since the product is *produced in-house*.

The Buy decision does not require a production capability to be set up. There is no fixed cost (or sometimes a nominal fixed cost). There is a higher variable cost per unit of output since the product is *purchased from an outside vendor* (who absorbs the fixed cost).

Example 2: (Make vs. Buy Decision). The EZ Chair Furniture Factory has decided to expand its output of chairs in order to meet the increased demand it has forecasted for the next few years. The expansion being considered by EZ Chair is in the range of up to 1500 additional chairs/month over its current output. The owner of EZ Chair must decide whether to Make or Buy the additional chairs required to expand output. The resources and costs required to expand output in the range of up to 1500 additional chairs/month for the actions Make and Buy are as follows:

Make	Lease more factory space and equipment ($2000/month) Hire an additional foreman ($7000/month) Hire two additional workers (each $1000/month) The unit cost per chair produced in-house will be $10
Buy	Contract with a nearby small furniture factory The unit cost per chair purchased from this factory will be $22

For any given choice by EZ Chair for the level to expand output (in the range of 0 to 1500 additional chairs/month), determine whether it is economical (minimum total monthly cost) for the EZ Chair Furniture Factory to Make or Buy chairs.

Answer: The decision is to determine which of the two actions Make or Buy (a simple deterministic decision) minimizes the total monthly cost TC(y) made up of a fixed cost FC(y) and a variable cost VC(y) for expanding output to a level y chairs over the current output. That is, the decision solves the expression:

Min TC(y)

or, equivalently,

Min FC(y) + VC(y)

for y in the range of up to 1500 additional chairs/month.

Examine the Make decision: The fixed cost $FC_{Make}(y)$ to Make the expanded output of y chairs/month (no matter how many or few) includes leasing more factory space ($2000/month) and hiring both a foreman ($7000/month) and two workers ($1000/month + $1000/month) for an overall total of $11,000/month. The variable cost $VC_{Make}(y)$ to Make the expanded output of y chairs/month in-house is $10y/month. The total monthly cost $TC_{Make}(y)$ to Make the expanded output of y chairs/month is:

$$TC_{Make}(y) = FC_{Make}(y) + VC_{Make}(y)$$

$$= \$11,000 + \$10y$$

Examine the Buy decision: The fixed cost $FC_{Buy}(y)$ to Buy the expanded output of y chairs/month is zero. The variable cost $VC_{Buy}(y)$ to Buy the y chairs/month from the nearby furniture factory is $22y/month. The total monthly cost $TC_{Buy}(y)$ to Buy y chairs/month is:

$$TC_{Buy}(y) = FC_{Buy}(y) + VC_{Buy}(y)$$

$$= \$0 + \$22y$$

The equations for the total monthly costs, $TC_{Make}(y)$ and $TC_{Buy}(y)$, associated with the Make and Buy actions, respectively, for EZ Chair are plotted in Figure 4.

Total Monthly Costs For EZ Chair Make vs. Buy Decision

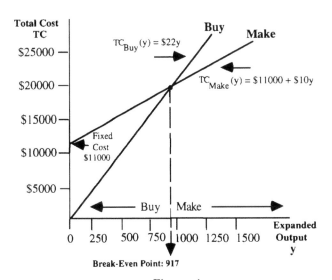

Figure 4

Notes: Assume that EZ Chair chooses to expand it monthly output by y chairs. From the total monthly cost lines for the Make and Buy actions in Figure 4, it can be observed that:

• If the chosen level of expanded monthly output y for chairs is low, it is economical to Buy due to the substantial fixed costs (\$11,000/month) associated with the Make decision.

• As the expanded monthly output for chairs increases, the savings from the lower variable costs of making the chairs (\$10 per chair to Make vs. \$22 per chair to Buy) start to become substantial and the total cost gap between Make and Buy narrows.

• At some higher level of expanded monthly output y, the most economical action will *switch* from Buy to Make. The level of output at which total costs associated with the Make and Buy actions are equal is called the <u>Break-Even Point</u>. EZ Chair is indifferent between the Make and Buy actions at this level of expanded monthly output.

Algebraically, the Break-Even Point (where the total monthly cost lines for Make and Buy intersect in Figure 4) can be determined by setting total monthly cost for the Make action $TC_{Make}(y)$ equal to the total monthly cost for Buy action $TC_{Buy}(y)$:

$$TC_{Make}(y) = TC_{Buy}(y)$$

$$\$11000 + \$10y = \$0 + \$22y$$

$$\$11000 = \$12y$$

$y \sim 917$ *additional chairs/months output is the Break–Even Point*

Notes:

• If the expanded output chosen by EZ Chair is below the Break-Even Point of 917 additional chairs/month, the action which minimizes total monthly cost is to Buy the additional chairs from the nearby furniture factory.

• If the expanded output chosen by EZ Chair is above the Break-Even Point of 917 additional chairs/month, the action which minimizes total monthly cost is to Make the additional chairs by leasing the factory space and hiring the additional foreman and workers.

• If the expanded output chosen by EZ Chair is close to the Break-Even Point of 917 additional chairs/month, other factors (e.g. strategic plan for expanding the company, quality control for the chairs, ability to control the production schedule, etc.) become more important than total monthly cost in the decision whether to Make or Buy.

3.4 Sensitivity Analysis

The best action for the Make vs. Buy decision given in Example 2 is dependent on the fixed values of the cost parameters:

- *Lease costs* (*$2000 / month*)
- *Hire additional foreman* (*$7000 / month*)
- *Hire two additional workers* (*each* $1000 / month)
- *Unit Make cost* (*$10 per chair*)
- *Unit Buy cost* (*$22 per chair*)

Sensitivity analysis examines the effect of changes in the values of the Make vs. Buy cost parameters on the Break-Even Point.

Example 3: (Make vs. Buy Decision Sensitivity Analysis). The $7000/month salary for the additional foreman in Example 2 is an *average* salary for a foreman at the EZ Chair Furniture Factory. Depending on the individual chosen and his or her qualifications, the salary for an additional foreman can range from $5000/month to $9000/month.

Examine the sensitivity of the Make vs. Buy decision in Example 2 to the \$7000/month salary for the additional foreman. Find the Break-Even Point for the Make vs. Buy decision when the foreman salary is at the low end of the salary range (\$5000/month) and at the high end of the salary range (\$9000/month).

Answer: Recall the Make vs. Buy decision from Example 2 where the foreman's salary was set at \$7000/month and the overall fixed cost for Make was \$11,000/month. For an expanded monthly output below the Break-Even Point of 917 chairs, the economical (minimum monthly cost) action is to Buy. For an expanded monthly output above 917 chairs, the economical action is to Make.

A change in the salary of the foreman (a fixed cost) will have no affect on the total monthly cost of the Buy action which remains the purchase price from the vendor of \$22 per chair:

$$TC_{Buy}(y) = FC_{Buy}(y) + VC_{Buy}(y)$$

$$= \$0 \ + \ \$22y$$

When the \$7000/month salary of the foreman is reduced to \$5000 or is increased to \$9000, the overall fixed cost of \$11000/month is reduced to \$9000 or is increased to \$13000, respectively. The total monthly cost of the Make action becomes:

$$TC_{Make}(y) = FC_{Make}(y) + VC_{Make}(y)$$

$$= \$9000 \ + \ \$10y \qquad (Foreman\ salary = \$5000\ /\ month)$$

$$TC_{Make}(y) = FC_{Make}(y) + VC_{Make}(y)$$

$$= \$13000 \ + \ \$10y \qquad (Foreman\ salary = \$9000\ /\ month)$$

The Break-Even Point for the Make vs. Buy decision when the foreman's monthly salary is reduced from \$7000/month to \$5000/month is calculated by setting the total monthly cost for the Make and Buy actions equal:

$$\$9,000 \ + \ \$10y \ = \ \$0 \ + \ \$22y$$

$$\$12y \ = \$9,000$$

$$y = 750\ additional\ chairs\ /\ month$$

Note: As the foreman's salary dropped from \$7000/month to \$5000/month, the fixed cost dropped from \$11000/month to \$9000/month. Due to the smaller fixed cost, it should be expected that a smaller expanded output will be required before it becomes economical to Make the chairs in-house. In fact, the Break-Even Point dropped from 917 to 750 additional chairs/month.

The Break-Even Point for the Make vs. Buy decision when the foreman's monthly salary is increased from $7000/month to $9000/month is also calculated by setting the total monthly costs for the Make and Buy actions equal:

$$\$13,000 + \$10y = \$0 + \$22y$$

$$\$12y = \$13,000$$

$$y \sim 1083 \; additional \; chairs \, / \, month$$

Note: As the foreman salary increased from $7000/month to $9000/month, the fixed cost increased from $11000/month to $13000/month. Due to the larger fixed cost, it should be expected that a larger expanded output will be required before it becomes economical to Make the chairs in-house. In fact, the Break-Even Point increased from 917 to 1083 additional chairs/month.

The equations for the total monthly cost (TC) for Make vs. Buy with foreman salaries of $5000/month and $9000/month as well as the Break-Even Points are plotted in Figure 5.

Sensitivity of Make vs. Buy Decision to the Foreman's Salary

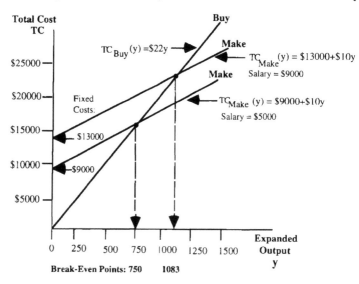

Figure 5

The sensitivity of the Break-Even Point for the Make vs. Buy decision to the foreman's salary can be summarized in Figure 6.

Break-Even Points For EZ Chair Make vs. Buy Decision

	Foreman Salary $5000/Month	Foreman Salary $7000/Month	Foreman Salary $9000/Month
Break-Even Point	Output of 750 Additional Chairs/Month	Output of 917 Additional Chairs/Month	Output of 1083 Additional Chairs/Month

Figure 6

Notes: The Break-Even Points in Figure 6 indicate the following sensitivity of the Make vs. Buy decision to the foreman's salary:

• When the expanded output level chosen by EZ Chair is quite low, the Buy action will be economical even when the fixed cost is nominal. For example, if the expanded output level chosen for chairs is less than 750/month, the Buy action is best even if the foreman's salary is as low as $5000/month (and the fixed cost is nominal at $9000/month). There is not enough additional output to offset even the lower fixed cost associated with the $5000 per month foreman salary.

• When the expanded output level chosen by EZ Chair is quite high, the Make action will be economical even when the fixed cost is substantial. For example, if the expanded output level chosen for chairs is greater than 1083/month, the Make action is economical even if the foreman's salary is as high as $9000/month (and the fixed cost is substantial at $13000/month). There is enough additional output to offset even the higher fixed cost associated with the $9000/month foreman salary.

• When the expanded output level chosen by EZ Chair is in a mid-range between low and high, the decision Make vs. Buy becomes *sensitive* to the value of the fixed costs. For example, if the expanded output level for chairs is between 750 and 1083 per month, the decision is sensitive to the foreman's salary. More information should be gathered by on the exact salary requirements of the intended foreman before deciding whether to Make or Buy to meet the chosen expanded output level for chairs/month.

3.5 Extensions to Make vs. Buy Decision

Often the fixed cost will not stay *constant* over the entire range of the product output which may be chosen by the decision maker to meet an increased demand forecasted. More realistically, the fixed cost will vary as a function of product output in an increasing *stepwise* fashion.

Fixed cost (FC) increasing with output in a stepwise manner is shown notionally in Figure 7.

Stepwise Fixed Cost vs. Output

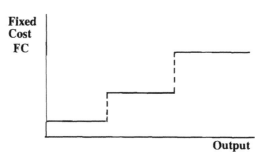

Figure 7

Example 4: (Make vs. Buy Decision With Stepwise Fixed Cost). In the EZ Chair Make vs. Buy decision given in Example 2, assume that the fixed cost associated with different ranges of expanded output required to meet the increased monthly demand forecasted for chairs are as given in following table:

Stepwise Fixed Cost vs. Range of Expanded Output

Range of Expanded Output	Additional Requirements	Fixed Cost
0-750 additional chairs/month	One full time foreman Two additional workers 2000 square feet factory space	$9,000 Per Month
750-1250 additional chairs/month	One full time foreman Three additional workers 3000 square feet factory space	$12,000 Per Month
1250-1500 additional chairs/month	One full time foreman One part time foreman Four additional workers 4000 square feet factory space	$16,000 Per Month

Also, assume that the EZ Chair Furniture Factory has lowered the cost of chairs purchased from the nearby furniture factory from $22 per chair to $20 per chair by providing a *guaranteed* payment of $1000 each month (regardless of the number of chairs purchased from the nearby furniture factory during the month).

For any given choice by EZ Chair for the level to expand output (in the range of 0 to 1500 additional chairs/month), determine whether it is economical (minimum total monthly cost) for the EZ Chair Furniture Factory to Make or Buy chairs.

Answer: Examine the Make action. The variable cost $VC_{Make}(y)$ remains at $10 per chair. The fixed cost $FC_{Make}(y)$ is a stepwise increasing function of the expanded output y. The total cost $TC_{Make}(y)$ including fixed and variable costs is given by:

$$TC_{Make}(y) = FC_{Make}(y) + VC_{Make}(y)$$

The variable cost (VC_{Make}) for the Make action for the expanded output level y is given by:

$$VC_{Make}(y) = \$10y$$

The fixed cost (FC_{Make}) for the ranges of expanded output for the Make action is given by:

$$FC_{Make}(y) = \$9,000 \qquad \qquad (Output\ 0 - 750)$$

$$= \$12,000 \qquad \qquad (Output\ 750 - 1250)$$

$$= \$16,000 \qquad \qquad (Output\ 1250 - 1500)$$

Hence, the total cost (TC_{Make}) of the Make action for the different ranges of expanded output is given by the following three line segments:

$$TC_{Make}(0 \leq y < 750) = \$9,000 + \$10y$$

      Total cost for 0 chairs: $9,000
      Total cost for 750 chairs: $16,500
      Total cost plots as a line segment: (0,$9000) to (750,$16500)

$$TC_{Make}(750 \leq y < 1250) = \$12,000 + \$10y$$

      Total cost for 750 chairs: $19,500
      Total cost for 1250 chairs: $24,500
      Total cost plots as a line segment: (750,$19500) to (1250,$24500)

$$TC_{Make}(1250 \leq y \leq 1500) = \$16,000 + \$10y$$

      Total cost for 1250 chairs: $28,500
      Total cost for 1500 chairs: $31,000
      Total cost plots as a line segment: (1250,$28500) to (1500,$31000)

Examine the Buy action. There is a now a nominal fixed cost $FC_{Buy}(y)$ of $1000 which EZ Chair pays the nearby furniture factory regardless of the output y of chairs purchased each month. However, the variable cost $VC_{Buy}(y)$ has been lowered to $20 per chair. The total cost $TC_{Buy}(y)$ including fixed and variable costs is given by:

$$TC_{Buy}(y) = FC_{Buy}(y) + VC_{Buy}(y)$$

The variable cost is given by:

$$VC_{Buy}(y) = \$20y$$

The fixed cost (the guaranteed payment each month) is given by:

$$FC_{Buy}(y) = \$1,000$$

Hence, the total cost of the Buy action for the expanded output y is given by:

$$TC_{Buy}(y) = \$1,000 + \$20y \qquad (Output\ 0\ -\ 1500)$$

Total cost for 0 chairs: $1000
Total cost for 1500 chairs: $31000
Total cost plots as a line segment: (0,$1000) to (1500,$31000)

The total monthly costs (TC) for the Make and Buy actions with the stepwise increasing fixed cost are plotted as a function of the expanded output y in Figure 8.

EZ Chair Make vs. Buy Total Costs With Stepwise Fixed Costs

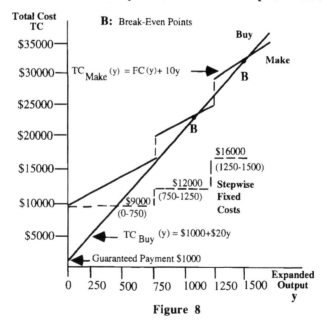

Figure 8

104

From the total monthly costs for Make and Buy (with stepwise fixed costs) in Figure 8, it can be observed that *more than one* Break-Even Point may exist for step-wise fixed cost Make vs. Buy decisions. More than one Break-Even Point is indicative that the cost differential between the Make and Buy actions is close over an extended output range.

In Figure 8, there are two Break-Even Points: at expanded monthly outputs of 1100, and 1500 chairs/month. These Break-Even Points can be derived by equating the TC_{Make} and TC_{Buy} for each step (range) for fixed costs individually. The Make and Buy cost lines will intersect at some point since they have different slopes (different costs per unit of output). If the intersection point falls within the range of the step for which the total costs have been equated, then the point is a Break-Even Point. If the intersection point falls outside the range of the step, there is no Break-Even Point in that step for fixed costs.

Once the Make and Buy costs are plotted as in Figure 8, one can observe in which steps the Make and Buy lines cross-over and derive Break-Even points algebraically only for those steps. Figure 8 seems to indicate a Break-Point two of the three steps for the fixed costs. Each step (range of outputs) will be checked for a possible Break-Even Point.

- Output of 0 - 750 chairs/month: Fixed Cost to Make = $9,000.

$$TC_{Make}(y) = TC_{Buy}(y)$$

$$\$9,000 + \$10y = \$1000 + \$20y$$

$$\$8,000 = \$10y$$

$$y = 800 \ additional \ chairs/month$$

Since 800 additional chairs/month is not in the range of this step (0 - 750) for fixed cost, this is not a Break-Even Point.

- Output of 750 - 1250 chairs/month: Fixed Cost to Make = $12,000.

$$TC_{Make}(y) = TC_{Buy}(y)$$

$$\$12,000 + \$10y = \$1000 + \$20y$$

$$\$11000 = \$10y$$

$$y = 1100 \ additional \ chairs/month$$

Since 1100 additional chairs/month is in the range of this step (750 -1250) for fixed cost, this is a Break-Even Point.

- Output of 1250 - 1500 chairs/month: Fixed Cost to Make = $16,000.

$$TC_{Make}(y) = TC_{Buy}(y)$$

$16,000 + $10y = $1000 + $20y$

$15,000 = $10y$

$y = 1500\ additional\ chairs\ /\ month$

Since 1500 additional chairs/month is in the range of this step (1250 - 1500) for fixed cost, this is a Break-Even Point.

Notes: From the total monthly costs for the EZ Chair Make vs. Buy decision with stepwise increasing fixed cost plotted in Figure 8, it can be observed that for Example 4:

• The Buy action (as expected) is best for a low level of expanded output (below 1100 chairs/month). Even though it is cheaper on a per unit (marginal) basis to Make the chairs in-house, it is still more economical to Buy due to the substantial fixed costs associated with the Make action.

• For larger levels of expanded output (between 1100 and 1500 chairs/month), the total monthly costs for Make and Buy are close. The decision for the EZ Chair Furniture Factory whether to Make or Buy will probably be made on *other* criteria besides cost (e.g. convenience, future growth plans, quality control, etc.).

3.6   Time Value of Money

Introduction

Two basic principles of the time value of money are:

- An given amount of money received today has *greater* value than the same amount of money received in the future.

- A smaller amount of money received today may have *equal or greater* worth than a larger amount of money received in the future.

Time value of money provides the mathematical relationship between money received today and money received in the future:

- Which has greater value?

- Under what conditions are they equivalent?

Note: The convention used in this presentation of the time value of money is that all revenues or costs occur (unless otherwise stated) at the *end* of the time periods (e.g. a revenue received in year 1 is received after 1 year has elapsed).

Example 5: An investment provides a guaranteed $600 in revenue paid out over the next four years. The investor is given the choice of the following two revenue flows:

Revenue Flow	Year 1	Year 2	Year 3	Year 4
#1	$400	$100	$50	$50
#2	$150	$150	$150	$150

Which $600 total revenue flow should be of greater value to the investor?

Answer: Both revenue flows pay out $600 over four years. However, Revenue Flow #1 returns a larger portion of the $600 total revenue in the early years than Revenue Flow #2. This money can then be *reinvested* to earn more money. The best case revenue flow for the investor would be to receive the total $600 revenue in Year 1.

Compound Interest

Compound interestcomputes the amount of revenue which accrues to the investor in the future if an amount of revenue is left in an interest bearing account over one or more interest (time) periods.

Example 6: If $1000 were invested today at 6% interest compounded annually, what revenue would an investor have accrued at the end of 5 years?

Answer: At the end of each year (the interest period), the investor would have accrued a revenue equal to 1 plus the yearly interest rate times the amount of revenue in the account at the start of the year. At the end of the first year, the investor would have:

$$\$1000 \cdot (1 + 0.06) = \$1000 \,(\textbf{1.06}) \qquad (\textit{See Table 7})$$
$$= \$1,060$$

At the end of the second year, the investor would have:

$$\$1060 \cdot (1 + 0.06) = \$1000 \cdot (1 + 0.06)^2 \qquad (\textit{See Table 7})$$
$$= \$1000 \cdot (\textbf{1.1236})$$
$$= \$1,123.60$$

Similarly, at the end of the fourth year, the investor would have accrued $1262. Hence, at the end of the fifth year, the investor would have:

$$\$1262 \cdot (1 + 0.06) = 1000 \cdot (1 + 0.06)^5 \qquad (\textit{See Table 7})$$
$$= \$1000 \cdot (\textbf{1.3382})$$
$$= \$1,338.20$$

The factors given in **bold**, the compound interest multiplication factors for the revenue accrued in future interest periods from an investment of $1 at 6% per period, are given in Table 7.

Example 7: If $1000 were invested today at 8% annual interest compounded *quarterly*, what revenue would an investor have accrued at the end of 5 years?

Answer: Over 5 years there are 20 quarters. At the end of each quarter, the investor will receive an additional revenue equal to the yearly interest rate divided by the number of quarters in a year times the amount of revenue in the account at the start of the quarter:

$$\frac{8\% \ per \ year}{4 \ quarters \ per \ year} = 2\% \ per \ quarter$$

At the end of year 5 (the 20th quarterly interest period), the investor would have:

$$\$1000 \cdot (1 + 0.02)^{20} = \$1000 \cdot (\mathbf{1.4859}) \qquad (See \ Table \ 7)$$
$$= \$1,485.90$$

Notes:

• If the $1000 had been invested at 8% *annual* interest, the amount of revenue the investor would have accrued in 5 years is given by:

$$\$1000 \cdot (1 + 0.08)^{5} = \$1000 \cdot (\mathbf{1.4693}) \qquad (See \ Table \ 7)$$
$$= \$1,469.30$$

The investor accrues more revenue when the $1000 is compounded (at the same interest rate) *quarterly* ($1485.90) rather than *yearly* ($1469.30) for 5 years.

• $1000 invested at 8% annual interest compounded annually, quarterly, and daily would result in an accrued revenue at the end of 1 year of $1,080.00, $1,082.40, and $1,083.28, respectively. The accrued revenue for $1000 invested for 1 year at 8% annual interest compounded *daily* is computed as follows:

$$\frac{8\% \ per \ year}{365 \ days \ per \ year} = \frac{0.08 \ per \ year}{365 \ days \ per \ year} = 0.000219178 \ per \ day$$

$$\$1000 \cdot (1 + 0.000219178)^{365} = \$1000 \cdot (\mathbf{1.0832775}) \qquad (Use \ calculator)$$
$$= \$1,083.28$$

• A fundamental property of compound interest revenue or cost is that more frequent compounding of interest is advantageous or disadvantageous, respectively, to the investor!

Present Value

Present value (PV) is the value *today* of a future revenue or a flow of future revenues.

Example 8: If it is possible to earn 10% interest yearly on money invested, what is the present value to an investor of $1000 in revenue 5 years from now?

Answer: If $1 were invested today, the revenue that would accrue in 5 years (at 10% yearly interest) is:

$1 \cdot (1 + 0.10)^5 = \$1 \cdot (1.61)$ (*See Table* 7)

$\qquad\qquad\qquad = \$1.61$

The present value of $1000 in 5 years (which is the money which must be invested *today* to return a revenue $1000 in 5 years at 10% yearly interest) is given by:

$PV = \$1000 / (1 + 0.10)^5$

$\quad = \$1000 / 1.61$

$\quad = \$1000 \cdot (0.621)$ (*See Table* 8)

$\quad = \$621$

This investor is indifferent between revenues of $621 today and $1000 in 5 years. The value 0.621 is called the single payment present value factor of $1 received in 5 interest periods from now at 10%. Single payment present value factors are provided in Table 8.

Example 9: If the return on investment is 10% yearly, determine the present value of the following revenue flow given in Example 5:

	Year 1	Year 2	Year 3	Year 4
Revenue Flow	$400	$100	$50	$50

Answer: The present value (PV) of the above revenue flow is computed by *discounting* (at 10%) the revenue in each year by the number of years into the future when the revenue is realized:

$PV = \$400/(1 + 0.10) + \$100/(1 + 0.10)^2 + \$50/(1 + 0.10)^3 + \$50/(1 + 0.10)^4$

$\quad = \$400 \cdot (0.909) + \$100 \cdot (0.826) + \$50 \cdot (0.751) + \$50 \cdot (0.683)$ (*See Table* 8)

$\quad = \$517.90$

Uniform Series Present Value

The uniform series present value is the present value (the value today) of a sequence of *equal* year end revenues.

Example 10: If the return on investment is 10% yearly, determine the uniform series present value of the following sequence of equal year end revenues given in Example 5:

	Year 1	Year 2	Year 3	Year 4
Revenue Flow	$150	$150	$150	$150

Answer: The present value of the equal yearly revenues of $150 for four years (at 10% yearly interest) is given by:

$$PV = \$150 \cdot \{ [1/(1 + 0.10)] + [1/(1 + 0.10)^2] + [1/(1 + 0.10)^3] + [1/(1 + 0.10)^4] \}$$

$$= \$150 \cdot \{0.909 + 0.826 + 0.751 + 0.683\} \qquad (See\ Table\ 8)$$

$$= \$150 \cdot (\mathbf{3.170}) \qquad (See\ Table\ 9)$$

$$= \$475.90$$

The value 3.170 is called the uniform series present value factor for equal $1 revenues for 4 consecutive interest periods at 10%. Uniform series present value factors are given in Table 9.

Notes:

• The uniform series present value of the revenue flow of four yearly $150 payments in Example 10 is *less* than $600 today. This occurs since an amount of money in the future is worth less than the same amount today.

• The present value (at 10% yearly interest) of the $600 total revenue flow in Example 9 of $517.90 is greater than the present value of the $600 total revenue flow in Example 10 or $475.90. This proves quantitatively the qualitative comparison of these two $600 revenue flows made in Example 5 which concluded that the first flow providing more money early is preferred.

• A fundamental property of present value is that it is advantageous to the investor to receive the larger portion of the revenue flow *early* (so that it can be reinvested).

Capital Recovery Factor

The capital recovery factor is used amortize a lump sum debt over N equal payments (e.g. paying off the cost of a house through a monthly mortgage payment for 15 years).

Example 11: An electronics store sells stereos for $1000. What should the store charge a customer in each of 5 equal yearly payments if the applicable interest rate is 18% per year?

Answer: The present value (PV) of the electronics store charging $1 (a revenue flow of $1) for each of 5 years discounted at 18% is:

$$PV = \$1 \cdot \{ [1/(1 + 0.18)] + [1/(1 + 0.18)^2] + \cdots + [1/(1 + .18)^5] \}$$

$$= \$1 \cdot \{0.847 + 0.718 + 0.609 + 0.516 + 0.437\} \qquad (See\ Table\ 8)$$

$$= \$1 \cdot (\mathbf{3.127}) \qquad (See\ Table\ 9)$$

$$= \$3.13$$

Hence, in order to earn $1000 present value at 18% interest over 5 years, the electronics store would have to charge annually:

$1000 / \{1 / [1/(1 + 0.18)] + [1/(1 + 0.18)^2] + \cdots + [1/(1 + 0.18)^5]\}$

$= \$1000 / (3.127)$

$= \$1000 \cdot (\textbf{0.320})$ (*See Table* 10)

$= \$320$

The electronics store must charge $320 for each of the next five years ($1600 total revenue over 5 years) in order to recoup the present value of the $1000 stereo. The amount 0.320, called the capital recovery factor, is charged in equal payments for 5 consecutive interest periods to recoup $1 at 18%. Capital recovery factors are provided in Table 10.

Example 12: What would be the annual payment *for 25 years* to buy a home with a $100,000 mortgage remaining (after the down payment) at 16% yearly interest?

Answer: From Table 10, a borrower must pay 16.4 cents every year for 25 years to pay off $1 loaned today at 16% interest. To pay off $100,000, the borrower must pay annually:

$100,000 \cdot (\textbf{0.164}) = \$16,400$ *per year*   (*See Table* 10)

Note: Over 25 years, the total payments by the borrower equal $16,400 x 25 = $410,000. Of this amount, $310,000 represents total interest incurred to pay off the principal of $100,000.

Example 13: What would be the annual payment *for 15 years* to buy a home with a $100,000 mortgage remaining (after the down payment) at 16% yearly interest?

Answer: From Table 10, the borrower must pay 17.9 cents every year for 15 years to pay off $1 loaned today at 16% interest. Thus to pay off $100,000, the borrower must pay annually:

$100,000 \cdot (\textbf{0.179}) = \$17,900$ *per year*   (*See Table* 10)

Notes:

• Over 15 years, the total payments by the borrower equal $17,900 x 15 = $268,500. Of this amount, $168,500 represents total interest payments to pay off the principal of $100,000.

• A 15 year mortgage will require a slightly higher annual (and thus monthly) mortgage payment than the 25 year mortgage ($17,900 vs. $16,400 per year in Examples 12 and 13, respectively). However, with this higher annual payment, it is possible to save a large amount in total interest payments ($310,000 - $168,500 = $141,500) if the loan is paid off in 15 years rather than 25 years!

• A fundamental property of capital recovery is that when a loan is paid off in a shorter time (sometimes with only a nominal increase in the payment each period), substantial savings in total interest payments over the life of the loan may result.

3.7 Net Present Value Method

Investments often involve an initial cost (e.g. a purchase of capital equipment) to the investor and then a series of revenues returning to the investor in future time periods.

The Net Present Value (NPV) Method discounts the future revenue flow (at the applicable interest rate) to obtain its present value. The initial cost is then subtracted from the present value of the future revenue flow to obtain the net present value of the investment.

• If NPV > 0, the investment is worthwhile from a *financial* standpoint. Other considerations (e.g. uncertainty in cash flows, lack of capital for the initial outlay, etc.) may cause the investment not to be implemented.

• If NPV < 0, the investment is not worthwhile from a *financial* standpoint. Other considerations (e. g. willingness to loss money in the short run to gain market share) may cause the investment to be implemented.

Example 14: (Net Present Value). The EZ Chair Furniture Factory has invested in a lathe machine costing $6000. The prevailing annual interest rate is 10%. Determine the Net Present Value (NPV) of the following expected revenue flow for the five year life of the lathe (there is no salvage value for the lathe at the end of the five years):

	Year 1	Year 2	Year 3	Year 4	Year 5
Revenue Flow	$1000	$1500	$3000	$2500	$2000

Answer: The calculation to determine the sum of the discounted revenue flows is given in Figure 9.

Discounted Revenue Flows For EZ Chair Lathe Machine

Year N	Revenue Flow	$[1/(1+.10)^N]$ (Table 8)	Discounted Revenue Flow
1	$1000	0.909	$909
2	$1500	0.826	$1240
3	$3000	0.751	$2254
4	$2500	0.683	$1708
5	$2000	0.621	$1242
			Sum = $7353

Figure 9

The Net Present Value NPV is computed as follows:

$$NPV = Discounted\ Revenue\ Flow\ Sum\ -\ Initial\ Cost$$

$$= \$7353 - \$6000$$

$$= + \$1353$$

Since NPV > 0, this investment is returning more than the prevailing 10% interest rate used to compute the NPV. This appears to be a good investment for the EZ Chair Furniture Factory from a strictly financial standpoint.

Equivalent Worth

Sometimes, the NPV is expressed in terms of its <u>equivalent worth</u> which are the *equal* revenues (for a positive NPV) or costs (for a negative NPV) in each interest period which have the same present value as the NPV.

Example 15: (Equivalent Worth). Compute the equivalent worth over 5 years (at 10% yearly interest) for the initial cost and revenue flow in Example 14 which has a *positive* NPV of $1353.

Answer: Since the NPV in Example 14 is positive, the equivalent worth will be expressed by a flow of equal revenues over the 5 years. From Table 9, the present value of a $1 revenue for 5 consecutive interest periods discounted at 10% is $3.791. The equivalent worth of the costs and revenues in Example 14 is computed as follows:

$$Equivalent\ Worth = \frac{NPV\ of\ \$1353}{PV\ of\ \$1\ over\ 5\ years\ at\ 10\%}$$

$$= \frac{\$1353}{\$3.791}$$

$$= \$357$$

Five equal yearly payments of $357 at 10% yearly interest would have the same present value as the Net Present Value of the initial cost and the revenue flow in Example 14.

Internal Rate of Return

The interest rate which makes NPV = 0 (> 10% in Example 14) is called the <u>Internal Rate of Return</u> (IRR or sometimes IROR).

- If the IRR is greater than the prevailing interest rate, the investment will appear attractive to the investor. The return of revenue to the investor is greater than if the initial outlay is left in the bank.

- If the IRR equals the prevailing interest rate, the investor is indifferent between investing and not investing. The investor could obtain the same return if the initial outlay is left in the bank.

- If the IRR is less than the prevailing interest rate, the investment doe not appear attractive to the investor. The return of revenue to the investor is less than if the initial outlay is left in the bank.

Example 16: (IRR) A new lathe at the Lazy Boy Furniture Factory costs $3000. Determine the Net Present Value NPV and the Internal Rate of Return IRR of the following expected revenue flow over the 4 year useful life (no salvage value) of the lathe if the prevailing yearly interest rate is 12%:

	Year 1	Year 2	Year 3	Year 4
Revenue Flow	$1000	$1000	$1000	$1000

Answer: The calculation to determine the sum of the discounted revenue flow is given in Figure 10:

Discounted Revenue Flows For Lazy Boy Lathe (12%)

Year N	Revenue Flow	$[1/(1+.12)^N]$ (Table 8)	Discounted Revenue Flow
1	$1000	0.893	$893
2	$1000	0.797	$797
3	$1000	0.712	$712
4	$1000	0.636	$636
		Sum = 3.037 (See Table 9)	**Sum = $3037**

Figure 10

The Net Present Value (NPV) to Lazy Boy is computed as follows:

$$NPV = Discounted\ Revenue\ Flow\ Sum\ -\ Initial\ Cost$$

$$= \$3037\ - \$3000$$

$$= +\$37$$

Note: There is a uniform series of $1000 revenues in Example 16. Table 9 can be used to determine that the present value of four yearly revenues of $1 at 12% interest is 3.037. Thus the present value of four yearly revenues of $1000 at 12% interest is $3037.

Since NPV = +$37 > 0, this investment is returning more than 12% on the initial outlay of $3000. The Internal Rate of Return IRR is the interest rate which makes the NPV = 0.

There is no simple closed form formula to find the IRR. The IRR is typically found by finding interest rates where NPV is slightly positive and slightly negative. The IRR is then *estimated* by linear interpolation.

Compute the NPV for a yearly interest rate of 14%. The calculation to determine the sum of the discounted revenue flow is given in Figure 11:

Discounted Revenue Flows For Lazy Boy Lathe (14%)

Year N	Revenue Flow	$[1/(1+.14)^N]$ (Table 8)	Discounted Revenue Flow
1	$1000	0.877	$877
2	$1000	0.769	$769
3	$1000	0.675	$675
4	$1000	0.592	$592
		Sum = 2.914 (See Table 9)	**Sum = $2914**

Figure 11

The Net Present Value (NPV) to Lazy Boy is computed as follows:

$$NPV = Discounted\ Revenue\ Flow\ Sum\ -\ Initial\ Cost$$

$$= \$2914 - \$3000$$

$$= -\$86$$

Note: There is a uniform series of $1000 revenues. Table 9 can be used to determine that the present value of four yearly revenues of $1 at 14% interest is 2.914. Thus the present value of four yearly revenues of $1000 at 14% interest is $2914.

Since NPV = -$86 < 0, this investment is returning less than 14% on the initial outlay of $3000. The Internal Rate of Return (IRR) is somewhere between 12% and 14%. The value of the IRR can be estimated using linear interpolation as follows:

• Two (Interest Rate, NPV) points given by (12%, +$37) and (14%, -$86) have been determined. The line segment connecting these two points will pass through NPV = 0 at the point (x, 0) where x is between 12% and 14%. The IRR is the ordinate of the point (x, 0) that lies on the line segment connecting (12%, +$37) and (14%, -$86).

• The NPV decreases by $37 - (-$86) = $123 as the interest rate increases 2% from 12% to 14%. Thus the NPV decreases by $61.5 for each one point rise in the interest rate.

• For the NPV to decrease by $37, the interest rate must rise by 37/61.5 = 0.61% to 12.61%.

115

• The IRR (which provides a NPV = 0 for this initial outlay of $3000 with 4 yearly revenues of $1000) is then *approximately* 12.6%.

The Internal Rate of Return (IRR) can also be determined graphically by plotting the Interest Rate vs. Net Present Value (NPV) as shown in Figure 12. The two computed interest rates (12% and 14%) and NPV's (one slightly positive at $37 and one slightly negative at -$86) are connected by a line segment. The interest rate at which the line segment crosses the x-axis (NPV = 0) is an estimate for the Internal Rate of Return to the Lazy Boy Furniture Factory.

Graphical Solution To Determine Lazy Boy Lathe IRR

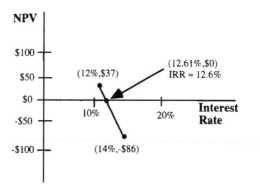

Figure 12

Exercise 3

Break-Even Analysis

Problem #1 (Make vs. Buy Decision With Stepwise Fixed Costs)

In an ongoing effort to re-invent itself and lower operating costs, Lake Woebegon General Hospital (LWGH) is reviewing two options for handling blood chemistry tests:

Option I: Contract Out (Buy). Trusty Test Tubes is willing to perform the blood chemistry tests for a $1.80/test fee. There is no significant fixed cost for this option. However, there would be a 24 hour turnaround time for test results.

Option II: Establish Lab at LWGH (Make). This option involves setting aside space, leasing an analyzer, and hiring lab technicians. The fixed cost is $3000/month for the space and the leasing of the analyzer plus $2500/month for each lab technician required. The variable cost would be only $1.50/test. Results of tests would be available very quickly.

The analyzer can handle up to 50,000 tests/month. The number of lab technicians that must be hired depends on the number of blood chemistry tests/month (output) performed at LWGH:

1 lab technician	Up to 15,000 tests/month
2 lab technicians	Between 15,000 and 30,000 tests/month
3 lab technicians	Between 30,000 and 40,000 tests/month
4 lab technicians	Between 40,000 and 50,000 tests/month

a) Plot the total (fixed plus variable) costs for the Make and Buy actions over the range of possible tests/month performed at LWGH up to the analyzer capacity of 50,000 tests/month.

b) Determine the Break-Even Point(s) over the range of possible tests/month.

c) Determine a reasonable Make vs. Buy decision strategy for LWGH over the range of possible tests/month.

d) (Sensitivity Analysis) What effect would an increase in the cost for the space and leasing the analyzer from $3000/month to $4500/month have upon the Make vs. Buy decision strategy?

Problem #2 (Compound Interest)

An investor places $1000 in a savings account earning 6% interest compounded annually. At the end of 10 years, how much will accrue in the savings account? Assume the investor makes no withdrawals.

Problem #3 (Present Value)

Assume that it is currently possible to earn 10% yearly on invested funds. Which is more advantageous (has the higher Present Value) to an investor: $3500 now or $1000 revenue at the end of each year for the next five years? Assume all revenues to the investor are net of taxes.

Problem #4 (Net Present Value)

The Furniture-R-Us company must purchase a lathe and is considering two options: Lathe X and Lathe Y. Lathe X requires an initial outlay of $8000 while Lathe Y requires an initial outlay of $10000. The prevailing interest rate is 10% yearly. Both lathes are expected to last 5 years and have no salvage value. The net (after tax) revenue flows for the output of products manufactured by the two lathes are forecasted as follows:

Year	Lathe X	Lathe Y
1	$2000	$2000
2	$2500	$3000
3	$2500	$3500
4	$2500	$3000
5	$2000	$2500

Determine the Net Present Value (NPV) for Lathe X and Lathe Y.

Problem #5 (Internal Rate of Return)

Determine the Internal Rate of Return (IRR) for Lathe X in Problem #4.

Chapter 3

Break-Even Analysis Additional Solved Problems

Problem #1 (Make vs. Buy)

St. Agony Hospital has the option of preparing meals for patients in-house or having Mariott bring prepared meals to the hospital. Each Mariott meal costs the hospital $10 on the average. Setting up and operating meal preparation facilities at St. Agony would have a fixed cost of $6000/month for demand less than 1000 meals/month and $10,000 per month for greater than 1000 meals/month. Once the meal preparation facilities at St. Agony have been set up, meals can be produced in-house for $5 on the average.

a) Identify the Fixed Cost (FC), Variable Cost (VC), and Total Cost as a function of demand y (meals/month) for the Make (St. Agony) and Buy (Mariott) actions.

Answer: The costs for the Make and Buy actions are as follows:

Make (St. Agony): Demand y= # meals/month

$$FC(Make) = \quad \$6,000 \qquad\qquad y < 1000 \text{ meals/month}$$
$$\$10,000 \qquad\qquad y \geq 1000 \text{ meals/month}$$

$$VC(Make) = \quad \$5y$$

$$TC(Make) = \quad FC(Make) + VC(Make)$$
$$TC(Make) = \quad \$6,000 + \$5y \qquad y < 1000 \text{ meals/month}$$
$$\$10,000 + \$5y \qquad y \geq 1000 \text{ meals/month}$$

Buy (Mariott): Demand y= # meals/month

$$FC(Buy) = 0$$

$$VC(Buy) = \$10y$$

$$TC(Buy) = FC(Buy) + VC(Buy)$$
$$TC(Buy) = \$10y$$

b) Plot the Total Cost per month versus demand (up to 3000 meals/month) for the Make and Buy actions.

Answer: The Total Cost (TC) Plot for Make vs. Buy is as follows:

119

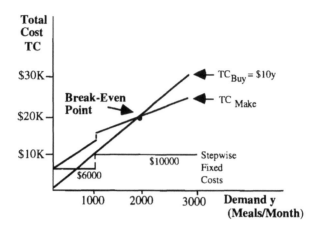

c) Solve algebraically for the Break-Even Point(s) and plot the Break Even Point(s) on the graph plotted for b).

Answer: Equate TC_{Buy} = TC_{Make} for the steps of the fixed costs. If the intersection points of the two lines occurs within the range of the step, the intersection point is a Break-Even point.

For the range 0 to 1000 meals/month:

TC_{Buy} = TC_{Make}
$10y = $6000 + $5y$
$5y = 6000
$y = 1200$ meals/month
1200 meals/month is not in the range (0, 1000) and is not a Break-Even Point.

For the range 1000 to 3000 meals/month:

TC_{Buy} = TC_{Make}
$10y = $10000 + $5y$
$5y = 10000
$y = 20000$ meals/month
2000 meals/month is in the range (1000, 3000) and is a Break-Even Point.

Problem #2 (Make vs. Buy)

Oliver Twist Hospital (OTH) provides bedside setups (including tissues, toiletries, water containers, etc.) for all its new patients (up to 2000 new patients per month). OTH can buy the patient setups from an outside vendor at $7 per setup. Alternatively, OTH can make the patient setups at the hospital. The fixed cost to OTH of space, storage, utilities, salaries, etc. required to make the patient setups is $4000 per month if the number of patient setups made is in the range of 0 up to 1000 per month and $5000 per month if the number of patient setups made is in the range of 1000 up to 2000 per month. Once the facilities are in place, OTH can make the patient setups for $4 each.

a) Identify the fixed and variable cost for the Make and Buy actions and plot the total cost for each action over the range of possible new patients per month at OTH.

Answer: The fixed and variable cost for the Make and Buy actions and plot the total cost for each action over the range of possible new patients per month at OTH are as follows:

Output of 0 – 1000 new patients (y):

Fixed Cost (Make) = $FC(Make)$ = $4000
Variable Cost (Make) = $VC(Make)$ = $4·y$
Total Cost (Make) = $TC(Make)$ = $4000 + $4·y$
Fixed Cost (Buy) = $FC(Buy)$ = 0
Variable Cost (Buy) = $VC(Buy)$ = $7·y$
Total Cost (Buy) = $FC(Buy) + VC(Buy)$ = $0 + $7·y$ = $7·y$

Output of 10000 – 2000 new patients (y):

Fixed Cost (Make) = $FC(Make)$ = $5000
Variable Cost (Make) = $VC(Make)$ = $4·y$
Total Cost (Make) = $TC(Make)$ = $5000 + $4·y$
Fixed Cost (Buy) = $FC(Buy)$ = 0
Variable Cost (Buy) = $VC(Buy)$ = $7·y$
Total Cost (Buy) = $FC(Buy) + VC(Buy)$ = $0 + $7·y$ = $7·y$

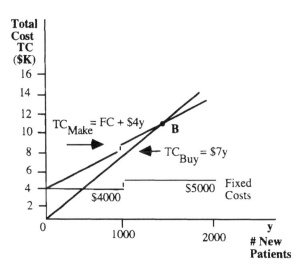

b) Determine the Break-Even Point(s) over the range of possible new patients per month at OTH and identify the Make vs. Buy strategy for patient setups.

Answer: The determination of the Break-Even Point(s) is as follows:

For the range 0 to 1000 new patients (y):

$TC(Buy) = TC(Make)$
$\$7 \cdot y = \$4000 + \$4 \cdot y$
$\$3 \cdot y = \$4000$
$y \approx 1333$
1333 *new patients is not in the range of 0 to 1000 (not a Break – Even Point)*

For the range 10000 to 2000 new patients (y):

$TC(Buy) = TC(Make)$
$\$7 \cdot y = \$5000 + \$4 \cdot y$
$\$3 \cdot y = \$5000$
$y \approx 1667$
1667 *new patients is in the range of 1000 to 2000 (Break – Even Point)*

Optimal Make vs. Buy strategy: less than 1667 new patients expected per month, Buy the patient setups; greater than 1667 new patients expected per month, Make the patient setups. The Break Even Point (cost to Make = cost to Buy) is approximately 1667 new patients per month.

Problem #3 (Investment Analysis)

Examine the following investment where the prevailing interest rate is 10%. An initial outlay of $10,000 returns to the investor the following revenue flow at the end of next 5 years.

Year	Revenue
1	$2000
2	$3000
3	$3500
4	$3000
5	$2500

a) Determine the Net Present Value (NPV) of the above investment.

Answer: The Net Present Value (NPV) of the investment is determined as follows:

Year	Revenue	$[1/(1+.10)^N$ (Table 8)	Discounted Revenue Flow
1	$2000	0.909	$1818
2	$3000	0.826	$2479
3	$3500	0.751	$2630
4	$3000	0.683	$2049
5	$2500	0.621	$1552
			Sum = $10528

Net Present Value (NPV) = *Discounted Revenue Flow* − *Initial Investment*

NPV = \$10528 − \$10000

$\quad$ = \$528

b) Determine the Equivalent Worth of the above investment.

Answer: The Equivalent Worth of the above investment is determined as followw:

$$Equivalent\ Worth = \frac{NPV}{Present\ Value\ of\ \$1\ for\ 5\ years\ at\ 10\%}$$

$$= \frac{\$528}{\$3.791} \quad (Table\ 9)$$

$$\approx \$139$$

"...but time and chance happenenth to them all."

4 | Inventory Decision Analysis

4.1 Deterministic Demand

Assume that a business has a *deterministic* (constant) demand (say 5000 units each year) for a particular item. The following cost considerations hold for the stockroom:

- Annual cost to buy enough inventory (5000 units per year)
- Annual cost to *order* (e.g. manpower, paperwork, etc.) inventory
- Annual cost to *hold* (e.g. storage space, capital, etc.) inventory

The stockroom could buy enough inventory for the entire year in one large order (small total ordering cost, large total holding cost) or in many small orders over the year (large total ordering cost, small total holding cost).

The inventory decision for a given item of inventory is stated as:

What is the optimum order size (called the Economic Order Quantity, EOQ) which will minimize the total cost of ordering and holding inventory?

Notes:

• The simplest inventory formulation is where the cost of an item is *independent* of the order size. That is, there is no reduction in the cost of an item when larger orders are placed. Relaxation of this assumption is examined later in this chapter.

• Over the course of each year, the business must buy the 5000 units required to meet demand. Whether the decision is to order inventory in large or small quantities, the yearly cost to buy the inventory itself is *constant* (5000 units times the cost per unit). Thus the cost of the inventory itself is *dropped* from the total cost function that will be minimized. The total cost function to be minimized consists only of the ordering cost and the holding cost (both of which vary according to the order size Q).

• It will be demonstrated (see Section 4.2) that the optimal average inventory may be expressed in terms of the Economic Order Quantity EOQ (in fact, it is simply EOQ/2). Thus, an *alternate* statement of the inventory decision for a given item may be stated as:

What average inventory should be carried in order to minimize the total cost of ordering and holding inventory?

Example 1: (Deterministic EOQ Formulation). The Acme Hardware Company which is open every day of the year buys its inventory of tires at an average cost of $60 per tire. The demand (assumed to be constant) for tires by its customers is 180 tires per day. There is an Ordering Cost of $40 independent of the tire order size. There is a Holding Cost (including both the cost of storage and the cost of money tied up in inventory) of $4.25 per tire per year. Formulate the inventory decision to determine the Economic Order Quantity (EOQ) which will minimize Acme's total annual cost for ordering and holding the inventory.

Answer: Define the Ordering Cost and the Holding Cost (sometimes call the Carrying Cost) for the Acme inventory decision:

- Ordering Cost: This includes the cost associated with placing the tire order (paperwork, taking delivery, etc.) but does not include the cost to buy the tires. The Ordering Cost is assumed to be a *fixed cost* associated with placing the order no matter how large or small the tire order size.

-- The Ordering Cost is $40 per order.

- Holding Cost: This includes the cost associated with storing the tires (e.g. storage space, insurance, etc.) plus the cost associated with capital tied up in tire inventory. The Holding Cost is assumed to be a *variable cost* which increases with the size of the tire order.

-- The Holding Cost is $4.25 per tire per year.

Notes: The deterministic EOQ formulation of the inventory decision in Example 1 will make the following assumptions:

• Only one item of stock is considered at a time. The Economic Order Quantity EOQ for tires is considered independent of all other items stocked by Acme Hardware.

• Demand is deterministic (180 tires/day) and constant over time. Each year Acme Hardware will buy 65,700 tires (180 tires/day x 365 days) to meet customer demand regardless of the size determined for each individual tire order.

• The cost of a tire to Acme does not depend on the size of the tire order (no quantity discounts). Since the cost c per tire is $60, Acme Hardware will spend $3,942,000 (65,700 tires x $60/tire) each year to buy tires. This constant yearly cost of tire inventory is independent of the tire order size and may be dropped from the total cost function.

• Lead time (LT) is the amount of time between placing an order for an item stocked by Acme and the arrival of that order. Initially, the lead time is considered to be zero. Relaxation of the zero lead time assumption will be examined later in this chapter.

For the deterministic inventory decision formulation, the single *decision variable* is Q, the size of the tire order each time an order is placed. A small number of variables will be defined as parameters of the decision. This is a simple deterministic decision.

The objective of the inventory decision is to minimize the Total Annual Cost TC(Q) of ordering and holding tires required to meet the demand (constant at 180 tires per day). If Q tires are ordered each time an order is placed, then the *objective function* becomes:

Min TC(Q) = Annual Ordering Cost (Q) + Annual Holding Cost (Q)

The only *constraint* on the decision variable Q is the largest order amount the tire supplier can deliver in a single order or the largest amount of tires Acme can hold. For this example, it is assumed that the constraint on the size of each order is given by:

$Q \leq 2000$ *tires*

For an order size Q, the Annual Ordering Cost is the cost of placing each order times the # of orders placed per year:

- Let k = cost to place an order (where cost is independent of order size) = \$40.

- Let d = demand = (180 tires/day)·(365 days/year) = 65,700 tires per year.

- For an order size Q, the number of orders per year is then equal to d/Q.

- Note: The optimum order size Q (denoted as the EOQ) is unknown at this point but will be derived analytically in Section 4.2.

- Then the Annual Ordering Cost (Q) = $k \cdot d/Q$ = (\$40)·(65,700/Q) .

For an order size Q, the Annual Holding Cost is the cost to store the inventory and to tie up capital in inventory times the *average inventory* in stock:

- Let h = cost to hold each tire for a year (cost associated with storage space and the use of capital) = \$4.25 per tire per year.

- Note: The Holding (Carrying) Cost for an item is sometimes given as a percentage i (e.g. 7.08%) of the total cost c. Then, h = i·c = 0.0708·\$60 = \$4.25.

- For an order size Q, the average inventory on hand is Q/2. This can be seen by considering Figure 1 which depicts a typical Inventory Level vs. Time graph.

- Note: An inventory cycle is defined as the time between the arrival of successive orders of size Q of an item.

- Q tires are ordered and are on hand at the start of each inventory cycle.

- Since the demand is constant (180 tires/day), it is possible to determine exactly (with certainty) the time when the level of inventory of the item will be zero.

- When the level of inventory diminishes to zero, the next order of EOQ units arrives immediately (lead time is initially zero days).

- Over the time between the arrival of successive orders (the cycle time), the inventory decreases linearly (at the constant demand rate d) from Q units to 0 units.

- The average inventory on hand is thus Q/2. That is, half the time there is more than Q/2 tires on hand and half the time there is less than Q/2 tires on hand.

- Then the Annual Holding Cost (Q) = $h \cdot Q/2$ = $4.25·Q/2 .

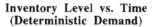

**Inventory Level vs. Time
(Deterministic Demand)**

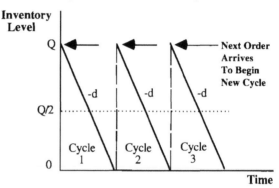

Figure 1

The formulation of the inventory decision to minimize Total Annual Cost TC(Q) for ordering and holding inventory is given by:

Min TC(Q)   = *Annual Ordering Cost (Q) + Annual Holding Cost (Q)*

          = $k \cdot d/Q \ + \ h \cdot Q/2$

subject to Q $\leq$ *2000 tires*

<u>Notes</u>:

• This inventory formulation assumes only enough space must be reserved in the stockroom for the *average* amount of inventory Q/2. The rationale is as follows:

    - Many different items are being ordered at different times of the year.
    - At any particular time, some items will be above their average inventory level while others will be below their average level.
    - Hence, it not typically necessary to reserve enough space for the entire order Q for each item stocked.

• If space must be reserved for the entire Q units (e.g. space is required in a special refrigeration unit not shared with other items), then the Annual Holding Cost equals $h \cdot Q$ instead of $h \cdot Q/2$.

Recall the inventory decision analysis cost parameters given in Example 1:

k *(Order Cost)* = $40 / *order*
d *(Demand)* = 65,700 *tires / year*
h *(Holding Cost)* = $4.25 / *tire / year*

Then the following formulation (objective function and constraints) is used to derive the order size Q which minimizes the Total Annual Cost TC(Q) to Acme Hardware:

Min $TC(Q)$ = *Ordering Cost(Q)* + *Holding Cost(Q)*

$$= k \cdot d / Q + h \cdot Q / 2$$

$$= (\$40) \cdot (65,700) / Q + (\$4.25) \cdot Q / 2$$

subject to $Q \leq 2000$ *tires*

The order size Q which minimizes the Total Annual Cost TC(Q) to Acme is denoted as the Economic Order Quantity (EOQ).

4.2 Economic Order Quantity

Two solution methods will be demonstrated to determine the Economic Order Quantity (EOQ) which minimizes the Total Annual Cost function TC(Q) consisting of the cost of ordering and holding inventory:

- Trial and Error Approach (a search method)
- Calculus Approach (an analytical method)

Trial and Error Approach

Example 2: Use the Trial and Error Approach to search for the Economic Order Quantity (EOQ) which minimizes the Total Annual Cost TC(Q) for tire inventory in the Acme Hardware Company inventory decision given in Example 1.

Answer: Pick a wide spectrum of possible discrete values for the order quantity Q whose range is expected to include the Economic Order Quantity EOQ. The value of Q with the minimum Total Annual Cost TC(Q) is an *estimate* for the Economic Order Quantity EOQ.

For Example 1, a spectrum of values for Q (≤ 2000) whose range is expected to include the Economic Order Quantity EOQ is chosen by Acme Hardware to be Q = 400, 700, 1000, 1300, 1600, and 1900 tires per order.

The Total Annual Cost function TC(Q) for Acme Hardware (derived in Example 1) is given by:

$$TC(Q) = k \cdot d/Q + h \cdot Q/2$$

$$= (\$40) \cdot (65,700)/Q + (\$4.25) \cdot Q/2$$

The values of the Total Annual Cost function TC(Q) for the range of possible values for Q chosen by Acme Hardware are given in Figure 2.

Trial and Error Approach Calculations For EOQ

Trial Value Q	Ordering Cost k·d/Q ($40·65,700/Q)	Holding Cost h·Q/2 $4.25·(Q/2)	Total Annual Cost TC(Q)
400	$6570	$850	$7420
700	$3754	$1487.5	$5241.5
1000	**$2628**	**$2125**	**$4753**
1300	$2021.5	$2762.5	$4784
1600	$1642.5	$3400	$5042.5
1900	$1383	$4037.5	$5420.5

Figure 2

The Economic Order Quantity EOQ which minimizes the Total Annual Cost for inventory is estimated to be 1000 tires per order (based on the range of discrete values for Q chosen by Acme Hardware).

Notes:

• The strength of the Trial and Error Approach is that it is *simple* and will work for *any* inventory decision problem formulation no matter how complicated.

• Consider a more complicated inventory formulation where the cost c per unit and the cost k to order are a function of the order size Q (i.e. c = c(Q) and k = k(Q)). A term for the cost to buy the annual demand of d units, d·c(Q), must be added to the Total Annual Cost function. Also, the term in the Total Annual Cost function for the cost to order k·d/Q now becomes k(Q)·d/Q. The Total Annual Cost function TC(Q) to be minimized becomes the more complicated function:

$$TC(Q) = d \cdot c(Q) + k(Q) \cdot d/Q + h \cdot Q/2$$

The Total Annual Cost function TC(Q) (regardless of how complicated it becomes) can still be easily numerically evaluated for the spectrum of order sizes. The order size Q with the minimum Total Annual Cost is an *estimate* for the Economic Order Quantity EOQ.

• The drawback of the Trial and Error Approach is that the *true* minimum cost order size will not usually be found. Typically, the value of the optimal order size EOQ will *not* be among the set of discrete values chosen for Q. However, a computer search using a large number of candidate order sizes (where the spectrum of order sizes is evaluated at small intervals) can come close enough to finding the EOQ which provides the true minimum cost solution.

Calculus Approach

The Calculus Approach will derive *analytically* a closed form (i. e. a function of the parameters of the decision) EOQ solution for a *continuous* Total Annual Cost function TC(Q). The Calculus Approach will *not* work if there are discontinuities in the Total Annual Cost objective function. Remember that the Trial and Error Approach may be used if the Total Annual Cost function has discontinuities or has a high level of complexity.

Figure 3 provides an example and a graph of a cost function discontinuity where the inventory cost c(Q) per tire is a discontinuous function of the order size Q.

Example of a Cost Function Discontinuity

Order Size Q For Tires	Cost Per Tire c(Q)
Q < 1000 Tires	$60
Q ≥ 1000 Tires	$50

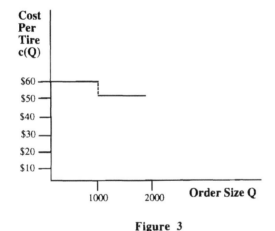

Figure 3

Recall from the field of calculus that for a *continuous* function:

   - The maximum or minimum of a function occurs at the value which solves the equation derived when the first derivative is set to zero.

   - The extreme is a minimum if the second derivative is *positive* at the value where the first derivative is zero.

Recall the inventory decision formulation derived in Example 1 where the cost per unit is not a function of order size and the cost to order is also not a function of order size. The

Total Annual Cost TC(Q) function and its first derivative with respect to the decision variable Q are given by:

$$TC(Q) = k \cdot d/Q + h \cdot Q/2$$

$$dTC(Q)/dQ = -k \cdot d/Q^2 + h/2$$

Setting the first derivative to zero implies that the optimal order size Q, the Economic Order Quantity (EOQ), can be determined by an expression involving the inventory parameters k, d, and h:

$$-k \cdot d/EOQ^2 + h/2 = 0$$

$$EOQ = (2k \cdot d/h)^{1/2}$$

The second derivative of the Total Annual Cost TC(Q) function with respect to the decision variable Q is given by:

$$d^2TC(Q)/dQ^2 = 2k \cdot d/Q^3$$

The second derivative will be greater than zero since all reasonable values for k, d, and Q are positive. This implies that the optimization of the Total Annual Cost TC(Q) function to determine the EOQ will provide the *minimum* total cost.

Notes: Examine the first derivative of the Total Annual Cost function TC(Q) given by $dTC(Q)/dQ = -k \cdot d/Q^2 + h/2$:

• The negative cost (payoff) $-k \cdot d/Q^2$ in the first derivative of the Total Annual Cost function TC(Q) is the *marginal benefit* defined as the decrease in total Annual Ordering Cost due to augmenting each order size by one unit and having to order less often.

• The positive cost h/2 in the first derivative of the Total Annual Cost TC(Q) function is the *marginal cost* defined as the increase in total Annual Holding Cost of augmenting the order size by one unit and holding an average of 1/2 more units through each inventory cycle.

• Setting the first derivative to zero is equivalent to equating marginal cost to marginal benefit. From the field of economics, this will derive the optimal order size EOQ.

• At optimality (at the value EOQ), the Annual Ordering Cost will *equal* the Annual Holding Cost. This can be verified by substituting $EOQ = (2k \cdot d/h)^{1/2}$ into the Total Annual Cost function $TC(EOQ)$:

$$Total\ Cost(EOQ) = Annual\ Ordering\ Cost(EOQ) + Annual\ Holding\ Cost(EOQ)$$

$$TC(EOQ) = k \cdot d/EOQ + h \cdot EOQ/2$$

$$TC(EOQ) = k \cdot d / (2k \cdot d/h)^{1/2} + h \cdot (2k \cdot d/h)^{1/2} / 2$$

$$TC(EOQ) = \frac{k^{1/2} \cdot d^{1/2} \cdot h^{1/2}}{2^{1/2}} + \frac{k^{1/2} \cdot d^{1/2} \cdot h^{1/2}}{2^{1/2}}$$

• The property that the Annual Ordering Cost equals the Annual Holding Cost at the EOQ will not typically hold for more general inventory formulations (e.g. where the cost of each unit or the cost to place an order is a function of the order size).

• However, for even the more general inventory formulations, the marginal holding cost will equal the marginal ordering benefit at optimality.

Recall the inventory decision analysis cost parameters given in Example 1:

k (*Order Cost*) = $40 / *order*
d (*Demand*) = 65,700 *tires / year*
h (*Holding Cost*) = $4.25 / *tire / year*

Then the analytical (closed form solution) for the EOQ for tires at Acme Hardware yields:

$$EOQ = (2k \cdot d/h)^{1/2}$$

$$= [(2)(40)(65700)/(4.25)]^{1/2}$$

$$= 1112.07$$

$$\approx 1113 \text{ (Round up to the next integer value!)}$$

The minimum Total Annual Cost can be determined by substituting EOQ = 1,113 into the Total Annual Cost function TC(Q):

$$TC(Q) = k \cdot d/EOQ \qquad + \qquad h \cdot EOQ/2$$

$$= (\$40) \cdot (65700/1113) + (\$4.25) \cdot (1113/2)$$

	Annual		*Annual*
	Ordering		*Holding*
	Cost		*Cost*
=	\$2361	+	\$2365
=	\$4726		

Many of the properties of the Acme Hardware inventory decision can be verified by looking at the graphical solution presentation given in Figure 4.

133

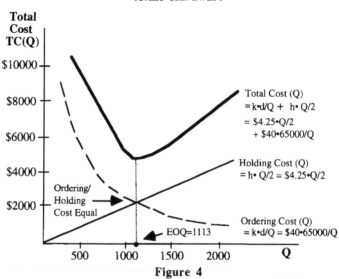

Figure 4

Notes: Figure 4 provides verification of the following:

• The Holding Cost curve h·Q/2 and the Ordering Cost curve k·d/Q are in classic *cost-benefit* form. As Q increases, the Ordering Cost decreases while Holding Cost increases. As Q decreases, the Ordering Cost increases while the Holding Cost decreases.

• The Annual Ordering Cost ($2361) is *approximately* equal to Annual Holding Cost ($2365) at the optimal Economic Order Quantity EOQ (= 1113 tires). They are not exactly equal since the optimal EOQ solution was rounded up to the next higher integer value.

• For more general inventory formulations (e.g. where the cost per unit is a function of the order size or the cost to order is a function of the order size), the optimal EOQ which minimizes TC(Q) may not provide equal Annual Ordering Cost and Annual Holding Cost. However, marginal ordering benefit will *always* equal marginal holding cost at the optimal EOQ.

• For the Acme Hardware inventory decision formulation, the Total Annual Cost curve TC(Q) changes slowly around the optimal order size EOQ. That is, using a reasonable estimate of the optimal analytical solution EOQ (e.g. the estimate derived using the Trial and Error Approach) will not greatly increase the value of the Total Annual Cost TC(Q) for inventory.

• The Trial and Error Approach (EOQ = 1000 tires/order) with only 6 values evaluated was quite close to the optimal value from the Calculus Approach (EOQ = 1113 tires/order)

• It is better to round off the value of the analytical EOQ to the next *higher* integer value since the Ordering Cost d·k/Q rises sharply (exponentially) as Q decreases.

Reorder Point

Recall that lead time (LT) is defined (see Section 4.1) to be the interval of time between a placement of an order and the arrival of that order. The inventory decision formulation examined up to this point has assumed a lead time of zero. For most applications, there is a *positive* lead time between the placement of an order and its arrival.

With a constant demand rate d, an order can be placed when current inventory on hand will run out exactly in the positive lead time it takes for the new order to arrive. This inventory level is called the reorder point (ROP). The reorder point occurs when inventory drops to the level of Lead Time (LT) times demand (d):

$$ROP = LT \cdot d$$

The lead time (LT) and the reorder point (ROP) for deterministic inventory cycles are displayed graphically in Figure 5.

- At the start of each inventory cycle (a new order has just arrived), EOQ units of inventory are on hand.

- During each cycle, the level of inventory decreases at the constant demand rate d. The average inventory level during each inventory cycle is EOQ/2.

- At the point where inventory decreases to the level ROP (the amount of inventory which will be used up during the lead time LT), a new order for EOQ units is placed.

- At the end of the lead time LT, the new order arrives just as the level of inventory drops to zero.

- A new inventory cycle begins with the arrival of the new order of EOQ units.

Deterministic EOQ Inventory Cycles With Positive Lead Time

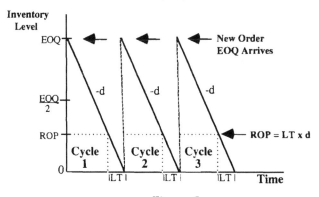

Figure 5

135

Example 3: (Reorder Point). If demand d for an item in inventory is 3650 units per year and the lead time (LT) before an order arrives is 5 days, calculate the reorder point (ROP).

Answer: The reorder point is calculated as lead time times demand per unit of time:

$$ROP = LT \cdot d$$

$$= 5 \; days \; \cdot \; \frac{3650 \; units/yr}{365 \; days/yr}$$

$$= 5 \; days \; \cdot \; 10 \; units \: / \: day$$

$$= 50 \; units$$

A new order is placed when the inventory level drops to the reorder point of 50 units. During the lead time of 5 days required for the new order to arrive, the inventory level will decrease from 50 units to exactly zero units.

4.3   Sensitivity Analysis

It has been demonstrated graphically (see Figure 4) that the Total Annual Cost function TC(Q) varies slowly as a function of the order size Q in the neighborhood around the optimal order size EOQ. One can also examine the sensitivity of the Total Annual Cost to changes in the other inventory formulation parameters:

- Demand (d)
- Ordering Cost (k)
- Holding Cost (h)

Example 4: For the tire inventory decision for the Acme Hardware Company given in Example 1, determine the sensitivity of the Total Annual Cost TC(Q) to a 20% difference between the forecasted annual demand d = 65,700 tires and the *actual* annual demand.

Answer: Consider the case where the actual demand is 20% less than the forecasted demand d (the actual demand $d_{-20\%}$ = 65,700 · 0.80 = 52,560 tires/year). Then the Economic Order Quantity for a 20% decrease in the level of demand is given by:

$$EOQ_{-20\%} = (2k \cdot d_{-20\%}/h)^{1/2}$$

$$= [(2)(40)(52560)/(4.25)]^{1/2}$$

$$= 994.67 \text{ or } 995 \text{ tires/order (round up)}$$

Note: Recall that EOQ = 1113 tires/order for d = 65,700 tires/year in Example 1.

The minimum Total Annual Cost for demand $d_{-20\%}$ equal to 52,560 tires/year can be determined by substituting $EOQ_{-20\%}$ = 995 tires into the Total Annual Cost function TC(Q):

$$TC(d_{-20\%}, EOQ_{-20\%}) = k \cdot d_{-20\%}/EOQ_{-20\%} + h \cdot EOQ_{-20\%}/2$$

$$= (40)(52560)/995 + (4.25)(995)/2$$

$$= Annual\ Ordering\ Cost\ +\ Annual\ Holding\ Cost$$

$$= \$2113 + \$2114$$

$$= \$4227$$

If the actual demand $d_{-20\%}$ = 52,560 tires per year had been known prior to the start of the year, each order would have been for 995 tires with an Total Annual Cost of $4227 made up of equal Annual Ordering Cost and Annual Holding Cost.

However, instead of $EOQ_{-20\%}$ = 995 tires, each order will consist of EOQ = 1113 tires based on a faulty forecasted demand of d = 65,700 tires per year. The non-optimal size order of EOQ = 1113 tires will cause the Total Annual Cost to increase over the minimum cost TC($d_{-20\%}$, $EOQ_{-20\%}$) = $4227 that would have resulted if the optimal order size of $EOQ_{-20\%}$ = 995 tires were used.

The actual Total Annual Cost (using the non-optimal order size EOQ = 1113 tires for the actual demand $d_{-20\%}$ = 52,560 tires/year) is given by:

$$TC(d_{-20\%}, EOQ) = k \cdot d_{-20\%}/EOQ + h \cdot EOQ/2$$

$$= (40)(52560)/1113 + (4.25)(1113)/2$$

$$= Annual\ Ordering\ Cost\ +\ Annual\ Holding\ Cost$$

$$= \$1889 + \$2365$$

$$= \$4254$$

Notes:

• The Total Annual Cost for tire inventory *rises* from $4227 (when the optimal $EOQ_{-20\%}$ = 995 tire order size which is compatible with the actual demand of $d_{-20\%}$ = 52,560 tires/year is used) to $4254 (when the non-optimal EOQ = 1113 tires which is not compatible with $d_{-20\%}$ = 52,560 tires/year is used).

• The Annual Ordering Cost ($1889) and the Annual Holding Cost ($2365) are not equal for EOQ = 1113 tires/order since this order size is not optimal for $d_{-20\%}$.

• There is a 1 percent error in Total Annual Cost for inventory due to the actual demand being 20% less than the forecasted annual demand for tires. The computation is as follows:

$$\frac{(Actual\ Total\ Cost - Optimal\ Total\ Cost)}{(Actual\ Total\ Cost)} = \frac{(\$4254 - \$4227)}{(\$4254)}$$

$$= 0.01$$

$$= 1\%$$

Figure 6 provides the sensitivity of Total Annual Cost for tire inventory to errors in demand forecasting. The percent error (increase) in Total Annual Cost is given when the order size EOQ = 1113 tires is used (based on the forecasted demand d = 65,700 tires/year) and the actual demand proves to be $d_{-40\%}$, $d_{-20\%}$, $d_{+20\%}$, and $d_{+40\%}$:

Sensitivity of Total Cost TC(Q) To Demand Forecasting Error

	$d_{-40\%}$	$d_{-20\%}$	d	$d_{+20\%}$	$d_{+40\%}$
Actual Demand	39420	52560	65700	78840	91980
Optimal Order Size Q	861	995	1113	1218	1316
TC(Optimal Q)	$3660	$4227	$4276	$5177	$5592
TC(EOQ = 1113)	$3784	$4254	$4276	$5199	$5672
Per Cent Increase	3%	1%	----	0.4%	1%

Figure 6

Notes:

• The Total Annual Cost TC(Q) function for Acme Hardware is not sensitive to errors in forecasting demand.

- If actual demand is 40% below the forecast, the error between the optimal Total Annual Cost (using the optimal order size 861 tires) and the actual Total Annual Cost (using the faulty EOQ of 1113 tires) is only 3%.

– If actual demand is 40% above the forecast, the error between the optimal Total Annual Cost (using the optimal order size 1316 tires) and the actual Total Annual Cost (using the faulty EOQ of 1113 tires) is only 1%.

• Similar sensitivity tables can be constructed for other inventory formulation parameters such as k (Ordering Cost) and h (Holding Cost).

• When the total inventory cost is sensitive to the value of an inventory decision parameter, close attention should be paid to the forecast or measurement of that parameter.

4.4 Production Order Quantity

Consider the deterministic <u>Production Order Quantity (POQ)</u> inventory decision formulation where:

- Instead of ordering an inventory item from an outside vendor, the business *produces the item in-house.*

- The business produces this inventory item by means of a short *production run* which produces the required inventory for the cycle quickly at a constant rate r.

- After the production run of POQ units, inventory that has been accumulated is used to meet demand d during the remainder of the cycle.

<u>Notes</u>:

• A cycle in the Production Order Quantity formulation is defined as the time between the start of successive production runs.

• The deterministic production rate r must be *strictly* greater than the deterministic demand rate d (that is, $r > d$). The business must be able to produce units of the item fast enough to *exceed* the demand during the production period in order to accumulate inventory.

In the POQ inventory decision formulation, a production run of POQ units is initiated at a cost k (the Set-up Cost of the production run). The period between time the production run is initiated and the time the first units are produced is the lead time LT. The usage and accumulation of inventory during the POQ cycles is displayed graphically in Figure 7.

- POQ units are produced by a production run of rate r at the start of each cycle. Demand for inventory decreases the inventory level at a constant at rate d during the entire cycle.

- Inventory is accumulated at a rate of r - d during the short period of time of the production run up to a maximum level. After the production run (where the lot of POQ units has been produced in the short period of time), inventory will decrease from the maximum level at a rate d.

- The set-up for a new production run (at a Set-up Cost k) is initiated at the reorder point (ROP) when inventory drops to the level of lead time times demand (LT x d). After the lead time, when the level of inventory decreases to exactly zero, a new production run begins to produce units of inventory.

- A new POQ inventory cycle begins with the start of the new production run (after the lead time for set-up).

- Inventory accumulated during a POQ inventory cycle must be held at a cost h (the Holding Cost) per unit time.

Deterministic POQ Inventory Cycles

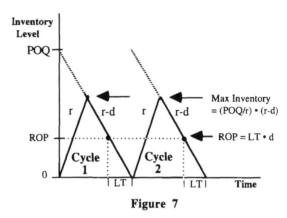

Figure 7

<u>Notes</u>: The following can be observed from Figure 7:

• The extended portion of the lines in Figure 7 display the theoretical case of an *infinite* production rate ($r = \infty$) where all POQ units would be available immediately at the start of the inventory cycle (as in the EOQ inventory formulation). Then the shapes of the EOQ inventory cycles (see Figure 5) and the POQ inventory cycles would be *identical*.

• *Less inventory* is accumulated during the POQ inventory cycles than during the EOQ inventory cycles (for a production run POQ and an order quantity EOQ of equal size). During a POQ cycle, not all the inventory arrives immediately. Also, current demand d during the period of the production run uses some of the production run inventory before it needs to be stored. Thus the entire production run of POQ units would never need to be stored (unlike the EOQ formulation).

• The production run of POQ units at a production rate r actually takes a *finite* amount of time (equal to POQ/r). During this time POQ/r at the start of each cycle, inventory is being accumulated up to a maximum inventory level (less than POQ units). The rate of accumulation of inventory during the time of the production run is r - d.

• The maximum and average inventory levels during a POQ inventory cycle are given by:

$$Max\ Inventory = (Time\ of\ Accumulation) \cdot (Rate\ of\ Accumulation)$$

$$= \frac{POQ}{r} \cdot (r - d)$$

$$= (POQ) \cdot \frac{(r - d)}{r} \qquad\qquad \left(vs.\ EOQ\right)$$

$$Ave\ Inventory\ =\ \frac{POQ}{2} \cdot \frac{(r-d)}{r} \qquad\qquad \left(vs.\ \frac{EOQ}{2}\right)$$

• The quantity $(r - d)/r$ is *less than 1* corresponding the fact that the maximum and average inventory levels for the POQ formulation will be *less* than that of the EOQ formulation when the inventory parameters k, d, and h have equal values. As the production rate r increases so that $(r - d)/r$ approaches 1 in value, the difference between the maximum and average inventory levels for the POQ and EOQ formulations approaches zero.

The Total Annual Cost to meet demand using production runs of size POQ includes the cost to set up the number of production runs required plus the cost to hold the average inventory. That is,

$$Total\ Annual\ Cost\ =\ (Set\text{-}up\ cost) \cdot (\#\ Production\ runs)$$
$$+\ (Holding\ cost) \cdot (Average\ inventory)$$

For a production run of size Q, the Total Annual Cost function TC(Q) becomes:

$$TC(Q)\ =\ k \cdot \left(\frac{d}{Q}\right) + h \cdot \frac{Q}{2} \cdot \left(\frac{r-d}{r}\right) \qquad \left(vs.\ k \cdot \frac{d}{Q} + h \cdot \frac{Q}{2}\right)$$

Using the Calculus Approach (see Section 4.2) to solve for the optimal Production Order Quantity (POQ), the first derivative of TC(Q) with respect to the decision variable Q (production run size) is set equal to zero:

$$\frac{dTC(Q)}{dQ}\ =\ -k \cdot \left(\frac{d}{Q^2}\right) + \frac{h}{2} \cdot \frac{r-d}{r}\ =\ 0$$

Then the value Q which makes the first derivative zero is the optimal production run size POQ:

$$POQ = [2k \cdot d/h]^{1/2} \cdot [r/(r-d)]^{1/2} \qquad \left(EOQ = [2k \cdot d/h]^{1/2}\right)$$

Note: The multiplier quantity $r/(r - d) > 1$ in the equation for the optimal production run size POQ is inverted from the multiplier $(r - d)/r < 1$ in the above equations for maximum and average inventory levels for the POQ formulation. The production run POQ is *greater* than the order quantity EOQ for equal values of k, d, and h.

When annual demand is met using production runs of size POQ, the Total Annual Cost TC(Q) is minimized. The extreme value of the TC(Q) curve is demonstrated to be a minimum since the second derivative is positive for all reasonable (positive) values of k, d, and Q:

$$\frac{dTC^2(Q)}{dQ^2}\ =\ 2k \cdot \left(\frac{d}{Q^3}\right) > 0$$

Notes:

• Figure 8 provides a comparison between the EOQ and POQ inventory decision formulations.

EOQ vs. POQ Inventory Decision Formulation Parameters

	EOQ	POQ
Cost to Order/Set Up	k	k
Holding Cost	h	h
Demand	d	d
Production Rate	Infinite (All EOQ at once)	r
Optimal EOQ/POQ	$[2k{\cdot}d/h]^{1/2}$	$[2k{\cdot}d/h]^{1/2}{\cdot}[r/(r{-}d)]^{1/2}$
Maximum Inventory	EOQ	POQ${\cdot}[(r{-}d)/r]$
Average Inventory	EOQ/2	POQ/2${\cdot}[(r{-}d)/r]$
Lead Time	LT	LT
Reorder Point	LT x d	LT x d

Figure 8

• At the optimal production run size POQ, the Annual Set-up Cost equals the Annual Holding Cost:

$$k \cdot \frac{d}{POQ} = h \cdot \left(\frac{POQ}{2} \cdot \frac{r-d}{r} \right)$$

$$k \cdot \frac{d}{\left\{ [2k \cdot d / h]^{1/2} \cdot [r/(r-d)]^{1/2} \right\}} = h \cdot \frac{\left\{ [2k \cdot d / h]^{1/2} \cdot [r/(r-d)]^{1/2} \right\}}{2} \cdot \frac{(r-d)}{r}$$

$$\frac{k^{1/2} \cdot d^{1/2} \cdot h^{1/2} \cdot (r-d)^{1/2}}{2^{1/2} \cdot r^{1/2}} = \frac{k^{1/2} \cdot d^{1/2} \cdot h^{1/2} \cdot (r-d)^{1/2}}{2^{1/2} \cdot r^{1/2}}$$

Annual Set–up Cost = Annual Holding Cost

• At the optimal production run size POQ, the marginal benefit (lower Annual Set-up Cost) of a one unit larger production run equals the marginal cost (higher Annual Holding Cost) of a one unit larger production run.

$POQ > EOQ$ since $r/(r - d) > 1$

Max Inventory POQ < *Max Inventory EOQ*   since $(r - d)/r < 1$

Ave Inventory POQ < *Ave Inventory EOQ*   since $(r - d)/r < 1$.

4.5   Stochastic Demand

The difference between the inventory decision with deterministic and stochastic demand focuses on the manner in which inventory decreases *after the number of units has decreased to the reorder point* quantity. To examine the difference in the inventory decision when demand is deterministic and stochastic, the definition of the reorder point must be expanded to account for the deterministic and stochastic demand formulations.

When demand is deterministic as in the EOQ and POQ inventory formulations presented earlier in this chapter, the demand is described by the constant rate d. Inventory decreases at a constant rate to the <u>deterministic reorder point</u> quantity ($ROP_D = LT \cdot d$) and a new order is placed or a new production run is initiated. During the lead time for the next order of EOQ units to arrive or the next production run of POQ units to begin, demand will be *exactly* the deterministic reorder point quantity ROP_D that is required to decrease the inventory level to zero.

For the case of stochastic demand, demand is described by a Random Variable with an *expected* rate d. Inventory decreases at a variable rate (see Figure 9) to the <u>stochastic reorder point</u> quantity ROP_S (quantified later in this section) and a new order is placed or a new production run is initiated. During the lead time for the next order of EOQ units to arrive or the next production run of POQ units to begin, lead time demand is described by a Random Variable with *expected value equal* to the deterministic reorder point quantity ($ROP_D = LT \cdot d$).

Typical EOQ Inventory Cycles With Stochastic Demand

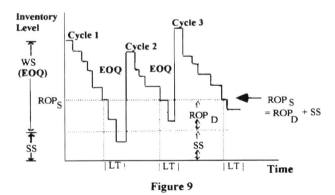

Figure 9

143

Note: In the stochastic demand inventory formulation, demand d is the only Random Variable (simple stochastic decision). The lead time is a fixed quantity. The stochastic reorder point (ROP$_S$) is also a fixed quantity (it will be shown that ROP$_S$ can be calculated as a function of the expected value and standard deviation of the demand during lead time).

Recall that for the deterministic demand formulation in Example 3, a lead time of 5 days with a constant demand of 10 units/day led to a deterministic reorder point quantity (LT · d) of 50 units. In a stochastic inventory formulation with *expected* demand rate 10 units/day, the demand during the 5 days lead time is a symmetric Random Variable with a expected value equal to the deterministic reorder point quantity of 50 units. Demand during the lead time will be *greater* than 50 units half the time and will be *less* than 50 units half the time.

Notes:

• A stockout will occur whenever the stochastic (variable) demand during the lead time is greater than the stochastic reorder point quantity of inventory on hand at the start of the lead time period.

• Demand not met leads to customer dissatisfaction (the cost of which is typically difficult for a business to quantify). The cost associated with customer dissatisfaction is often quantified either through customer surveys or though expert opinion. Experts can be either those who work in a similar business and have direct experience with customer reactions to stockouts of a given item or those who study general consumer reaction to stockouts.

• High costs associated with consumer dissatisfaction will cause a decision maker to take steps (e.g. carry more inventory) to try to avoid frequent stockouts.

The following definitions are useful in analyzing inventory decisions with stochastic demand:

Working stock (WS): The general term in inventory analysis for the EOQ order size or the POQ production run size. The quantities EOQ and POQ are calculated using the *previous deterministic formulas* derived in Sections 4.2 and 4.4, respectively (with the expected rate d substituted for the constant rate d). During an inventory cycle, the working stock will typically be used up meeting the demand.

Safety stock (SS): An *additional* amount of stock carried over *all* the inventory cycles to help meet instances of high demand (greater than the expected demand quantity ROP$_D$) during the lead time. During an inventory cycle, the safety stock will typically not be totally used up meeting demand during the lead time.

Typical inventory cycles with stochastic demand for the EOQ formulation displayed in Figure 9 specify the deterministic reorder point ROP$_D$ and the stochastic reorder point ROP$_S$. The deterministic reorder point ROP$_D$ is defined as LT·d. The stochastic reorder point (ROP$_S$) is equal to the deterministic reorder point (ROP$_D$) plus the safety stock (SS):

$$ROP_s = ROP_D + SS$$

Notes: More complete observations on Figure 9:

• An order the size of the working stock (EOQ) plus the safety stock (SS) is ordered in the initial inventory cycle. The safety stock is ordered (at no additional cost since the cost to order is independent of order size) in the initial cycle only and is carried through the remaining cycles. The cost to buy the safety stock is a *one-time* cost $c \cdot SS$ in Cycle 1. The annual cost to hold the safety stock is $h \cdot SS$.

• An order is placed when inventory drops to the stochastic reorder point ROP_S. In the first inventory cycle, demand during the lead time is greater than its expected value ROP_D. However, there is enough safety stock to meet the excess demand over the expected value of demand during lead time. If not, a stockout would have occurred.

• At the beginning of the second inventory cycle (when the next order of EOQ units arrive), there will be a current inventory of slightly *below* the level of working stock (EOQ) plus safety stock. Some of the EOQ units have been used to replenish safety stock used in the first inventory cycle.

• In the second inventory cycle, an order of EOQ units is again placed when the inventory level drops to the stochastic reorder point level ROP_S. However, in this inventory cycle, the demand during the lead time is less than the expected value ROP_D. The excess stock is added to the working stock (EOQ) available at the start of the third inventory cycle.

• The inventory cycles continue on in this manner. When the inventory drops to ROP_S, a new order of EOQ units is placed. Whenever demand during the lead time is greater than an amount equal to the stochastic reorder point ROP_S (= ROP_D + SS), a stockout occurs in that cycle.

Figure 10 summarizes the similarities and differences between the deterministic and stochastic EOQ inventory decision formulations.

EOQ Inventory Decision For Deterministic vs. Stochastic Demand

	Deterministic	Stochastic
Order Size Cycle 1	EOQ	EOQ + SS
Order Size Cycles 2 - N	EOQ	EOQ
Reorder Point	$ROP_D = LT \times d$	$ROP_S = ROP_D + SS$
Lead Time Demand	ROP_D	Expected Value = ROP_D

Figure 10

In an inventory decision analysis with stochastic demand, two decisions are required:

- What is the working stock (EOQ or POQ) which minimizes Total Annual Cost consisting of Annual Ordering Cost and Annual Holding Cost?

- What amount of safety stock should be carried over all the inventory cycles?

The working stock (EOQ or POQ) are determined by simply substituting the expected demand rate d into the EOQ or POQ formulas (see Sections 4.2 and 4.3) derived for a deterministic demand rate d.

Safety Stock

Consider the safety stock to be carried over all the inventory cycles. The two extreme amounts of safety stock (SS) are unattractive options:

- *None*: Demand is met in half the inventory cycles. This approach may cause intolerable dissatisfaction to the customer.

   -- $ROP_S = ROP_D + SS = ROP_D$ when $SS = 0$. Since the demand during lead time is a symmetric RV with expected value ROP_D, a stockout (lead time demand greater than ROP_S) will occur in half the inventory cycles.

- *Very large*: Demand is met in almost every inventory cycle. This approach may cause unbearable inventory capital costs and holding costs to the business.

   -- $ROP_S = ROP_D + SS$ becomes very large. A stockout (lead time demand greater than ROP_S) will occur very rarely.

The amount of safety stock required is determined by the desired customer service level which is defined as the probability that demand during lead time will be met in any inventory cycle.

Assume that the stochastic (variable) demand during lead time is a given by the symmetric Normal Random Variable (see Figure 11) specified as follows:

$\mu = $ *Expected Value* $= ROP_D = LT \cdot d$

$\sigma = $ *Standard Deviation* (*Must be measured or forecasted*)

Then the safety stock is specified as $K\sigma$ (that is, in terms of a multiplier K times the standard deviation σ of the demand during lead time):

- The inventory on hand at the time inventory drops to the stochastic reorder point $ROP_S = ROP_D + SS$ or, equivalently, $ROP_D + K\sigma$.

- When demand during the lead time is greater than $ROP_D + K\sigma$, a stockout will occur.

- The service level (the probability that the demand during lead time is less than $ROP_S = $ the Expected Value ROP_D plus the safety stock $K\sigma$) is given by the area under the Normal curve in Figure 11 that is *unhatched*.

- The stockout probability (the probability that demand during lead time exceeds $ROP_S = ROP_D + K\sigma$) is given by the area under the Normal curve in Figure 11 that is *hatched*.

146

Normally Distributed Demand During Lead Time

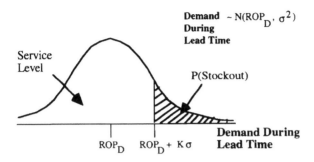

Figure 11

When the safety stock is specified in terms of in terms of K times the standard deviation of demand during lead time σ, Figure 12 provides the correlation between K and the service level.

Safety Stock Multiplier K vs. Service Level

K	Safety Stock SS	Stochastic Reorder Point ROP$_D$ + SS	Service Level
0	0σ	ROP$_D$	50%
1	1σ	ROP$_D$ + 1σ	84.1%
2	2σ	ROP$_D$ + 2σ	97.7%
3	3σ	ROP$_D$ + 3σ	99.87%

Figure 12

The service level (the area under the Normal curve to the left of the expected value of lead time demand ROP$_D$ plus K = 0, 1, 2, and 3 times the standard deviation σ) in Figure 12 can be determined using Table 1b.

For the initial inventory cycles, the standard deviation σ of demand during lead time typically may have to be *forecasted* using data from other businesses stocking similar items. After a sufficient number of inventory cycles, the standard deviation of demand during lead time for the business can be *measured* using actual customer demand.

Often the standard deviation of demand has been previously measured for a *different* length of time than that of the lead time required for the current safety stock decision. However, standard deviations for demand may be manipulated *statistically* to provide the standard deviation for the desired lead time demand.

Suppose there is a measurement of the standard deviation σ_{OT} of demand for some other time (OT) than the standard deviation σ_{LT} of the desired lead time (LT). Then the following relationship holds between σ_{LT} and σ_{OT}:

147

$$\sigma_{LT}^2 / \sigma_{OT}^2 = LT / OT$$

$$\sigma_{LT} = (LT/OT)^{1/2} \cdot \sigma_{OT}$$

Example 6: (Lead Time Demand Calculation). The standard deviation of demand for 16 days has been previously measured to be 200 units. Derive the standard deviation of the demand for the desired lead time of 4 days.

Answer: For this example, OT = 16 days and $\sigma_{OT} = \sigma_{16}$ = 200 units. The desired lead time (LT) is 4 days. For LT = 4 days, $\sigma_{LT} = \sigma_4$ is derived as follows:

$$\sigma_{LT} = (LT/OT)^{1/2} \cdot \sigma_{OT}$$

$$\sigma_4 = (4/16)^{1/2} \cdot \sigma_{16}$$

$$\sigma_4 = (4/16)^{1/2} \cdot 200$$

$$= 100 \ units$$

The stochastic POQ inventory decision is analogous to the stochastic EOQ inventory decision described above. The calculations for the stochastic POQ inventory decision are illustrated in Example 7.

Example 7: (Stochastic POQ Inventory Decision). Same Day Tire Company estimates the expected customer demand for tires to be 10,000 tires per year. The average cost to Same Day to produce each tire is $50. The in-house production rate of tires is 25,000 tires per year. The cost to set-up a tire production run is $2000. The lead time for set up is 15 days. The cost to hold each tire in inventory is $1.20 per tire per year. The standard deviation of demand for tires has been measured to be 5 tires per day. The policy of Same Day is to provide a 95% customer service level for all items stocked.

a) What is the working stock (POQ)?

Answer: This calculation is the same as in the deterministic POQ formulation except the expected demand rate is substituted for the constant demand rate:

> d (Expected demand) = 10,000 tires/year
> r (Production rate) = 25,000 tires/year
> k (Set-up Cost) = $2000
> h (Holding Cost) = $1.20/tire/year

Then,

$$POQ = [2k \cdot d/h]^{1/2} \cdot [r/(r-d)]^{1/2}$$

$$= [(2)(2000)(10000) / (1.20)]^{1/2} \cdot [(25000) / (25000 - 10000)]^{1/2}$$

$$\approx 7454 \ tires \ per \ production \ run \ (round \ up)$$

148

b) What is the expected Annual Set-up Cost for the working stock (POQ)?

Answer: The expected Annual Set-up Cost is the cost per production run set-up times the expected number of production runs per year:

> k *(Set-up Cost per production run)* = $2000
> d *(Expected demand)* = 10,000 *tires/year*
> *POQ (Production run size)* = 7454 *tires*

Then,

> *Annual Set-up Cost* = $k \cdot (d/POQ)$
>
> $\qquad = (\$2000) \cdot (10000 \: / \: 7454)$
>
> $\qquad \approx \$2683$

c) What is the expected Annual Holding Cost for the working stock (POQ)?

Answer: The expected Annual Holding Cost equals the average inventory times the holding cost per tire per year:

> $Ave \; inventory = \dfrac{(POQ)}{2} \cdot \dfrac{(r-d)}{r}$
>
> $\qquad = \dfrac{(7454)}{2} \cdot \dfrac{(25000 - 10000)}{25000}$
>
> $\qquad \approx 2236 \; tires$
>
> *Holding Cost h* = $1.20 /*tire/year*
>
> *Annual Holding Cost* $= h \cdot \dfrac{(POQ)}{2} \cdot \dfrac{(r-d)}{r}$
>
> $\qquad = (\$1.20) \cdot (2236)$
>
> $\qquad \approx \$2684$

Notes:

• At the optimal POQ of 7454 tires, the expected Annual Set-up Cost ($\approx$ $2683) approximately equals the expected Annual Holding Cost ($\approx$ $2684). Remember that the POQ was rounded up to the next integer value.

• The expected Total Annual Cost for the working stock (POQ) is given by the Total Annual Cost function $TC(POQ) = k \cdot d/POQ + h \cdot (POQ/2) \cdot (r-d)/r \approx \$2683 + \$2684 \approx \$5367$.

• The cost of the safety stock includes producing the tires and holding the additional tires over the inventory cycles. The production cost of each safety stock tire is $50 (a one-time cost in inventory cycle 1). The cost to hold a safety stock tire is $1.20 per tire per year. The additional Set-up Cost per safety stock tire is zero since the Set-up Cost k is assumed to be *independent* of the production run size (i.e. no change in the Set-up Cost whether POQ or POQ plus safety stock tires are produced in inventory cycle 1).

d) What is the deterministic reorder point ROP$_D$?

Answer: The deterministic reorder point equals the lead time times the expected demand d:

$$LT = 15 \, days$$
$$d = 10{,}000 \, tires \, per \, year \, / \, 365 \ days \, per \, year = 27.4 \ tires/day$$

Then,

$$ROP_D = LT \cdot d$$

$$= (15 \ days) \cdot (27.4 \ tires/day)$$

$$\approx 411 \ tires$$

e) What are the expected value and standard deviation of the Normal RV describing lead time demand?

Answer: The expected value of the Normal RV describing the stochastic demand during lead time equals ROP$_D$ (computed above):

$$Expected \ Value = ROP_D = 411 \ tires$$

The standard deviation of the Normal RV is computed for desired lead time LT (15 days) by statistically manipulating the measured standard deviation of 5 tires for 1 day given in the problem description.

$$LT = 15 \ days$$
$$OT = 1 \ day$$
$$\sigma_{OT} = \sigma_1 = 5 \ tires$$

Then, σ_{LT} is computed as follows:

$$\sigma_{LT} = (LT/OT)^{1/2} \cdot \sigma_{OT}$$

$$\sigma_{15} = (15/1)^{1/2} \cdot \sigma_1$$

$\sigma_{15} = (15/1)^{1/2} \cdot 5 \ tires$

$= 19.36 \ tires$

f) What is the safety stock required at the stochastic reorder point?

Answer: Table 1b for the Normal distribution is used to determine that the 95% service level in terms equates to a K multiplier of 1.64 standard deviations above the expected value:

$95th \ percentile = 1.64 \ \sigma$

$K = 1.64$

$Safety \ Stock = K \ \sigma$

$= (1.64) \cdot (19.36 \ tires)$

$= 31.75 \ tires$

The safety stock required for a 95% service level is approximately 32 tires. The cost in inventory cycle 1 to produce 32 tires of safety stock is $50 \cdot 32 = \$1600$. The cost to hold the safety stock tires is $\$1.20 \cdot 32 = \$38.40$ per year.

g) What is the stochastic reorder point ROP$_S$?

Answer: The stochastic reorder point equals the deterministic reorder point plus the safety stock:

$ROP_S = ROP_D + Safety \ Stock$

$= 411 + 32$

$= 443 \ tires$

The set up for the next production run is initiated when the inventory level drops to the stochastic reorder point ROP$_S$ of 443 tires. This 32 tires of safety stock will provide a customer service level of 95%.

Exercise 4

Inventory Decision Analysis

Problem #1 (Deterministic EOQ Formulation)

The Ball Four novelty company orders coffee mugs with assorted college logos at a price of $2.00 per mug. Demand has been almost constant over the last few years at approximately 13,000 mugs per year. The Ordering Cost is estimated to be $50 and the Lead Time for an order of mugs to arrive is approximately 5 days. The Holding Cost is estimated to be $0.45 per mug per year.

a) What is the Economic Order Quantity (EOQ) for coffee mugs?

b) What is the Total Annual Cost (including Ordering Cost and Holding Cost) TC(EOQ)?

c) Determine the sensitivity of the EOQ and the Total Annual Cost to a 30% error in the estimated Ordering Cost (above and below $50).

d) What is the (deterministic) Reorder Point (ROP)?

Problem #2 (Stochastic EOQ Formulation)

The Super Market chain stocks many items. Some are purchased from vendors and others are produced in-house. Consider cases of baby formula which is a purchased inventory item with an expected annual demand of 12,300 cases at a cost of $19.00 per case. The cost of processing each order is $200. The Annual Holding Cost for inventory have been determined to be 5% of the purchase cost of the item on the average. The Lead Time is 20 days and the standard deviation of demand is 11 cases per day. The policy of the Super Market chain is to provide a 99% service level for all items stocked.

a) What is the Economic Order Quantity (EOQ) for cases of baby formula?

b) What is the Total Annual Cost TC(EOQ)?

c) What is the Deterministic Reorder Point (ROP$_D$)?

d) What is the Safety Stock (SS) needed to the 99% service level?

e) What is the Stochastic Reorder Point (ROP$_S$)?

Problem #3 (Stochastic POQ Formulation)

The Forever Young Toy Company sells many items. Some are purchased from small vendors and others are produced in-house. Consider high powered squirt guns which is an inventory item produced in-house to meet an expected annual demand of 36,800 units. The manufacturing cost is $1.00 per squirt gun. Squirt guns can be manufactured at a rate of 120 units/day (the manufacturing plant is operational seven days a week). The cost of setting up the production run for squirt guns is $240 and the Lead Time is 2 days. The Annual Holding Cost for inventory at the toy manufacturing plant has been determined to be 5% of the manufacturing cost of the item on the average. The standard deviation of demand is 100 squirt guns per day. The policy of the toy manufacturer is to provide a 95% service level for all items stocked.

a) What is the Production Order Quantity (POQ) for high powered squirt guns?

b) What is the Total Annual Cost TC(POQ)?

c) What is the Deterministic Reorder Point (ROP$_D$)?

d) What is the Safety Stock (SS) needed to provide a 95% service level?

e) What is the Stochastic Reorder Point (ROP$_S$)?

Chapter 4

Inventory Decision Analysis Additional Solved Problems

Problem #1 (Inventory)

Annual demand for IV solutions at Hertz Hospital averages 1250 bottles. The cost per bottle is $7. The cost to order IV bottles is $200 per order. Holding costs per year at Hertz average 5% of the cost per unit of inventory. The Lead Time is 10 days and the standard deviation of demand has been measured to be 8 bottles per day. Hertz Hospital wishes to meet a customer service level of 90%. Of course, Hertz Hospital is open every day of the year.

a) What is the EOQ for IV solution bottles?

Answer: The EOQ for IV solution bottles is computed as follows:

$$d = 1250 \text{ bottles/year (expected value)}$$
$$k = \$200/\text{order}$$
$$h = (\$7)\cdot(.05) = \$0.35/\text{year}$$

$$EOQ = [2dk / h]^{1/2}$$

$$EOQ = [2(1250)(200)/(0.35)]^{1/2}$$

$$EOQ = 1195.23 \text{ bottles}$$

$$EOQ \approx 1196 \text{ bottles (round up)}$$

b) What is the expected annual Total Cost to order and hold IV solution inventory?

Answer: The expected annual Total Cost (TC) of Ordering and Holding IV solution bottles is computed as follows:

$$TC = k \cdot d / EOQ + h \cdot EOQ / 2$$

$$TC = (200)\cdot(1250)/1196 + (0.35)\cdot(1196)/2$$

$$TC = \$209.03 + \$209.30$$

$$TC = \$418$$

c) What is the Deterministic Reorder Point?

Answer: The Deterministic Reorder Point is computed as Lead Time times expected demand:

$$ROP_{\text{Deterministic}} = LT \cdot d$$

$$= (10 \text{ days})\cdot(1250 \text{ bottles/year} / 365 \text{ days/year})$$

$$= 34.25 \text{ bottles}$$

$$\approx 34 \text{ bottles}$$

d) What is the Safety Stock required?

Answer: To compute the Safety Stock, one must first determine the standard deviation of demand during the 10 day lead time:

$$\sigma_{LT} = (LT \, / \, OT)^{\frac{1}{2}} \cdot \sigma_{OT}$$

$$\sigma_{10} = (10 \, / \, 1)^{\frac{1}{2}} \cdot \sigma_1$$

$$\sigma_{10} = (10 \, / \, 1)^{\frac{1}{2}} \cdot 8$$

$$\sigma_{10} = 25.3 \text{ } bottles$$

A 90% service level corresponds to a K-factor of 1.28 (the 90th percentile of the $N(0,1)$ distribution from Table 1b) for σ_{10}.

$$\text{Safety Stock} = K \cdot \sigma_{10}$$

$$= 1.28 \cdot (25.3 \text{ bottles})$$

$$= 32.38 \text{ bottles}$$

$$\approx 32 \text{ bottles}$$

e) What is the Stochastic Reorder Point?

Answer: The Stochastic Reorder Point is the Deterministic Reorder Point plus the Safety Stock:

$$\text{Reorder Point}_{\text{Stochastic}} = \text{Reorder Point}_{\text{Deterministic}} + \text{Safety Stock}$$

$$= 34.25 \text{ bottles} + 32.38 \text{ bottles}$$

$$= 66.63 \text{ bottles}$$

$$\approx 67 \text{ bottles}$$

Problem #2 (Inventory)

The Exotic Boot Company (EBC) is open every day of the year and has an expected demand of 40 pair per day for its boots made from the skins of ostriches. EBC produces its ostrich skin boots in-house and is capable of producing 29,200 pair per year. The cost to produce a pair of these boots is an exorbitant $1200. The set-up cost of a production run is $800 while the cost to hold a pair of ostrich boots is 5% of its cost per year. The lead time for a new production run is 5 days while the standard deviation of demand for ostrich boots has been measured to be 8 pair of boots per day.

a) Determine the Production Order Quantity (POQ) for ostrich boots at the Exotic Boot Company.

Answer: Determine the Production Order Quantity (POQ) for ostrich boots at the Exotic Boot Company.

d = Expected demand = 40 pair per day = 14,600 pair per year
r = Production rate = 29,200 pair per year
k = Setup cost = $800
c = Unit cost of pair of boots = $1200
h = Holding cost = $(0.05) \cdot (\$1200)$ = $60 per pair per year
LT = Lead time = 5 days
σ_1 = Standard deviation of demand = 8 pair per day
Service level = 85%

$$POQ = \sqrt{\frac{2kd}{h}} \cdot \sqrt{\frac{r}{r-d}}$$

$$= \sqrt{\frac{2(\$800)(14600/yr)}{\$60/yr}} \cdot \sqrt{\frac{29200/yr}{(29200 - 14600)/yr}}$$

$$= \sqrt{389,333} \cdot \sqrt{2}$$

$$\approx 883 \ (Round \ up)$$

b) Determine the optimum annual set-up cost and the annual holding cost for ostrich boots.

Answers: The optimum annual set-up cost and holding cost are as follows:

$$Annual \ Set\text{-}up \ Cost = k \cdot \frac{d}{POQ}$$

$$= (\$800) \cdot \frac{14600/yr}{883}$$

$$\approx \$13228$$

$$Annual \ Holding \ Cost = h \cdot \frac{POQ}{2} \cdot \frac{r-d}{r}$$

$$= (\$60) \cdot \frac{883}{2} \cdot \frac{(29200 - 14600)/yr}{29200/yr}$$

$$= (\$60) \cdot \frac{883}{2} \cdot \frac{1}{2}$$

$$\approx \$13245$$

c) What is the Safety Stock required to meet an 85% service level for customers of the Exotic Boot Company?

Answer: The Safety Stock required to meet an 85% service level is determined as follows:

$K = Z_{.85} = 1.04$

$OT = 1 \, day$

$\sigma_{OT} = \sigma_1 = 8 \, pair$

$LT = 5 \, days$

$\sigma_{LT} = (^{LT}/_{OT})^{1/2} \cdot \sigma_{OT}$

$\sigma_s = (^5/_1)^{1/2} \cdot 8 = 17.89 \, pair$

$Safety \, Stock = K \cdot \sigma_s = (1.04) \cdot (17.89) \approx 19 \, pair$

d)What is the Stochastic Reorder Point?

Answer: The Stochastic Reorder Points is determined as follows:

$ROP_s = ROP_D + Safety \, Stock$

$ROP_D = LT \cdot d = (5 \, days) \cdot (40 \, pair/day) = 200 \, pair$

$ROP_s = 200 + 19 \approx 219 \, pair$

e) How much inventory is produced in the initial inventory cycle? In all subsequent inventory cycles?

Answer: The inventory in the initial and subsequent inventory cycles is given as follows:

- Inventory period 1: Produce POQ + SS = 883 + 19 = 902 pair of ostrich boots.

- All subsequent inventory periods: Produce POQ = 883 pair of ostrich boots.

f) For the Safety Stock carried, what are the annual inventory cost (cost to produce), annual set-up cost, and the annual holding cost?

Answer: The costs for the Safety Stock are determined as follows:

- Annual inventory cost = c·SS = ($1200)·(19 pair) = $22,800.

- Annual setup cost = $0 (no additional setup cost for extra 19 pair of boots produced).

- Annual holding cost = h·SS = ($60)·(19 pair) = $1140

"Practice without theory is blind, theory with practice is sterile."
Immanuel Kant, as amended by Karl Marx

5 | Project Managment (PERT/CPM)

5.1 Introduction

A *network* is a sequence of interrelated tasks to be undertaken in a particular sequence. The term network is often used interchangeably with the term *project*. Performance of the set of interrelated tasks in a specified sequence will lead to the completion of the project.

The Program Evaluation and Review Technique (PERT), sometimes referred to as the Critical Path Method (CPM), is used to:

> - View the required sequencing of the tasks of the project.

> - Determine the earliest completion time for the project (the longest or critical path through the network).

> - Identify the critical tasks which must be completed on schedule in order not to delay the project beyond its earliest completion time.

The deterministic PERT technique will be demonstrated in Example 1. The stochastic PERT technique will be demonstrated in Section 5.5.

Example 1: (Deterministic PERT). Consider the start-up of the Narcissus Health Club (the project). A list of tasks (with task names in **bold**) required to open the health club for business is given as follows:

> 1. Find and lease **Space** for the health club.

> 2. Hire both a **Director** and an Assistant Director.

3. Purchase **Equipment** for the health club.

4. **Remodel** the space.

5. **Install** the equipment.

6. Hire **Fitness** personnel.

7. **Train** all health club personnel.

8. Design and implement the **Publicity** campaign.

9. Perform **Final** preparations.

Note: The task list for this project has been simplified to account for only the top level tasks. However, the application of PERT can be generalized to networks with a large number of tasks. There are many computer software packages which enable PERT to be applied to large networks.

Build the PERT network (show the task sequencing) for the start-up of the Narcissus Health Club. Determine the earliest completion time and critical tasks for the project.

Answer: The initial step in the application of PERT to a project requires the specification of two characteristics for each task:

1) The task completion time.

2) The predecessor tasks. A task can not begin until all its predecessor tasks are completed. After its predecessors are completed, a given task can start at any time.

Estimates have been made for the completion time for each task required to accomplish the start-up of the Narcissus Health Club. Also, the sequencing of the tasks (the identification of each task's predecessors) has been established. The completion time and the predecessors for each of the 9 tasks required to start-up the Narcissus Health Club are given in the Task Time/Precedence Table (see Figure 1).

Narcissus Health Club Task Time/Precedence Table

Task	Completion Time	Predecessors
1. Space	7 Weeks	None
2. Director	8 Weeks	None
3. Equipment	6 Weeks	2
4. Remodel	3 Weeks	1,2
5. Install	1 Weeks	3,4
6. Fitness	2.5 Weeks	None
7. Train	2 Weeks	6
8. Publicity	5.5 Weeks	None
9. Final	1.5 Weeks	5,7,8

Figure 1

Example 1 examines the simplest PERT formulation which assumes *deterministic* (constant) task completion times (stochastic task completion times will be examined in Section 5.5).

The application of PERT to a large project involves many decisions. The *decisions* for PERT are the appropriate starting times for each task in the project. PERT with deterministic task completion times is a complex deterministic decision technique.

The *objective* of PERT is to determine the earliest completion time of the project.

The next step is to build a PERT network (see Section 5.2) which graphically shows the sequencing of the tasks. The Activity-on-Node (AON) method will be used in this presentation to build the PERT network. Activities in this application of PERT are tasks.

Notes:

• Some presentations use the Activity-on-Arrow (AOA) method (which is sometimes called Activity-on-Arc method). The AON and AOA methods are completely equivalent and will lead to the same starting times for each task and the same earliest completion time of the project.

• Building the PERT network is more straightforward using the Activity-on-Node (AON) method than the Activity-on-Arrow (AOA) method. For example, consider the case where Task 1 is a predecessor for both Task 3 and Task 4 while Task 2 is a predecessor for Task 4. A comparison of the display of these precedence relationships using the Activity-on-Node method and the Activity-on-Arrow method is given in Figure 2. A *dummy task* must be introduced in the Activity-on-Arrow method (see the dotted arrow in Figure 2) to indicate that Task 1 is a predecessor to both Task 3 and Task 4.

Comparison of AON vs. AOA Methods

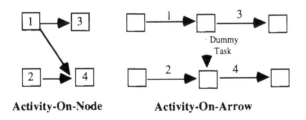

Activity-On-Node   **Activity-On-Arrow**

Figure 2

• The Activity-on-Node method is used in this presentation since it is initially easier to implement (the introduction of dummy tasks to display graphically the precedence among the tasks is not required).

5.2 Building The PERT Network

The process of building the PERT network starts with establishing the precedence network which displays graphically the precedences among the tasks. In establishing the precedence network, it is convenient to start at the final task and work *backwards* toward the initial task (the start).

From the Task Time/Precedence Table (see Figure 1), Tasks 5, 7 and 8 precede the final Task 9. This relationship is indicated by arrows leading from the nodes for Tasks 5, 7, & 8 into the node for Task 9 (see Figure 3).

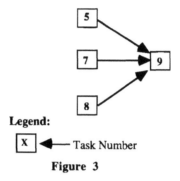

Legend:

| X | ◄——— Task Number

Figure 3

Notes:

• The Tasks (activities) 5, 7, 8, and 9 are placed on the *nodes* using the Activity-on-Node method.

• Figure 3 displays graphically the fact that Task 9 can not begin until *all three* of its predecessor tasks (Tasks 5, 7, and 8) are completed.

From the Task Time/Precedence Table, Tasks 6 precedes Task 7. The node for Task 6 has an arrow leading into the node for Task 7 (see Figure 4).

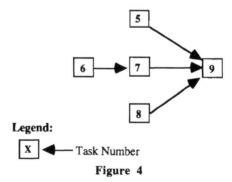

Legend:

| X | ◄——— Task Number

Figure 4

From the Task Time/Precedence Table, Tasks 3 and 4 precede Task 5. The nodes for Task 3 and Task 4 each have an arrow leading into the node for Task 5 (see Figure 5).

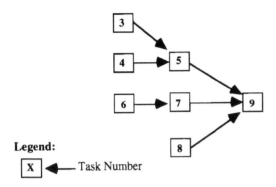

Legend:

Figure 5

From the Task Time/Precedence Table, Tasks 1 and 2 precede Task 4. The nodes for Task 1 and Task 2 each have an arrow leading into the node for Task 4 (see Figure 6).

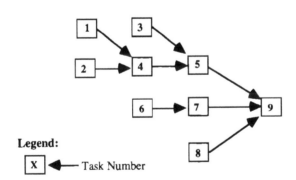

Legend:

Figure 6

From the Task Time/Precedence Table, Task 2 precedes Task 3. The node for Task 2 has an arrow leading into the Node for Task 3 (see Figure 7).

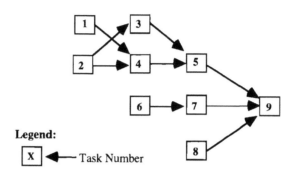

Figure 7

Since no initial or "Start" Task was given, it is added into this network of tasks. The Start Task, designated Task 0, takes *zero* time to complete and has *no* predecessor tasks. Connect the node for the Start Task 0 to the nodes for all tasks which have no predecessors (i.e. Tasks 1, 2, 6, and 8). Figure 8 provides the precedence network for the Narcissus Health Club described by the Task Time/Precedence Table (see Figure 1).

Precedence Network For Narcissus Health Club

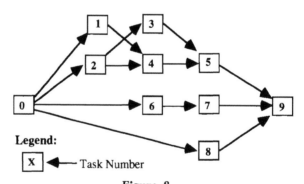

Figure 8

After completing the precedence network, check the network to ensure that all the precedences listed in the Task Time/Precedence Table are preserved.

The Activity-on-Node PERT network is completed by adding the task completion times from the Task Time/Precedence Table onto the precedence network (see Figure 9).

Activity-on-Node PERT Network For Narcissus Health Club

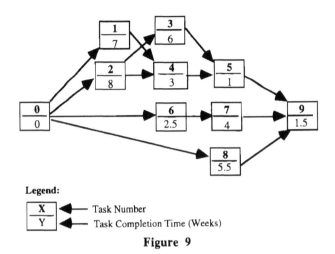

Figure 9

5.3 Solving the PERT Network

Recall that the decisions for PERT are the appropriate starting times for each task and that the objective of PERT is to determine the earliest completion time of the project (see Section 5.1).

The following definitions for each task are useful in solving the PERT network:

- Earliest Completion Time (ECT): The *earliest* time a task can be completed. For a task to be completed at it Earliest Completion Time, all preceding tasks must be at their Earliest Completion Time.

- Latest Completion Time (LCT): The *latest* time a task can be completed without causing a delay in the project Earliest Completion Time.

Earliest Completion Time (ECT)

When a given task has only a *single* predecessor, the ECT for the given task is the ECT for the predecessor plus the completion time for the given task.

Task ECT = Predecessor ECT + Task Completion Time

To calculate the Earliest Completion Time (ECT) for each task in the network, it is convenient to work *forward* through the PERT network from the initial node (Task 0) to the final node (Task 9).

The Earliest Completion Time (ECT) for Task 0 is 0 weeks since Task 0 has no predecessors and Task 0 itself takes zero time to complete. The time 0 is placed above the Task 0 node in the PERT network (see Figure 10).

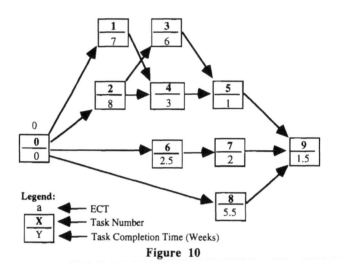

Figure 10

The Earliest Completion Time for Task 1 is 7 weeks. The derivation of this ECT is as follows:

- Task 1 has only Task 0 for a predecessor
- The ECT for Task 0 is 0 weeks (previous ECT calculation)
- Task 1 itself takes 7 weeks to complete
- The ECT for Task 1 is 0 weeks + 7 weeks = 7 weeks

The Earliest Completion Time for Task 2 is 8 weeks. The derivation of this ECT is as follows:

- Task 2 has only Task 0 for a predecessor
- The ECT for Task 0 is 0 weeks (previous ECT calculation)
- Task 2 itself takes 8 weeks to complete
- The ECT for Task 2 is 0 weeks + 8 weeks = 8 weeks

The Earliest Completion Time for Task 6 is 2.5 weeks. The derivation of this ECT is as follows:

- Task 6 has only Task 0 for a predecessor
- The ECT for Task 0 is 0 weeks (previous ECT calculation)
- Task 6 itself takes 2.5 weeks to complete
- The ECT for Task 6 is 0 weeks + 2.5 weeks = 2.5 weeks

The Earliest Completion Time for Task 8 is 5.5 weeks. The derivation of this ECT is as follows:

- Task 8 has only Task 0 for a predecessor
- The ECT for Task 0 is 0 weeks (previous ECT calculation)
- Task 8 itself takes 5.5 weeks to complete
- The ECT for Task 8 is 0 weeks + 5.5 weeks = 5.5 weeks

The Earliest Completion Time for Task 3 is 14 weeks. The derivation of this ECT is as follows:

- Task 3 has only Task 2 for a predecessor
- The ECT for Task 2 is 8 weeks (previous ECT calculation)
- Task 3 itself takes 6 weeks to complete
- The ECT for Task 3 is 8 weeks + 6 weeks = 14 weeks

The Earliest Completion Times for Tasks 1, 2, 3, 6, and 8 are placed above their nodes in the PERT network (see Figure 11).

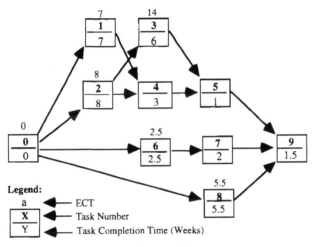

Figure 11

Notes: Examine Task 4:

• Task 4 has two predecessor tasks (Task 1 and Task 2). A task with *multiple* predecessor tasks can not begin until *all* predecessor tasks have finished.

• If it is assumed that all predecessors of a task finish at their earliest possible time (given by their ECT), then the task can begin at a time equal to the maximum of the ECT's for the predecessor tasks (the earliest time when all the predecessors tasks will be complete).

• The ECT for a task with predecessors is equal to the *maximum* of the ECT's for its predecessors plus the time to complete the task itself.

Task ECT = Max (Predecessor ECT's) + Task Completion Time

The Earliest Completion Time for Task 4 is 11 weeks. The derivation of this ECT is as follows:

- Task 4 has Task 1 and Task 2 for predecessors
- The ECT for Task 1 is 7 weeks (previous ECT calculation)
- The ECT for Task 2 is 8 weeks (previous ECT calculation)
- The Max (latest) ECT for any predecessor is 8 weeks (Task 2)
- Task 4 itself takes 3 weeks to complete
- The ECT for Task 4 is Max (7 weeks, 8 weeks) + 3 weeks = 11 weeks

The Earliest Completion Time for Task 5 is 15 weeks. The derivation of this ECT is as follows:

- Task 5 has Task 3 and Task 4 for predecessors
- The ECT for Task 3 is 14 weeks (previous ECT computation)
- The ECT for Task 4 is 11 weeks (previous ECT computation)
- The Max (latest) ECT for any predecessor is 14 weeks (Task 3)
- Task 5 itself takes 1 week to complete
- The ECT for Task 5 is Max (14 weeks, 11 weeks) + 1 = 15 weeks

The Earliest Completion Time for Task 7 is 4.5 weeks. The derivation of this ECT is as follows:

- Task 7 has only Task 6 for a predecessor
- The ECT for Task 6 is 2.5 weeks (previous ECT calculation)
- Task 7 itself takes 2 weeks to complete
- The ECT for Task 7 is 2.5 weeks + 2 weeks = 4.5 weeks

The Earliest Completion Time for Task 9 (Final Task) is 16.5 weeks. The derivation of this ECT is as follows:

- Task 9 has Task 5, Task 7, and Task 8 for predecessors
- The ECT for Task 5 is 15 weeks (previous ECT calculation)
- The ECT for Task 7 is 4.5 weeks (previous ECT calculation)
- The ECT for Task 8 is 5.5 weeks (previous ECT calculation)
- The Max (latest) ECT for any predecessor is 15 weeks (Task 5)
- Task 9 itself takes 1.5 weeks to complete
- The ECT for Task 9 is Max (15 weeks, 4.5 weeks, 5.5 weeks) + 1 week = 16.5 weeks

The Earliest Completion Times for Tasks 4, 5, 7, and 9 are placed above their nodes in the PERT network (see Figure 12).

Earliest Completion Times (ECT's) For Health Club Tasks

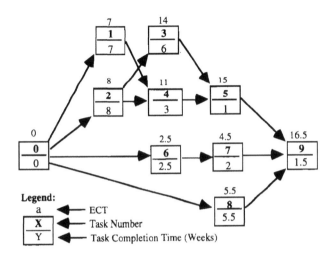

Figure 12

Notes:

• The ECT of 16.5 weeks for the *final task* (Task 9) is the ECT for the *project* and represents the length of the *longest path* through the Narcissus Health Club PERT network.

• The Earliest Completion Time (ECT) technique is a structured method for finding the *length* only of the longest path. Identification of the actual path that is longest can not be determined at this point in the presentation.

• One could enumerate all paths (there are five) through the Narcissus Health Club PERT network and determine the longest path. Path 4 is determined to be the longest path (16.5 weeks) as follows:

Path	Path Length
1. Task 0→Task 8→Task 9	0+5.5+1.5 = 7 weeks
2. Task 0→Task 6→Task 7→Task 9	0+2.5+2+1.5 = 6 weeks
3. Task 0→Task 1→Task 4→Task 5→Task 9	0+7+3+1+1.5 = 12.5 weeks
4. Task 0→Task 2→Task 3→Task 5→Task 9	0+8+6+1+1.5= **16.5 weeks**
5. Task 0→Task 2→Task 4→Task 5→Task 9	0+8+3+1+1.5= 13.5 weeks

• Enumeration of paths is not a satisfactory method for determination of the longest path. For a large network with hundreds of task nodes where there might be many thousands of individual paths through the network. Enumerating and then evaluating the length of each of the thousands of paths through this network is a very difficult undertaking.

• The ECT technique finds the length of the longest path (the earliest completion time) for the network using the simple calculation *Max (Predecessor ECT) + task completion time* at each of the task *nodes* . The actual paths through the PERT network are never enumerated.

• After the Latest Completion Time (LCT) presentation, it will be possible to identify the longest path through the network and the critical tasks on the longest path whose delay will delay the project beyond its Earliest Completion Time. Again the actual paths through the network are never enumerated.

Latest Completion Time (LCT)

Recall that the Latest Completion Time (LCT) for a task is the *latest* time that task can be completed without causing a delay in the project Earliest Completion Time.

Note: For each task, the Latest Completion Time (the latest time a task can complete without delaying the project) is always greater than or equal to the Earliest Completion Time (the earliest time it is possible to complete a task).

$$ECT \leq LCT$$

For a given task, a <u>successor</u> is defined as a task which can be started *immediately after* completing the given task. When a given task has only a *single* successor, the LCT for the given task is the LCT for the successor minus the successor completion time.

Task LCT = Successor LCT − Successor Completion Time

To calculate the Latest Completion Time (LCT), it is convenient to work *backwards* through the network from the final node (Task 9) to the initial node (Task 0). The LCT will be placed above the task node to the right of the ECT in the form ECT|LCT (see Figure 13).

The Latest Completion Time for Task 9 (and the project) is 16.5 weeks. The derivation of this LCT is as follows:

- Task 9 (the final task) has no successor tasks
- Task 9 must be complete by the project ECT = 16.5 weeks not to delay the project beyond its ECT

The Latest Completion Time for Task 5 is 15 weeks. The derivation of this LCT is as follows:

- Task 9 is the single successor for Task 5
- The LCT for Task 9 is 16.5 weeks (previous LCT calculation)
- Task 9 takes 1.5 weeks to complete
- The LCT for Task 5 is 16.5 weeks - 1.5 week = 15 weeks

The Latest Completion Time for Task 7 is 15 weeks. The derivation of this LCT is as follows:

- Task 9 is the single successor for Task 7
- The LCT for Task 9 is 16.5 weeks (previous LCT calculation)
- Task 9 takes 1.5 weeks to complete

- The LCT for Task 7 is 16.5 weeks - 1.5 weeks = 15 weeks not to delay Task 9 beyond its ECT

The Latest Completion Time for Task 8 is 15 weeks. The derivation of this LCT is as follows:
- Task 9 is the single successor for Task 8
- The LCT for Task 9 is 16.5 weeks (previous LCT calculation)
- Task 9 takes 1.5 weeks to complete
- The LCT for Task 8 is 6.5 weeks - 1.5 weeks = 15 weeks not to delay Task 9 beyond its ECT

The Latest Completion Time for Task 3 is 14 weeks. The derivation of this LCT is as follows:

- Task 5 is the single successor for Task 3
- The LCT for Task 5 is 15 weeks (previous LCT calculation)
- Task 5 takes 1 week to complete
- The LCT for Task 3 is 15 weeks - 1 week = 14 weeks not to delay Task 5 beyond its ECT

The Latest Completion Time for Task 4 is 14 weeks. The derivation of this LCT is as follows:

- Task 5 is the single successor for Task 4
- The LCT for Task 5 is 15 weeks (previous LCT calculation)
- Task 5 takes 1 week to complete
- The LCT for Task 4 is 15 weeks - 1 week = 14 weeks not to delay Task 5 beyond its ECT

The Latest Completion Time for Task 6 is 13 weeks. The derivation of this LCT is as follows:

- Task 7 is the single successor for Task 6
- The LCT for Task 7 is 15 weeks (previous LCT calculation)
- Task 7 takes 2 weeks to complete
- The LCT for Task 6 is 15 weeks - 2 weeks = 13 weeks not to delay Task 7 beyond its ECT

The Latest Completion Time for Task 1 is 11 weeks. The derivation of this LCT is as follows:

- Task 4 is the single successor for Task 1
- The LCT for Task 4 is 14 weeks (previous LCT calculation)
- Task 4 takes 3 weeks to complete
- The LCT for Task 1 is 14 weeks - 3 weeks = 11 weeks not to delay Task 4 beyond its ECT

Notes:

• Task 2 has two successor tasks (Task 3 and Task 4). A task with multiple successor tasks must finish in time so as not to delay the start of *all* successor tasks beyond their LCT's.

• It is assumed that all successors of a task finish at their latest completion time (given by their LCT). Then for any successor task, the task must finish by a time equal to the successor's LCT minus the successor's completion time in order not to delay that successor task beyond its LCT.

• The LCT for a task with *multiple* successors is then the *minimum* (over the set of all successors) of the successor LCT's minus the corresponding successor completion time.

   Task LCT = Min (Successor LCT – Successor Completion Time)

The Latest Completion Time for Task 2 is 8 weeks. The derivation of this LCT is as follows:

   - Tasks 3 and 4 are successor tasks for Task 2
   - The LCT for Task 3 is 14 weeks (previous LCT calculation)
   - Task 3 takes 6 weeks to complete
   - Task 2 must be complete by 14 weeks - 6 weeks = 8 weeks not to delay Task 3
   - The LCT for Task 4 is 14 weeks (previous LCT calculation)
   - Task 4 takes 3 weeks to complete
   - Task 2 must be complete by 14 weeks - 3 weeks = 11 weeks not to delay Task 4
   - The LCT for Task 2 is Min (14 - 6, 14 - 3) = Min (8, 11) = 8 weeks
   - Task 2 must be complete by 8 weeks not to delay Tasks 3 or 4 beyond their LCT's

The Latest Completion Time for Task 0 is 0 weeks. The derivation of this LCT is as follows:

   - Tasks 1, 2, 6, and 8 are successor tasks of Task 0
   - The LCT for Task 1 is 11 weeks (previous LCT calculation)
   - Task 1 takes 7 weeks to complete
   - Task 0 must be complete by 11 weeks - 7 weeks = 4 weeks not to delay Task 1
   - The LCT for Task 2 is 8 weeks (previous LCT calculation)
   - Task 2 takes 8 weeks to complete
   - Task 0 must be complete by 8 weeks - 8 weeks = 0 weeks not to delay Task 2
   - The LCT for Task 6 is 13 weeks (previous LCT calculation)
   - Task 6 takes 2.5 weeks to complete
   - Task 0 must be complete by 13 weeks - 2.5 weeks = 10.5 weeks not to delay Task 6
   - The LCT for Task 8 is 15 weeks (previous LCT calculation)
   - Task 8 takes 5.5 weeks to complete
   - Task 0 must be complete by 15 weeks - 5.5 weeks = 9.5 weeks not to delay Task 8
   - The LCT for Task 0 is Min (11 - 7, 8 - 8, 13 - 2.5, 15 - 5.5) = Min (4, 0, 10.5, 9.5) = 0 weeks
   - Task 0 must be complete by 0 weeks not to delay Tasks 1, 2, 6, or 8 beyond their LCT's

The Activity-on-Node PERT Network with ECT's and LCT's for each task node is given in Figure 13.

PERT Network For Narcissus Health Club

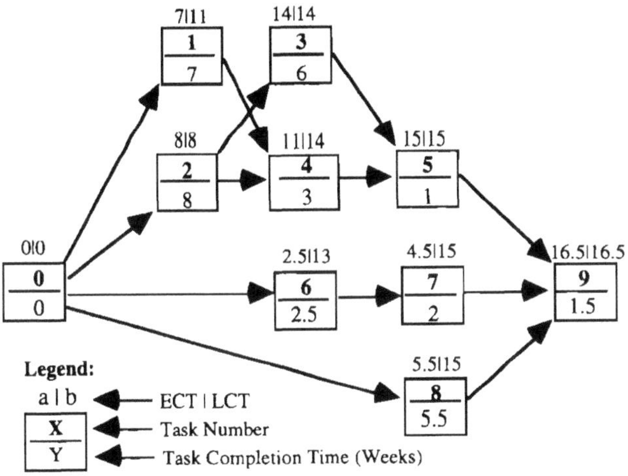

Figure 13

Notes:

• When the paths for Example 1 were enumerated, the longest path (16.5 weeks) was determined to be Task 0 → Task 2 → Task 3 → Task 5 → Task 9. In the PERT Network for the Narcissus Health Club (see Figure 13), the tasks on this path all have the property that ECT = LCT.

• When the Earliest Completion Time (ECT) equals the Latest Completion Time (LCT) for a task, the task is on the critical path (another name for the longest path). Thus, computation of the ECT's and LCT's for a network is a structured technique to determine the length of the critical path and the tasks on the critical path (again without having to enumerate all the paths through the network). For a large project, the ECT and LCT may have to be computed for hundreds of task nodes. But this technique is much easier than having to enumerate possibly thousands of paths in order to find the tasks on the critical (longest) path.

• The most important feature of tasks on the critical path is that *any delay* in the completion time of these tasks beyond their ECT (which equals their LCT) will cause the project to be delayed beyond its Earliest Completion Time. This occurs because the longest path (the ECT for the project) becomes longer.

• There need not be a unique critical path through the network (though all critical paths will have the same length equal to the project Earliest Completion Time). A delay in any task on any critical path will cause a delay in completion of the project beyond its project Earliest Completion Time.

Slack Time

For any task not on the critical path, its ECT will be *strictly less* than its LCT. The slack time (sometimes called float) for any task is defined as follows:

$$Slack\ Time = LCT - ECT$$

By definition, a task delayed in completion beyond its Latest Completion Time will delay the project Earliest Completion Time. However, when a task has positive *slack* (such that LCT - ECT > 0), the completion of the task may be delayed beyond its Earliest Completion Time up to its amount of its slack time (LCT - ECT) without delaying the project Earliest Completion Time. Though the task will complete after its Earliest Completion Time, the task can still complete by its Latest Completion Time (and thus the project will not be delayed).

The calculation of the slack time for the nine Narcissus Health Club tasks of Example 1 is given in Figure 14. The starred (*) tasks (Task 0 → Task 2 → Task 3 → Task 5 → Task 9) make up the critical (longest) path through the PERT network.

Slack Time Table For Narcissus Health Club

Task	Earliest Completion Time	Latest Completion Time	Slack Time
0*	0	0	0
1	7	11	4
2*	8	8	0
3*	14	14	0
4	11	14	3
5*	15	15	0
6	2.5	13	10.5
7	4.5	15	10.5
8	5.5	15	9.5
9*	16.5	16.5	0

Figure 14

Notes:

• Tasks on the critical path have *zero* slack time (LCT - ECT = 0). Any delay to these tasks will cause a delay in the project Earliest Completion Time.

• If the completion time for a task not on the critical path is delayed beyond its Earliest Completion Time by more than its slack time, a *longer* critical path for the project will emerge and the project will be delayed beyond its Earliest Completion Time.

5.4 Gantt Charts

While a PERT network displays visually the spatial relationship (precedences) among the tasks, it does not display the *time* relationship among the tasks in a network. There is no mechanism in PERT to display the relative lengths of the tasks or the slack time for each task.

A Gantt Chart (see Figure 15), sometimes called a *Critical Path Schedule*, displays the following task information visually:

- The tasks of the project
- The task lengths (completion times)
- The task ECT's
- The task LCT's
- The critical path for the project

Gantt Chart For The Narcissus Health Club Project

Week 0 1 2 3 4 5 6 7 8 9 10 11 12 13 14 15 16 17

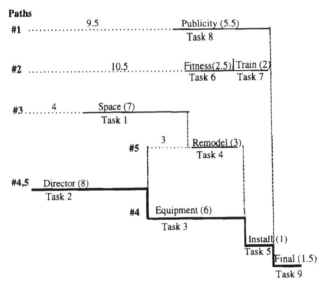

Week 0 1 2 3 4 5 6 7 8 9 10 11 12 13 14 15 16 17

Figure 15

Consider the 5 paths through the Narcissus Health Club PERT network from Example 1:

1. *Publicity (Task 8) → Final (Task 9) = 5.5 + 1.5 = 7 weeks*

2. *Fitness (Task 6) → Train (Task 7) → Final (Task 9) = 2.5 + 2 + 1.5 = 6 weeks*

175

3. *Space (Task 1) → Remodel (Task 4) → Install (Task 5) → Final (Task 9)*
= 7 + 3 + 1 + 1.5 = 12.5 *weeks*

*4. *Director (Task 2) → Equipment (Task 3) → Install (Task 5) → Final (Task 9)*
= 8 + 6 + 1 + 8 = 16.5 *weeks*

5. *Director (Task 2) → Remodel (Task 4) → Install (Task 5) → Final (Task 9)*
= 8 + 3 + 1 + 1.5 = 13.5 *weeks*

These 5 paths are numbered on the Gantt Chart for the Narcissus Health Club in Example 1 (see Figure 15). Path 4 (16.5 weeks) is the longest path through the network.

Notes: Observations on Figure 15:

• The time (in weeks) to complete each task has been provided in parenthesis after the task name. The length of the horizontal line below the task is the task completion time.

• A path through the Gantt Chart without any horizontal dashed lines (which indicates slack time) is a critical path. In the Gantt Chart given in Figure 14, the **bolded** Path #4: Director (Task 2) → Equipment (Task 3) → Install (Task 5) → Final (Task 9) through the network is without any slack time and is a critical path.

• The numbers on the horizontal dashed lines show how much a task or tasks extending from the dotted lines can be delayed. For example, the start time of the Fitness (Task 6) task or the start time of the Train (Task 7) task (or the combination of the start times of the two tasks) can be delayed 10.5 weeks without delaying the project Earliest Completion Time.

Given limited resources (manpower, equipment, money, etc.), a Gantt Chart allows a decision maker to visualize how slack times can be used to delay the start of tasks not on the critical path beyond their Earliest Completion Time without delaying the project beyond its Earliest Completion Time. Such delays may ease the peak allocation of personnel, equipment, payroll, etc. and thus save resources without delaying the project completion schedule.

5.5   PERT With Stochastic Time Estimates

Introduction

The PERT presentation up to this point has assumed deterministic (constant) task completion times. However, a network can also be viewed stochastically with task completion times which are Random Variables.

Task completion times in a stochastic network (project) are often specified by a Beta distribution which is a continuous, bounded, and usually skewed probability distribution (see Figure 16). A Beta distribution for the completion time of each task is defined by three parameters determined by the answers to the following three questions:

- What is an optimistic (o) estimate of the task completion time?
- What is the most likely (m) estimate of the task completion time?
- What is a pessimistic (p) estimate of the task completion time?

Typical Beta Distribution For Task Completion Times

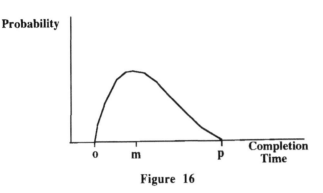

Figure 16

<u>Notes</u>

• The Beta distribution (like all continuous probability distributions) has an area under the curve of 1 for any set of stochastic task completion time estimates o, m, and p.

• In general, a Beta distribution used to specify a stochastic task completion time will not be symmetric around its most likely estimate m. The parameter m represents the *mode* (most likely value) and not the mean or median for the stochastic task completion time.

Beta distribution estimates for the optimistic (o), most likely (m), and pessimistic (p) task completion times for the nine Narcissus Health Club tasks in Example 1 are given in Figure 17. The most likely task completion time estimates m are the same as the deterministic task completion times (see Figure 1) though this is not true in general.

Beta Distribution Estimates For Task Completion Times

Tasks	Stochastic Estimates		
	o	m	p
0. Start	0	0	0
1. Space	4	7	10
2. Director	6	8	9
3. Equipment	3	6	7.5
4. Remodel	2.5	3	4
5. Install	0.5	1	2.5
6. Fitness	1.5	2.5	3.5
7. Train	1	2	2.5
8. Publicity	5	5.5	7.5
9. Final	1	1.5	2.5

Figure 17

The length of a path through a deterministic PERT network is the sum of the constant completion times for the tasks on the path. The length of the longest path is the earliest completion time for the project.

For a stochastic network, each task completion times is a Random Variable (RV) and thus the following attributes apply:

1) The length of each path is the sum of task RV's and is itself a Random Variable.

2) Which path is the longest path will vary based on the actual stochastic values realized for the task completion times. In fact, each path has some probability (perhaps very small) of being the longest path.

3) The length of the varying longest path (the earliest completion time for the project) is described by the *maximum* of the RV's associated with the length of each path. For example, for 5 paths whose lengths are described by the RV's Path 1 through Path 5, the length of the longest path through the network is given by the statistic Max (Path 1, ⋯, Path 5).

4) This statistic (the maximum of the RV's associated with each path) is itself a Random Variable and its distribution can be described by a sample mean $\bar{X}$ and a sample standard deviation s.

There are two options to determine the mean and standard deviation of the Random Variable denoting the length of the longest path (earliest project completion time) for a stochastic network:

- **Simulation Method:** Takes random draws from the given completion time Beta distribution (see Figure 17) for each of the tasks in the network.

-- The completion time random draws for the tasks on each path are summed to obtain the length of each path for this one set of random draws.

-- The path with the largest sum is the critical path for this one set of random draws. The longest path may vary with each set of random draws.

-- This process of taking random draws for each task in the stochastic network is repeated many times to obtain:

1) The sample mean and standard deviation for the length of each individual path.

2) The probability (proportion) that each path is the longest path.

3) The sample mean $\bar{X}$ and the sample standard deviation s of the RV for the longest path (the maximum of the RV's associated with the lengths of the individual paths).

- **Stochastic PERT:** Takes advantage of the Central Limit Theorem which states that if one samples from even a small set of Random Variables (e.g. less than 30 tasks on a path), the sum of the samples will be approximately a Normal Random Variable.

-- The fixed expected values for the Beta distribution task completion times are used to determine the *single* critical path.

-- The expected values of the completion times of the tasks on the single critical path are added to obtain the expected value of the earliest completion time for the project.

-- The variances of the task completion times of the tasks on the critical path are added to obtain the variance of the earliest completion time for the project. The square root of the single critical path variance provides the standard deviation of the earliest completion time for the project.

The expected value T_{exp}, variance σ^2, and standard deviation σ of the Beta distribution task completion times with estimates o (optimistic), m (most likely), and p (pessimistic) are defined as follows:

$$T_{exp} = (o + 4m + p)/6 \qquad \text{(Expected Value)}$$

$$\sigma^2 = [(p - o)/6]^2 \qquad \text{(Variance)}$$

$$\sigma = (p - o)/6 \qquad \text{(Standard Deviation)}$$

Figure 18 provides the expected value T_{exp}, standard deviation σ, and variance σ^2 for the nine Narcissus Health Club tasks given in Example 1 based on the Beta distribution stochastic task completion time estimates o, m, and p.

Beta Distribution Stochastic Time Estimate Table

	Tasks	Stochastic Estimates			Beta Distribution		
		o	m	p	T_{exp}	σ	σ^2
*	0. Start	0	0	0	0	0	0
	1. Space	4	7	10	7.00	1.00	1.00
*	2. Director	6	8	9	7.83	0.50	0.25
*	3. Equipment	3	6	7.5	5.75	0.75	0.56
	4. Remodel	2.5	3	4	3.08	0.25	0.06
*	5. Install	0.5	1	2.5	1.15	0.33	0.11
	6. Fitness	1.5	2.5	3.5	2.50	0.33	0.11
	7. Train	1	2	2.5	1.90	0.25	0.06
	8. Publicity	5	5.5	7.5	5.75	0.42	0.17
*	9. Final	1	1.5	2.5	1.60	0.25	0.06

Figure 18

Stochastic PERT

A key assumption of _stochastic PERT_ is that a *single* critical path for a stochastic network can be computed using the *fixed* expected values (T_{exp}) of the Beta distribution task completion times.

The expected values (T_{exp}) of the task completion times are used to determine the Earliest Completion Time (ECT) and Latest Completion Time (LCT) for each task in the stochastic network. The critical path is determined in the same manner as the deterministic network (i.e. those tasks where ECT = LCT).

Using the fixed expected values (T_{exp}) from Figure 18, the critical path derived for the stochastic Narcissus Health Club PERT network turns out to be the same critical path derived for the deterministic PERT network in Example 1:

$$Critical\ Path\ =\ Task\ 0\ \rightarrow\ Task\ 2\ \rightarrow\ Task\ 3\ \rightarrow\ Task\ 5\ \rightarrow\ Task\ 9$$

Note: The critical path determined for the stochastic PERT network using the estimates o, m, and p for each task and computing the expected value T_{exp} for each task need not be the same as that determined for the deterministic PERT network where a single (constant) completion time is assigned to each task.

For a stochastic network, the project earliest completion time, $T_{project}$, is assumed (invoking the Central Limit Theorem) to be a Normal Random Variable. The expected value of the project earliest completion time ($T_{exp, project}$) is assumed to be the *sum* of the expected values (T_{exp}) of the completion times (see Figure 18) for the tasks on the critical path:

$$T_{exp,\ project} = T_{exp}\ (Task\ 0) + T_{exp}(Task\ 2) + T_{exp}(Task\ 3) + T_{exp}(Task\ 5) + T_{exp}(Task\ 9)$$

$$= 0\ + 7.83\ + 5.75\ + 1.15\ + 1.60$$

$$= 16.33$$

The variance, $\sigma^2_{project}$, of the project earliest completion time is assumed to be the *sum* of the variances of the completion times (see Figure 18) for the tasks on the critical path:

$$\sigma^2_{project}\ =\ \sigma^2\ (Task\ 0) + \sigma^2\ (Task\ 2) + \sigma^2\ (Task\ 3) + \sigma^2\ (Task\ 5) + \sigma^2\ (Task\ 9)$$

$$= 0\ + 0.25\ + 0.56\ + 0.11\ + 0.06$$

$$= 0.98$$

The standard deviation, $\sigma_{project}$, of the project earliest completion time is then the *square root* of the variance of the project earliest completion time:

$$\sigma_{project}\ = (\sigma^2_{project})^{1/2}$$

$$= (0.98)^{1/2}$$

$= 0.99$

Note: Recall from the field of statistics:

• It is not valid statistically to obtain the standard deviation of the critical path length (the project earliest completion time) by adding the individual standard deviations of the completion times for the tasks on the critical path. Statistically, the addition of the standard deviations of a set Random Variables will provide an *incorrect* estimate of the standard deviation of the sum of these Random Variables.

• The valid statistical method is to add the variances of the completion times for the tasks on the critical path. The standard deviation of the critical path length (the project earliest completion time) is then computed as the square root of the sum of the variances of the completion times for the tasks on the critical path.

Using the stochastic PERT technique, it is assumed that the distribution of the earliest completion time $T_{project}$ for the Narcissus Health Club can be specified by a Normal Random Variable of expected value ($T_{exp, project}$) equal to 16.33 weeks and of standard deviation ($\sigma_{project}$) equal to 0.99 weeks. That is,

$$T_{project} \sim N (T_{exp, project}, \sigma^2_{project})$$

$$\sim N (16.33, 0.98)$$

Example 2: (Completion Time Probability). Consider the stochastic task completion time estimates given in the Stochastic Time Estimate Table (see Figure 18) for the Narcissus Health Club in Example 1. What is the probability that the Narcissus Health Club start-up will be completed in 18 weeks or less?

Answer: The stochastic network earliest project completion time $T_{project}$ is assumed to be a Normal Random Variable with an expected value of 16.33 weeks and a standard deviation of 0.99 weeks (i.e. $T_{project} \sim N(16.33, 0.98)$). Then the probability of completing the project in 18 weeks or less is given by:

$$P(T_{project} \leq 18 \ weeks) = P[(T_{project} - 16.33)/ 0.99 \leq (18 - 16.33)/ 0.99]$$

$$= P(Z \leq 1.67) \quad where \ Z = (T_{project} - 16.33/ 0.99) \sim N(0, 1)$$

$$= 0.953 \qquad (Table \ 1b)$$

There is 95.3% confidence that the Narcissus Health Club start-up will be completed in 18 weeks or less.

Expediting the Earliest Project Completion Time

If one had resources (manpower, equipment, etc.) available to expedite a task on the critical path, then it may be possible to reduce (called "crashing") the earliest project completion time (the longest path). The earliest project completion time should be reduced *one* unit at a time since the longest path could change and further reduction in the completion time for that task if it is no longer on the critical path would not improve the earliest project completion time.

Under what conditions would a decision maker spend resources to reduce the earliest project completion time by one, two, three, etc. units? In general, one would spend to reduce the earliest project completion time until the *cost* of the next unit of project completion time reduction exceeds the *benefit* (e.g. profit) of the next unit of project time reduction. This is the traditional marginal cost - marginal benefit approach.

Which task to "crash" next is determined as the task on a critical path that requires the *least* resources to speed up a unit of time. This is the traditional highest marginal return approach.

Exercise 5

Project Management (PERT/CPM)

Problem #1 (Deterministic PERT).

In order to build a new house, the Habitat Contracting Company has identified 19 key tasks with the completion times (in days) and predecessor tasks given by the following Time/Task Precedence Table. This house is complete when Task 19 is successfully completed.

Tasks	Completion Time (Days)	Predecessor Tasks
1	1	None
2	7	None
3	5	1,2
4	2	3
5	2	4
6	10	5
7	3	6
8	3	6
9	1	4
10	3	6
11	4	4
12	7	4
13	10	12
14	2	13
15	4	14
16	1	10
17	1	7,8
18	5	6,9,11
19	5	15,16,17,18

a) Build the PERT network for building the new house.

b) Find the house (project) earliest completion time (ECT).

c) Identify the critical path(s) for the project.

d) Identify a qualitative "crashing" strategy to speed up the house earliest completion time by three days.

Problem #2 (PERT With Stochastic Time Estimates)

The Fairbanks City Hall wishes to install a new air conditioning unit. Its architect has identified the following 11 key tasks and their immediate predecessor tasks. The installation is complete when Task K is successfully completed.

Task	Predecessor
A	None
B	None
C	A,B
D	None
E	D
F	C,E
G	D
H	G
I	H,F
J	I
K	J

The optimistic (o), most probable (m), and pessimistic (p) times for each activity are given in the following Stochastic Time Estimate Table:

Task	Time Estimates (Weeks)		
	Optimistic (o)	Most Likely (m)	Pessimistic (p)
A	0.5	1	2
B	1	2	3
C	1	2	3
D	0.5	0.5	0.5
E	4	6	11
F	1	2	3
G	2	3	4
H	8	10	16
I	2	3	6
J	2	3	4
K	1	2	4

a) Build the PERT network for the air conditioning installation project.

b) Find the tasks on the critical path.

c) Find the distribution of the project completion time $(T_{project})$.

d) Find the probability the air conditioning installation project can be completed in 20 weeks or less.

Chapter 5

Project Management (PERT/CPM) Additional Solved Problems

Problem #1 (PERT)

Consider the following task completion times and precedences for a project:

Task	Precedences
A	None
B	None
C	A
D	B,C
E	C,D

a) Formulate the PERT precedence network for the project.

Answer: The PERT precedence network is given below. Note that a starting task with zero completion time and no predecessors has been added to the network and it is connected to all task (A and B) with no predecessors.

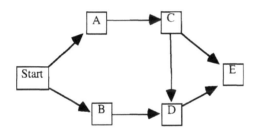

Consider the following optimistic, most likely, and pessimistic times (in years) for each task:

Task	Optimistic o	Most Likely m	Pessimistic p
A	1	2	3
B	2	4	6
C	1	4	7
D	4	8	12
E	3	4	5

b) Use the most likely time for each task to determine the critical path (determine the ECT and LCT for each task).

Answer: Determine the ECT and LCT for each task. The tasks with ECT = LCT are on the critical path:

Task ECT = Max (Predecessor ECT's) + Task Completion Time

Task LCT = Min (Successor LCT's - Successor Completion Time)

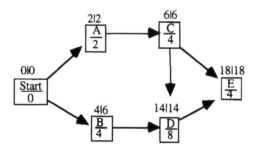

The critical path for the project is Task Start → Task A → Task C → Task D → Task E.

c) What is the distribution of the project completion time?

Answer: The estimates for the expected completion time, standard deviation, and variance are computed using the triangular distribution parameters o, m, and p:

Texp = (o + 4m + p)/6

$\sigma = (p - o) / 6$

$\sigma^2 = [(p - o) / 6]^2$

Task	Optimistic o	Most Likely m	Pessimistic p	Texp	σ	σ^2
A	1	2	3	2	1/3	1/9
B	2	4	6	4	2/3	4/9
C	1	4	7	4	1	1
D	4	8	12	8	4/3	16/9
E	3	4	5	4	1/3	1/9

The expected value of the project completion time Task Start → Task A → Task C → Task D → Task E is the sum of the expected completion times for the tasks on the critical path:

$T_{exp.project}$ = T_{exp}(Task Start) + T_{exp}(Task A) + T_{exp}(Task C) + T_{exp}(Task D) + T_{exp}(Task E)
= 0 + 2 + 4 + 8 + 4
= 18

Project Management (PERT/CPM) Additional Solved Problems

The variance for the project completion time is the sum of the variances of the completion times for the tasks on the critical path:

$$\sigma^2_{project} = \sigma^2(Task\ Start) + \sigma^2(Task\ A) + \sigma^2(Task\ C) + \sigma^2(Task\ D) + \sigma^2(Task\ E)$$
$$= 0 + 1/9 + 1 + 16/9 + 1/9$$
$$= 27/9$$
$$= 3$$

The standard deviation of the project completion time is then the square root of the variance of the project completion time of 3 years:

$$\sigma_{project} = 1.73$$

The distribution of the project completion time $T_{project}$ is then Normal with expected value $(T_{exp,project})$ 18 years and variance $(\sigma^2_{project})$ 3 years.

$$T_{project} \sim N(T_{exp,project}, \sigma^2_{project})$$

$$T_{project} \sim N(18, 3)$$

d) What is the probability that the project will finish sometime between year 15 and year 22?

Answer: The probability of finishing the project between the 15th and 22th year is computed as follows:

$$P(15\ years \leq T_{project} \leq 22\ years)$$
$$= P\left(\frac{15-18}{1.73} \leq \frac{T_{project}-18}{1.73} \leq \frac{22-18}{1.73}\right)$$
$$= P(-1.73 \leq Z \leq 2.31)$$
$$= P(Z \leq 2.31) - P(Z \leq -1.73)$$
$$= 0.9896 - 0.0418 \qquad\qquad (Table\ 1b)$$
$$= 0.9478$$

$$where\ Z = \frac{T_{Project} - 18}{1.73} \sim N(0,1)$$

Problem #2 (PERT)

Six major tasks (designated A through F) will be required to build the new hospital in honor of the philanthropist Ms. Trisha Fortune. The task precedence relationships are given as follows:

Task	Precedences
A	--
B	A
C	A
D	A
E	B,C,D
F	E

a) Formulate the PERT network for Ms. Fortune Hospital.

Answer: The Ms. Fortune Hospital PERT precedence network is given below. Note that a starting task "Start" with zero completion time and no predecessors has been added to the network and it is connected to all tasks (i.e. Task A) with no predecessors.

Ms. Fortune Hospital Precedence Network

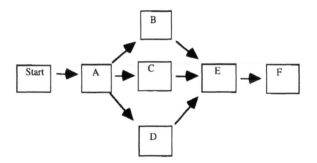

Consider the following optimistic, most likely, and pessimistic completion times (in months) for each task:

Task	Optimistic o	Most Likely m	Pessimistic p
A	3	6	9
B	2	4	12
C	1	4	19
D	2	8	14
E	3	6	9
F	2	11	20

b) Using the most likely time m as the *deterministic* time for each task, determine the critical path (use the ECT/LCT Method) for the Ms. Fortune Hospital project.

Answer: Using the ECT|LCT Method to determine the critical path (where LCT = ECT):

Project Management (PERT/CPM) Additional Solved Problems

Ms. Fortune Hospital PERT Network

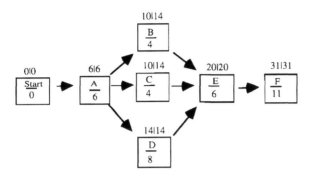

The Earliest Completion Time for the Ms. Fortune Hospital project is 31 months. The critical path (ECT = LCT) is Task Start → Task A → Task D → Task E → Task F.

c) Using the optimistic, most likely, and pessimistic task completion time estimates, determine the mean, standard deviation and the probability distribution for the *stochastic* completion time for the Ms. Fortune Hospital project?

Answer: The mean (T_{exp}) given by $(o + 4 \cdot m + p)/6$ and the variance σ^2 given by $[(p - o)/6]^2$ for each task is given below.

Task	Optimistic o	Most Likely m	Pessimistic p	T_{exp} $(o + 4 \cdot m + p)/6$	σ^2 $[(p - o)/6]^2$
Start	0	0	0	0	0
A	3	6	9	6	1
B	2	4	12	5	100/36
C	1	4	19	6	9
D	2	8	14	8	4
E	3	6	9	6	1
F	2	11	20	11	9

The expected completion time for the project $(T_{exp,project})$ of 31 months is determined by the sum of the expected completion times for the tasks on the critical path:

$$T_{exp,project} = T_{exp}(Task\ Start) + T_{exp}(Task\ A) + T_{exp}(Task\ D) + T_{exp}(Task\ E) + T_{exp}(Task\ F)$$
$$= 0 + 6 + 8 + 6 + 11$$
$$= 31\ months$$

The variance for the project $(\sigma^2_{project})$ of 15 months is determined by the sum of the variances of the completion times for the tasks on the critical path:

189

$$\sigma^2_{project} = \sigma^2(Task\ Start) + \sigma^2(Task\ A) + \sigma^2(Task\ D) + \sigma^2(Task\ E) + \sigma^2(Task\ F)$$
$$= 0 + 1 + 4 + 1 + 9$$
$$= 15\ months$$

The completion time for the Ms. Fortune Hospital project is assumed to be Normal with expected value ($T_{exp,project}$) of 31 months and variance ($\sigma^2_{project}$) of 15 months:

$$T_{Project} \sim N(T_{exp,project}, \sigma^2_{project})$$
$$T_{Project} \sim N(31,15)$$

d) What is the probability that the Ms. Fortune Hospital project will finish between the 28th and 34th month from the start of the project?

Answer: The probability of finishing the Ms. Fortune Hospital project between month 28 and month 34 is computed as follows:

$$P(28\ months \le T_{Project} \le 34\ months)$$
$$= P(\frac{28 - 31}{\sqrt{15}} \le \frac{T_{Project} - 31}{\sqrt{15}} \le \frac{34 - 31}{\sqrt{15}})$$
$$= P(-0.77 \le Z \le 0.77)$$
$$= P(Z \le 0.77) - P(Z \le -0.77)$$
$$= 0.7794 - 0.2206$$
$$= 0.5588$$

$$Note: Z = \frac{T_{Project} - 31}{\sqrt{15}} \sim N(0,1)$$

Problem #3 (PERT)

Consider the following 7 tasks which are required to bring Wham-O Company's latest new product to market. The Task/Time Precedence Table with stochastic task completion time estimates (in weeks) for this project is given below.

Task	Precedence	o	m	p
1. Design Product	None	12	24	30
2. Build Prototype	1	3	6	9
3. Evaluate Prototype	1	6	8	16
4. Test Prototype	2	3	4	5
5. Initial Report	3,4	3	7	11
6. Market Forecast	3,4	9	10	11
7. Deliver Final Product	5,6	4	4	4

a) Build the PERT Network showing the precedences and the expected completion time for each task in the project.

Answer: The expected value and variance of the task completion time for each task is given in the following table.

Task	Precedence	o	m	p	T_{exp}	σ^2
0. Start	None	0	0	0	0	0
1. Design Product	None	12	24	30	23	9
2. Build Prototype	1	3	6	9	6	1
3. Evaluate Prototype	1	6	8	16	9	25/9
4. Test Prototype	2	3	4	5	4	1/9
5. Initial Report	3,4	3	7	11	7	16/9
6. Market Forecast	3,4	9	10	11	10	1/9
7. Deliver Final Product	5,6	4	4	4	4	0

where:

$$T_{exp} = \frac{o + 4 \cdot m + p}{6}$$

$$\sigma^2 = \left(\frac{p - o}{6}\right)^2$$

Note: A Start Task 0 has been added to the network as a predecessor to all tasks without a predecessor. Since there is only one task without a predecessor (Task 1), this step of adding an initial task is optional. That is, Task 1 could be considered as the Start task. The earliest completion time for the project as well as the ECT's and LCT'S for the tasks are unchanged whether the initial Start Task 0 is added or Task 1 is considered the initial task.

The PERT network showing task precedences and expected completion times is as follows:

PERT Network

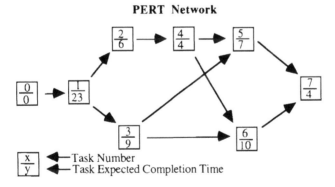

b) Use the Stochastic PERT technique to find the earliest completion time and the critical path for the project. Label the Earliest Completion Time (ECT) and the Latest Completion Time (LCT) for each task in the project.

Answer: The T_{exp} for each task is used to determine the project earliest completion time and the critical path. The ECT and LCT for each task in the project must be determined.

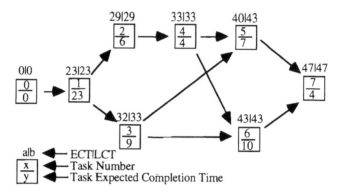

The critical path is Task 0 → Task 1 → Task 2 → Task 4 → Task 6 → Task 7. All tasks on this path have ECT = LCT.

c) What is the distribution of the earliest completion time for the project derived using the Stochastic PERT technique? Provide the name of the distribution and its defining parameters. Show all work.

Answer: The distribution of the stochastic PERT network is as follows:

$$T_{exp,project} = T_{exp}(Task\ 0) + T_{exp}(Task\ 1) + T_{exp}(Task\ 2) +$$
$$T_{exp}(Task\ 4) + T_{exp}(Task\ 6) + T_{exp}(Task\ 7)$$
$$= 0 + 23 + 6 + 4 + 10 + 4$$
$$= 47\ weeks$$

$$\sigma^2_{project} = \sigma^2(Task\ 0) + \sigma^2(Task\ 1) + \sigma^2(Task\ 2) +$$
$$\sigma^2(Task\ 4) + \sigma^2(Task\ 6) + \sigma^2(Task\ 7)$$
$$= 0 + 9 + 1 + \tfrac{1}{9} + \tfrac{1}{9} + 0$$
$$= 10.22\ weeks$$

$$T_{Project} \sim N(T_{exp.project}, \sigma^2_{project}) \sim N(47, 10.22)$$

d) Determine the 90% confidence interval for the earliest project completion time.

Answer: The 90% confidence interval for the earliest project completion time is determined as follows:

$Z_{.05} = -1.645$

$Z_{.95} = 1.645$

$T_{exp,project} = 47\ weeks$

$\sigma_{project} = \left(\sigma^2_{project}\right)^{\frac{1}{2}} = (10.22)^{\frac{1}{2}} = 3.20\ weeks$

$90\%\ Confidence\ Interval = (T_{exp,project} + Z_{.05}\cdot\sigma_{project}\ ,\ T_{exp,project} + Z_{.95}\cdot\sigma_{project})$

$$= (47 - 1.645\cdot(3.20)\ ,\ 47 + 1.645\cdot(3.20))$$

$$\approx (41.75\ weeks\ ,\ 52.25\ weeks)$$

e) Determine the probability that the project will be completed in 42 weeks or less.

Answer: The probability that the project will be completed in 42 weeks or less is determined as follows:

$$T_{Project} \sim N(T_{exp,project}, \sigma^2_{project}) \sim N(47, 10.22)$$

$$P(T_{Project} \le 42\ weeks) = P(\frac{T_{Project} - 47}{3.2} \le \frac{42 - 47}{3.2})$$

$$= P(Z \le -1.56)\quad where\ Z = \frac{T_{Project} - 47}{3.2} \sim N(0,1)$$

$$= 0.06$$

"Statistical thinking will one day be as necessary for efficient citizenship as the ability to read and write."

H. G. Wells

6 | Simple Regression

6.1 Introduction

The decision whether to expand a business facility, to hire additional people, to start new programs, etc. often is made on the basis of meeting future *forecasted* demand.

Forecasting future demand for a business is a multi-stage process:

- Collection of past demand data for the business (if available).

- Identification of the trend in the past demand data.

- Projection of the past demand trend into the future.

Forecasts of *future* demand are typically based on collection of *past* demand data for a given system. If no past demand data exists (e.g. a business facility has not yet been built), demand data of a *similar* system (e.g. a business in a similar location with a similar market) should be collected and examined.

Forecasting future demand based on past demand requires two assumptions:

1) Past demand has exhibited some measurable trend.

2) This trend will continue into the future.

Notes:

• Whether past demand has exhibited some measurable trend is a mathematical proposition which can be tested statistically. This chapter provides statistics to test whether a trend exists in the past demand data.

195

• There is no statistical test to determine whether future demand will follow the trend of past demand. Whether the future will follow the past is a *logical* proposition rather than a *mathematical* proposition. The decision maker must decide based on logic and intuition whether it is valid to assume that future demand will follow the trend of the past demand.

• Careful consideration must also be given to how far into the future a past trend is projected. Projection of past trends into the future assumes "all things remain the same". That is, all the factors that affected the past demand (product excitement, market potential, lack of competition, etc.) will continue without significant change into the future. The validity of assuming all things remain the same should be more closely examined the further into the future demand is forecasted.

- Consider the demand for VCR's between 1980 and 1990. There was an inclination to expect that the explosive growth in the early 1980's would continue into the late 1980's (see Figure 1). However, such growth could not be sustained over the entire decade. The environment changed in that the market for VCR's became saturated (see Figure 2). Things did not remain the same over the entire decade.

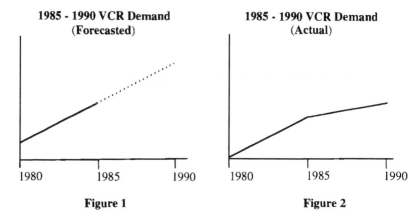

| **1985 - 1990 VCR Demand** | **1985 - 1990 VCR Demand** |
| (Forecasted) | (Actual) |

Figure 1   **Figure 2**

• Examining demand data graphically is always the best place to start the fitting of a trend to past demand data. Visual inspection ("eyeballing") of the demand data is sometimes more definitive than a statistical test in determining whether a line will be a good fit. For example, a small number of outliers (data points well off the linear trend) will quickly degrade statistical confidence in the linear trend. Visual inspection can establish that except for the anomalous outliers, the linear trend characterizes the data quite well.

• While visual inspection can help indicate whether a line will be a good fit to the demand data, it can not answer the question of which line will *best* fit the demand data.

• Regression analysis is the fitting of a *line* to past demand data in order to: 1) characterize the trend in past demand; and 2) forecast future demand. While regression analysis can characterize *only linear trends*, we will see later on in this chapter that there is a great deal of application of linear regression to non-linear functions (which can be transformed mathematically into linear functions of the trend parameters).

• A <u>trend</u> is a discernible pattern (see Figure 3) in the change in demand data over time. There will be random fluctuations (noise) about the trend over time since demand is a Random Variable. Random fluctuations about a constant mean value (see Figure 4) does not constitute a trend in demand.

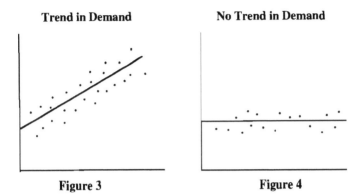

Trend in Demand	**No Trend in Demand**
Figure 3	**Figure 4**

• Regression analysis uses the <u>Method of Least Squares</u> as the mathematical technique to find the line with the "best" fit to a linear trend in the past demand data. The line derived is typically called the Least Squares line.

6.2 Method of Least Squares

Example 1: (Method of Least Squares to Characterize Past Demand). The Round Table Furniture Company has been in business for eight years and has the capacity to produce up to 400 tables per year. In recent years, demand for tables has exceeded capacity and orders above 400 tables per year have been lost. The demand for the tables produced by Round Table over the past eight years is given as follows:

Year t	Demand Y_t
1	180
2	324
3	360
4	378
5	540
6	504
7	490
8	648

Prior to making the investment for expansion, the company CEO needs to know if the changes in the yearly demand for tables represent a trend of increasing demand (see Figure 3) or merely random fluctuations around a constant mean level of demand (see Figure 4).

Use the Method of Least Squares to determine the line which best fits the demand data over the past three years for tables produced by Round Table over past eight years.

Answer: The past demand data for tables produced by Round Table when plotted and examined visually seems to increase in a linear fashion (see the dotted line in Figure 5).

Actual Past Round Table Demand Data

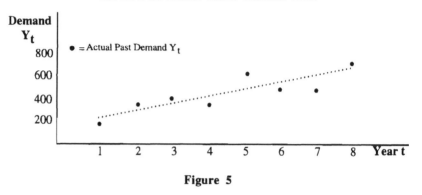

Figure 5

It will be hypothesis tested that the demand data for the past eight years represents a trend (linear increase). However, even if there is a linear trend in demand, many factors affecting demand (most with small positive or negative effect and which are assumed to cancel each other out) have not been included in the regression. This omission of such factors plus possible measurement errors in collecting the past demand leads to a random component (noise) in the demand data (see Figure 3):

Past Demand Data = Trend + Randomness

Hence, the past demand data is actually a Random Variable consisting of a fixed component (the trend) and a random component (the fluctuations about the trend). The random component will vary each time a set of demand data is collected. The eight outcomes of past demand data collected by Round Table represent but one of an infinite set of possible outcomes for the demand for tables over the past eight years.

One may find that statistically there is no trend in the demand data. The demand is then assumed to be constant over time (estimated by the sample mean of the past demand data). However, even if there is no trend in demand, there still will be random fluctuations over time around the sample mean (see Figure 4) due to the same effects of small factors being omitted and any measurement errors in collecting data.

Regression analysis in characterizing a trend in the data will examine the change in the demand data over time due to the underlying trend and due to the random effects. Statistical confidence in a trend in the demand data will increase as the amount of the change in demand over time that can be explained by the trend (referred to as the *explanatory power* of the regression) increases and the amount of change in demand which must be assigned to randomness decreases.

Define the following Random Variables for demand over time:

Y_t = Actual (collected) past demand for tables in year t ($t = 1, 2, \cdots, 8$)

Y_t^* = Estimated past demand for tables in year t ($t = 1, 2, \cdots, 8$)

The Least Squares line is derived based on the actual past demand collected. The Least Squares line which provides the best *estimate* of past demand may then be used to provide a *forecast* for future demand. In this presentation, the term "estimate" is applied to past demand while the term "forecast" is applied to future demand.

Recall that a line is *completely* defined by two parameters: its slope b and its y-intercept a. Everything about a line is known once its slope b and y-intercept a are defined. The line used to the estimate past demand can be expressed as:

$$Y_t^* = a + b \cdot t \qquad\qquad t = 1, 2, \cdots, 8$$

where t (time) is the single <u>independent variable</u> and Y_t^* (estimated demand) is the <u>dependent variable</u>. Y_t^* is a function of t; once the year t (the independent variable) is chosen, an estimate of past demand Y_t^* (the dependent variable) is specified by the linear trend $a + b \cdot t$.

<u>Notes:</u>

• The regression of a dependent variable (e.g. demand Y_t) on a single independent variable (e.g. time t) is called <u>simple regression</u>.

• It is important to specify correctly which variable is the independent variable and which variable is the dependent variable. There are no mathematical criteria for choosing which variable is the independent variable; it is chosen using logic. The demand for tables is logically a function of the year specified (while the year is not logically a function of the demand for tables).

• When the independent variable is an economic parameter (e.g. price, advertising budget, etc.) or a natural parameter (e.g. weather, consumer confidence), the regression is a <u>causal model</u> which hypothesizes a direct cause and effect relationship between the independent and dependent variables.

• When the independent variable is time (e.g. the change in the demand for tables produced by the Round Table Furniture Company in Example 1 varies over time), the regression is a <u>time series model</u> (see Chapter 7) which does not hypothesize a direct cause and effect relationship between the independent variable time and the dependent variable.

• The change in time itself is not the cause of the change in the dependent variable demand. In different instances, time can be associated with an increase, a decrease, or no change in demand for a product. Time is often used in regression analysis as a <u>surrogate variable</u>. A surrogate variable aggregates the effects of other *known* factors (e.g. advertising budget, new housing starts in the local community, etc.) or *unknown* factors which change as the surrogate variable itself (e.g. time) changes.

• The points on the Least Squares line $a + b \cdot t$ (see Figure 6) corresponding to years 1 through 8 are the estimated past demand data Y_t^*. In general, there will be deviation between the actual past demand Y_t and the estimated past demand Y_t^* in each year t.

Deviations Between Actual and Estimated Past Round Table Demand Data

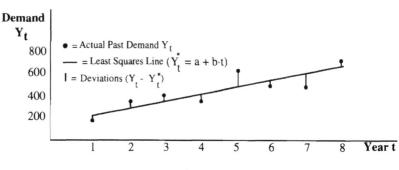

Figure 6

• The smaller the deviation between the actual past demand and the estimated past demand data, the greater is our confidence that this line provides a good characterization of the past demand data trend (in the sense that this line explains much of the change in demand over time). The Method of Least Squares derives the line which minimizes the deviation between the actual and the estimated past demand data.

• If the Least Squares line satisfactorily characterizes the past demand data trend, the line can then be extended (see Figure 8) to forecast future demand Y_t^* for tables over the next few years (t = 9, 10, 11, ···). This future forecast of demand for tables is the information necessary to the CEO considering expansion at the Round Table Furniture Company.

Application of the Method of Least Squares

The Method of Least Squares derives the line that is the "best" fit to the actual (collected) past demand data Y_t. The Least Squares line is best among all lines that can be hypothesized in the sense that it *minimizes the sum of the squares of the deviations* between the actual past demand data Y_t (which is known) and the estimated past demand data Y_t^* (which is found on the Least Square line).

This optimization is expressed mathematically as follows:

$$\text{Min } \Sigma \, (\, Y_t - Y_t^* \,)^2 \qquad t = 1, 2, \cdots, N$$

or, equivalently, since the estimated past demand $Y_t^* = a + b{\cdot}t$, determine the parameters a and b which:

$$\text{Min } \Sigma \, (Y_t - a - b{\cdot}t)^2 \qquad t = 1, 2, \cdots, N$$

200

For the Round Table Furniture Company, the Least Squares line minimizes the sum of the squares of the deviations between the actual past demand (given in the problem statement for Example 1) and the estimated past demand for tables for the N = 8 years (see Figure 6).

Calculus is used to solve for the parameters a and b which minimize the sum of the squares of the deviations between the actual and estimated demand data:

- The first derivatives of the sum of the squares of the deviations with respect to the slope b and with respect to the y-intercept a are each set equal to zero. This will result in two simultaneous linear equations for the two unknowns a and b.

- This system of linear equations can be solved algebraically to derive expressions for both a and b which are functions of the actual data collected.

The mathematical expressions for the slope b and the y-intercept a for the Least Squares line which minimize the sum of the squares of the deviations between the actual past demand data (Y_t) and the estimated past demand data (Y_t^*) are given by:

$$b = \{N(\Sigma Y_t \cdot t) - (\Sigma t) \cdot (\Sigma Y_t)\} / \{N(\Sigma t^2) - (\Sigma t)^2\} \qquad t = 1, 2, \cdots, N$$

$$a = \{\Sigma Y/N\} - b \cdot \{\Sigma t/N\} \qquad t = 1, 2, \cdots, N$$

$$= \bar{Y} - b \cdot \bar{t}$$

where $\bar{Y} (= \Sigma Y/N)$ is the sample mean of the actual past demand data, $\bar{t} (= \Sigma t/N)$ is the sample mean of the time data, and N is the number of actual past demand data points available.

Notes:

• The mathematical expressions for the Least Squares line parameters a and b are a function of the actual past demand data Y_t, the time data t, and the number of actual past demand data points N. All this data was provided in the problem statement for Example 1.

• The Least Squares line $Y_t^* = a + b \cdot t$ *always* passes through the point $(\bar{t}, \bar{Y})$ which represents the sample means of the time data and the past demand data. In statistics, the point $(\bar{t}, \bar{Y})$ is called the *centroid* or most likely point for the time vs. demand data. Hence, once the slope b of the Least Squares line is derived, the y-intercept a of the Least Squares line can be determined by solving the equation $\bar{Y} = a + b \cdot \bar{t}$.

For the Round Table actual past demand data Y_t and time data t given in Example 1, the Linear Regression Calculation Table (see Figure 7) provides the quantities required to determine the slope b and the y-intercept a of the Least Squares line which best fits the data for past demand for tables.

Round Table Linear Regression Calculation Table

t	Y_t	$Y_t \cdot t$	t^2	Y_t^2
1	180	180	1	32400
2	324	648	4	104976
3	360	1080	9	129600
4	378	1512	16	142884
5	540	2700	25	291600
6	504	3024	36	254016
7	490	3430	49	240100
8	648	5184	64	419904
Σt $= 36$	ΣY_t $= 3424$	$\Sigma Y_t \cdot t$ $= 17758$	Σt^2 $= 204$	ΣY_t^2 $= 1615480$

Figure 7

where:

$$\bar{t} \ (= \Sigma t/N) \ = 36/8 = 4.5$$

$$\bar{Y} \ (= \Sigma Y/N) \ = 3424/8 = 428$$

Substituting the data from the Linear Regression Calculation Table into the Least Squares expressions for the slope b and y-intercept a gives:

$$b = \{N(\Sigma Y_t \cdot t) - (\Sigma t) \cdot (\Sigma Y_t)\} / \{N(\Sigma t^2) - (\Sigma t)^2\}$$

$$= \{(8)(17758) - (36)(3424)\} / \{(8)(204) - (36)^2\}$$

$$= 18800 / 336$$

$$= 55.95$$

$$a = \bar{Y} - b \cdot \bar{t}$$

$$= 428 - (55.95) \cdot (4.5)$$

$$= 176.2$$

indicating that the best fit Least Squares line for estimating the past demand trend and for forecasting future demand for tables (see Figure 8) is given by:

$$Y_t^* = a + b \cdot t = 176.2 + 55.95 \cdot t$$

• The slope b (approximately 56 tables per year) of the Least Squares line represents the marginal change in the dependent variable (demand for tables) per unit change in the independent variable (time in years).

Least Squares Line For the Round Table Demand Data

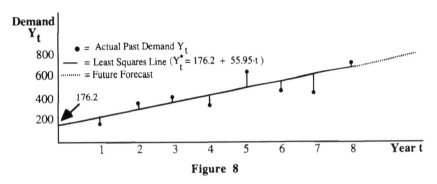

Figure 8

The Least Squares line $Y_t^* = 176.2 + 55.95 \cdot t$ represents the best *linear* fit to the actual demand data over the past 8 years. Before the Least Square line is used to forecast demand over the next few years (which is necessary to help the CEO of the Round Table Furniture Company to determine whether or not to expand), two questions must be addressed:

- How likely is it that the change in past demand data over time reflects a linear trend or reflects merely random fluctuations around the sample mean of the data)? The F-statistic will be used to test the hypothesis of no trend (b=0) versus the hypothesis of a linear trend (b≠0).

- How good is the fit of the Least Squares line to the past demand data? The R^2-statistic will be used to measure the goodness of the fit of the linear trend described by Least Squares line.

6.3 Testing the Linear Trend (F-Statistic)

Examine the deviations between the actual past demand data Y_t (t = 1, 2, 3, ⋯, 8) and the sample mean of the past demand data, $\bar{Y} = 428$ (see Figure 9).

Deviations Of The Round Table Demand Data From The Sample Mean

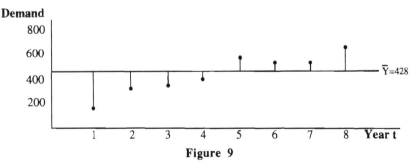

Figure 9

The F-statistic will determine whether it is better (in a statistical sense) to use a constant value (the sample mean) or a linear trend (a line) to characterize the past demand data.

The sum of the differences between each actual past demand data point Y_t and the sample mean $\bar{Y}$, given by $\Sigma (Y_t - \bar{Y})$, is defined as the <u>total deviation</u> from the mean of the data. Each deviation between an actual past demand data point and the sample mean given by $(Y_t - \bar{Y})$ for t = 1, 2, $\cdots$, N has two parts:

- The <u>explained</u> deviation in the regression given by $(Y_t^* - \bar{Y})$, the difference between the estimated data on the Least Squares line and the sample mean. This is the portion of the total deviation from the mean explained by the linear *trend* (change) in the data over time.

- The <u>unexplained</u> <u>deviation</u> in the regression given by $(Y_t - Y_t^*)$, the difference between the actual demand data and the estimated data on the Least Squares line. This is the portion of the total deviation from the mean unexplained by the linear trend and attributed to *randomness*.

<u>Note:</u> Recall (see Section 6.2) that the past demand data consists of a trend component and a random component (i. e. Past Demand Data = Trend + Randomness). The *numerical* relationships between the total deviation, the explained deviation, and the unexplained deviation are given in Figure 10.

Relationship Between Types of Deviation in the Regression

Total Deviation =	Explained Deviation +	Unexplained Deviation
(Actual Data - Mean) =	(Estimated Data - Mean) +	(Actual Data - Estimated Data)
$(Y_t - \bar{Y})$ =	$(Y_t^* - \bar{Y})$ +	$(Y_t - Y_t^*)$
$\Sigma(Y_t - \bar{Y})^2$ =	$\Sigma(Y_t^* - \bar{Y})^2$ +	$\Sigma(Y_t - Y_t^*)^2$

Figure 10

The total deviation (between the actual past demand data Y_t and the sample mean $\bar{Y}$) sum of squares given by $\Sigma(Y_t - \bar{Y})^2$ includes both a linear trend (i.e. change) in demand Y_t over time t = 1, 2, $\cdots$, 8 and randomness.

The explained deviation (between the estimated past demand Y_t^* and the sample mean $\bar{Y}$) sum of squares given by $\Sigma(Y_t^* - \bar{Y})^2$ is the deviation from the sample mean caused by the linear trend (i.e. change) in the demand Y_t^* over time.

The unexplained deviation (between the actual past demand data Y_t and the estimated past demand Y_t^*) sum of squares given by $\Sigma (Y_t - Y_t^*)^2$ is the deviation from the sample mean due to randomness.

Example 2: (F-Statistic). Test the hypothesis that the change in the past Round Table Furniture Company demand data over time in Example 1 exhibits a linear trend (rather than just randomness around the sample mean of the past demand data).

Answer: The *null hypothesis* is that the deviations in the past demand data from the sample mean $\bar{Y}$ represent just random fluctuations about the sample mean (i.e. b=0) and thus the best estimate is the sample mean $\bar{Y}$ (i. e. a=$\bar{Y}$). The *alternative hypothesis* is that the deviations in the demand data from the sample mean $\bar{Y}$ (the total deviation) are due to a linear trend in the data (i. e. b≠0).

Notes:

• If all the actual past demand data Y_t were to lie *exactly* on the Least Squares line $Y_t^* =$ 176.2 + 55.95·t (i.e. the actual past demand data and the estimated past demand data are identical), the unexplained deviation would be *zero*. Then all the deviation from the sample mean $\bar{Y}$ in the demand data would be explained by the linear trend.

• To the extent that the actual past demand data Y_t deviates from the estimated past demand data Y_t^*, the unexplained deviation (due to randomness) increases and the statistical confidence in a linear trend in the past demand data decreases.

The F-statistic, the ratio of the explained *variance* to the unexplained *variance*, is used to test the null hypothesis of no trend (b=0). The F-statistic is defined as:

$$F = \text{Explained Variance} \ / \ \text{Unexplained Variance}$$

$$= [\Sigma(Y_t^* - \bar{Y})^2/1] \ / \ [\Sigma(Y_t - Y_t^*)^2/(N - 2)]$$

where the F-statistic follows the F Distribution with $v_1 = 1$ degrees of freedom associated with the explained deviation and $v_2 = N - 2$ degrees of freedom associated with the unexplained deviation (where N is the # past demand data points).

Notes:

• In general, the explained and unexplained variances are determined by dividing the explained and unexplained deviation sum of squares, respectively, by their applicable number of degrees of freedom (see Section 1.10). For a simple regression:

Total Deviation = Explained Deviation + Unexplained Deviation
$$(N - 1) \, df \quad = \quad 1 \, df \quad + \quad (N - 2) \, df$$

• Overall, the degrees of freedom (df) for the total deviation in the N past demand data points is equal to N - 1 (the sample means being calculated and known account for a loss of 1 degree of freedom in the data).

• The degrees of freedom for explained deviation is equal to the # *independent variable* (equal to 1 in Example 1 since time is the only independent variable). For a simple regression, the explained deviation is divided by 1 to provide an estimate of the explained *variance* in the data.

• The degrees of freedom for the unexplained deviation is equal to (N-2), which is (N-1) minus the # independent variables. For a simple regression, the unexplained deviation divided by N-2 to provide an estimate of the unexplained *variance* in the data.

• A *larger* F-statistic provides a higher statistical confidence that the data reflects a linear trend. More of the total deviation in the model is explained deviation (in the numerator of the F-statistic) and less of the total deviation is unexplained deviation (in the denominator of the F-statistic).

A more convenient formula than dividing sums of squares for the calculation of the F statistic uses the data in the Linear Regression Calculation Table (see Figure 7):

$$F = [e/1] / [u/(N-2)]$$

where:

$$e = [\Sigma Y_t \cdot t - (\Sigma t)(\Sigma Y_t)/N]^2 / [\Sigma t^2 - (\Sigma t)^2/N] \qquad \textit{(Explained Deviation)}$$

$$u = \Sigma Y_t^2 - (\Sigma Y_t)^2/N - e \qquad \textit{(Unexplained Deviation)}$$

Using the data from the Linear Regression Calculation Table (see Figure 7), the explained deviation e and the unexplained deviation u sum of squares are given by:

$$e = [\Sigma Y_t \cdot t - (\Sigma t)(\Sigma Y_t)/N]^2 / [\Sigma t^2 - (\Sigma t)^2/N]$$

$$= [17758 - (36)(3424)/8]^2 / [204 - (36)^2/8]$$

$$= [2350]^2 / 42$$

$$= 131,488.1$$

$$u = \Sigma Y_t^2 - (\Sigma Y_t)^2/N - e$$

$$= 1,615,480 - (3424)^2/8 - 131,488.1$$

$$= 18,519.9$$

Hence, the F-statistic for the null hypothesis of a no trend in past demand for the Round Table Furniture Company tables versus time is given by:

$$F = [e/1] / [u/(N-2)]$$

$$= [131,488.1 / 1] / [18,519.9 / 6]$$

$$= 42.6$$

For this simple regression, the value for the F distribution with $v_1 = 1$ and $v_2 = N-2 = 6$ degrees of freedom at the 99th percentile is $F_{1,6,.99} = 13.75$ (see Table 2a). Since the F-statistic = 42.6 is *larger* than the 99th percentile value 13.75, the null hypothesis that the past demand data exhibits no trend is rejected with 99% statistical confidence. We accept the alternative hypothesis of a linear trend in the demand data.

Notes:

• The hypotheses that the change in the demand data over time exhibits a linear trend, or an exponential trend, or a polynomial trend, etc. are tested *individually* versus the null hypothesis of random fluctuations about a constant demand given by the sample mean (the most likely value for the data). Different trends are *never* tested versus each other (e.g. the hypothesis of a exponential trend in the past demand data is never tested *directly* versus the hypothesis of an exponential trend in the past demand data).

• The trend (among all the trends considered) with the highest F-statistic provides the best characterization of the past demand data in a statistical sense. The most appropriate trend for the data should be chosen by the decision maker based on *logic*. The F-statistic for the trends considered only provides valuable information to the decision maker. If no logical trend provides a statistically significant F-statistic, the null hypothesis (the data represents only random fluctuations around the sample mean) should not be rejected. Remember in hypothesis testing, the null hypothesis is *never accepted*.

Analysis of Variance (ANOVA) Table

The procedure of decomposing deviation and variance for a regression and analyzing its components is called analysis of variance. An Analysis of Variance (ANOVA) Table (see Figure 11) is the common vehicle used in statistics to display at a glance the total deviation in the data, the explained and unexplained deviation in the data, the unexplained and unexplained variance in the data, as well as the calculation of the F-statistic.

Round Table Analysis of Variance (ANOVA) Table

Deviation Source	Deviation Sum of Squares	Degrees of Freedom	Variance
Explained	e = 131,488.1	1	$e/1$ = 131,488.1
Unexplained	u = 18,519.9	6	$u/6$ = 3,086.65
Total	150,008	7	F-statistic = 42.6

Figure 11

Notes:

• The larger the F-statistic, the greater is the statistical confidence that the hypothesized trend characterizes the past demand data well.

• The *large* F-statistic (42.6) for the Round Table Furniture Company data allows us to reject the hypothesis of no trend (b=0) at the 99th percentile (since $F_{1,6,.99} = 13.75$). Smaller F-statistics would result in being able to reject the hypothesis of no trend only at lower percentiles such as the 95th percentile (where $F_{1,6,.95} = 5.99$).

• If the null hypothesis of no trend (b=0) can not be rejected and the alternative of a linear trend (b≠0) accepted at the 95th percentile or higher, other *logical* trends should be tested.

• It may be the case that some non-linear trend (e.g. exponential, polynomial, geometric, etc.) characterizes the past demand data *better* than a linear trend (as evidenced by a larger F-statistic). The Method of Least Squares can be used to derive the best fit line for several non-linear functions. The transformation of non-linear functions into linear functions of their trend parameters (as required to apply regression analysis which fits a line to data) is addressed in Section 6.5.

6.4 Measuring the Goodness of the Linear Fit (R^2-Statistic)

The R^2-statistic, defined as the ratio of the explained deviation to the total deviation from the sample mean $\bar{Y}$ of the data, is used to determine the goodness of fit of the Least Squares line to the past demand data. The R^2-statistic is defined as:

$$R^2 = \text{[Explained Deviation] / [Total Deviation]}$$

$$= [\Sigma(Y_t^* - \bar{Y})^2] / [\Sigma(Y_t - \bar{Y})^2]$$

Thus the R^2-statistic for Example 1 indicates the *proportion* of the total deviation from the sample mean $\bar{Y}$ in the Round Table Furniture Company past demand data that is explained by the Least Squares linear trend $Y_t^* = 176.2 + 55.95 \cdot t$.

The square root of R^2 is defined as the correlation coefficient R. The correlation coefficient R is a measure of the *statistical association* between the independent variable time (e.g. time t) and the dependent variable demand (e.g. demand Y_t). The correlation coefficient R takes on values between -1 and 1, inclusive, and has the following properties:

    - R has the same sign as b (the slope) of the Least Squares line $Y_t = a + b \cdot t$

    - If R is positive, then as t increases, Y_t generally increases (*positive correlation*)

    - If R is negative, then as t increases, Y_t generally decreases (*negative correlation*)

Values of R close to -1 or 1 indicate a strong negative or positive *linear* correlation, respectively, between the independent variable and the dependent variable. Changes in the independent variable (e.g. time t) would then explain much of the changes in the dependent variable (e.g. demand Y_t). If R = 0, the independent variable and dependent variable are observed to have no *linear* correlation to each other (a linear regression using the independent variable explains none of the change in the dependent variable). Values of R close to 0 indicate a weak linear correlation between the independent variable and the dependent variable.

Note: Since the Method of Least Squares only examines the *linear* relationship between the independent and dependent variable, it may be that the two variables are strongly correlated by means of a non-linear trend.

A more convenient formula than dividing sums of squares for the calculation of the correlation coefficient R uses the data in the Linear Regression Calculation Table (see Figure 7):

$$R = [N(\Sigma Y_t t) - (\Sigma t)(\Sigma Y_t)] / \{[N(\Sigma t^2) - (\Sigma t)^2][N\Sigma Y_t^2 - (\Sigma Y_t)^2]\}^{1/2}$$

The value of the correlation coefficient R must be squared to obtain the R^2-statistic. R^2 takes on values between 0 and 1, inclusive, which is appropriate for a statistic which defines the *proportion* of the total deviation in the data explained by the regression.

Example 3: (R^2-Statistic). For the Least Squares line for the past demand data for the Round Table Furniture Company in Example 1, determine and interpret the R^2-statistic.

Answer: Using the data from the Linear Regression Calculation Table (see Figure 7):

$$R = [N(\Sigma Y_t t) - (\Sigma t)(\Sigma Y_t)] / \{[N(\Sigma t^2) - (\Sigma t)^2][N\Sigma Y_t^2 - (\Sigma Y_t)^2]\}^{1/2}$$

$$= [8(17,758) - (36)(3424] / \{[8(204) - (36)^2][8(1,615,480) - (3424)^2]\}^{1/2}$$

$$= 0.936$$

$$R^2 = 0.877$$

Notes:

• The large positive coefficient of correlation R (well above 0.90) indicates that demand for tables has strong *positive* linear correlation with time (see Figure 6 to verify the positive slope b for the Least Squares line).

• The R^2-statistic of 0.877 indicates that 87.7% of the total deviation from the sample mean $\bar{Y}$ of the past demand data can be explained by a linear trend for demand data over time given by the Least Squares line $Y_t^* = 176.2 + 55.95 \cdot t$.

• An R^2-statistic close to 0.90 or higher is indicative of a good fit of the estimated past demand given by the Least Squares line to the actual past demand data.

209

• A large R^2-statistic (and thus a large correlation coefficient R) indicates a strong statistical association between the independent variable and the dependent variable. However, a large correlation coefficient R does not imply that the change in the dependent variable is *caused* by a change in the independent variable. Causation is not a statistical proposition but a *logical* proposition.

> - For example, one can perform a regression analysis of the price of diapers against the intensity of the sunspot activity over many years. One would find some deviation in diaper price from its sample mean explained by sunspot activity (i.e. correlation coefficient R would be non-zero indicating some correlation). But logic would tell us that changes in sunspot activity are not correlated to changes in diaper prices (in fact, they are uncorrelated logically).

• One method to increase the magnitude of the R^2-statistic is to *add additional independent variables* to the regression (e.g. yearly national GNP, population growth, advertising spending, etc.) which may have contributed to the change in demand over time. Since each additional independent variable is likely to have some explanatory power (some correlation with the dependent variable), the R^2-statistic for the regression will increase.

• The large F-statistic provides confidence that a linear trend is a good characterization of the past demand data. The large R^2-statistic provides confidence in the fit of the demand data to the Least Squares line $Y_t^* = 176.2 + 55.95 \cdot t$.

• Together, the large F-statistic ($42.6 > F_{1,6,.99}$) and the large R^2-statistic (0.877) provide confidence that the past demand (over the last 8 years) for tables produced by the Round Table Furniture Company is well characterized by a linear trend described by the Least Squares line $Y_t^* = 176.2 + 55.95 \cdot t$.

6.5 Error Terms for the Regression

Up to this point in the presentation of simple regression, it has been assumed that the regression model has been properly specified (i. e. an appropriate set of independent variables affecting the dependent variable has been defined). The presentation has focused on: 1) the derivation of the best fit line to the past demand data (Method of Least Squares); 2) the appropriateness of hypothesizing a linear trend (F-statistic); and 3) the goodness of fit of the Least Squares line (R^2-statistic).

In the formal presentation of multiple regression (see Chapter 8), considerations in specifying the regression model (i.e. have all significant independent variables been included) will be addressed. Possible errors in specifying the regression model will be cited and approaches to mitigating these specification errors will be presented.

A final consideration for the regression of demand data over time is the examination of the error terms (the deviations between the actual past demand and the estimated past demand):

$$\varepsilon_t = Y_t - Y_t^* \qquad\qquad t = 1, 2, \cdots, N$$

The error terms (sometimes called the residuals) for a *correctly specified regression* (i.e. all significant independent variables are included) will have the following properties:

1. The error terms are independent of each other. That is, the value of an error term is uncorrelated with (has no effect on) the value of any other error term.

- The error terms should be uncorrelated and exhibit a random pattern when plotted over the values of the independent variable (e. g. time).

2. The expected value of the error terms is zero (i.e. $E\varepsilon_t = 0$). This implies that the expected value of the estimated past demand Y_t^* should always equal the expected value of the actual past demand Y_t (the regression is said to be *unbiased*). Although many factors that have some effect on the demand may have been left out of the regression, their effects are small and tend to cancel each other out.

- The sample mean (ε_t / N) for the error terms when calculated should be quite small compared to the magnitude of the data.

3. The variance σ of the error terms is constant over time. The relative magnitude of the error terms ε_t should not increase or decrease significantly as the value of the independent variable (e.g. time) increases.

- The error terms ε_t when plotted over the values of the independent variable (e.g. time) should remain in a horizontal band of fixed size above and below the expected value zero. The size of this horizontal band should not be increasing or decreasing nor should it have a rainbow shape.

4. The error terms have a Normal distribution. This occurs due to the fact that although many factors affecting demand may have been left out of the regression, these factors tend to be small so that large error terms occur infrequently. This is the pattern of the Normal distribution. If the error terms do not have a Normal distribution, it is likely that independent variables with significant effects on the dependent variable have been left out of the regression.

- The error terms when analyzed should have the properties of the Normal distribution with an expected value of zero. Most error terms should be near zero with infrequent large deviations. The difference between the largest positive deviations and the largest negative deviations when normalized by the sample standard deviation s should be about 6 (± 3 standard deviations).

Note: After completing a regression, it is always instructive to calculate and plot the error terms over the values of the independent variable. Typically, a mistake in a hand calculation will be apparent since the error terms will not be independent (randomly scattered) and the sample mean (ε_t / N) will not be close to zero. Inspection o r "eyeballing" the plotted error terms for independence and zero mean is often *more sensitive* to mistakes than statistical techniques used to test for independence and zero mean.

Example 4: (Error Terms). Calculate and plot the error terms for the regression of demand versus time for tables at the Round Table Furniture Company given in Example 1.

Answer: The Error Term Table (see Figure 12) provides the actual past demand, the estimated past demand, and the error term for the each of the eight years of demand data.

Round Table Regression Error Term Table

Time t	Actual Demand Y_t	Estimated Demand $Y_t^* = 176.2 + 55.95 \cdot t$	Error Term $\varepsilon_t = Y_t - Y_t^*$
1	180	232.17	-52.17
2	324	287.12	+35.88
3	360	344.07	+15.93
4	378	400.02	-22.02
5	540	455.98	+84.02
6	504	511.93	-7.93
7	490	567.88	-77.88
8	648	623.83	+24.17
			$\Sigma \varepsilon_t = +0.001$
			$(\Sigma \varepsilon_t / 8) \approx 0$

Figure 12

The plot of the error terms (ε_t) versus the independent variable time (t) for the Round Table Furniture Company regression is given in Figure 13.

Plot of Error Terms For Round Table Regression

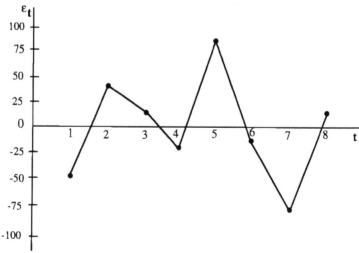

Figure 13

The sum of the 8 error terms is given by $\Sigma \varepsilon_t = +0.001$ (see Figure 12). The sample mean $(\Sigma \varepsilon_t / 8)$ for the eight error terms is very close to the ideal of zero sample mean. There

212

seems to be no pattern in the error terms (e.g. multiple consecutive positive or negative error terms in a non-random fashion) that would preclude independence (randomness).

6.6   General Regression

Regression analysis is limited to finding the best fit *line* to the past demand data. The limitation of regression to fitting lines to data is not as restrictive as it may seem in that many non-linear functions may be transformed into linear functions of their parameters. This section examines several non-linear functions whose parameters can be derived by the linear Method of Least Squares.

Log-Linear (Exponential) Regression

It may be possible to obtain a better fit (i. e. a larger F-statistic and R^2-statistic) by using a non-linear function to explain the trend in the past demand data.

Consider an exponential growth trend in past demand which is described mathematically as:

$$Y_t^* = e^{a + b \cdot t}$$

Over time, demand with an exponential growth trend appears to be increasing at an ever increasing rate (see Figure 14).

Log-Linear (Exponential) Demand Data Trend

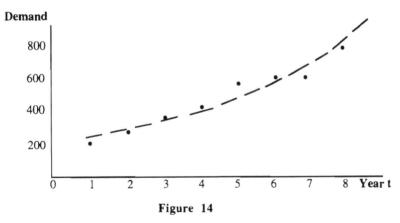

Figure 14

Then, taking the natural logarithm of both sides of the exponential growth trend equation yields:

$$lnY_t^* = a + b \cdot t$$

This equation is *log-linear* (i.e. linear in the parameters a and b after applying the natural logarithm function to both sides of the exponential growth trend equation). The values lnY_t

213

and t are known and fixed; the only variables in the linear equation $lnY_t^* = a + b \cdot t$ are the parameters a and b (which are raised to the first power).

All the previous Method of Least Squares expressions used to derive the slope b and the y-intercept a are applicable when the term lnY_t is substituted for the term Y_t. The Log-Linear Regression Calculation Table (with lnY_t substituted for Y_t in the previous Linear Regression Calculation Table given in Figure 7) used to calculate the Least Squares line parameters a and b is given in Figure 15.

Log-Linear Regression Calculation Table

t	lnY_t	$lnY_t \cdot t$	t^2	$(lnY_t)^2$
1	5.1929	5.1929	1	26.9662
2	5.7807	11.5614	4	33.4165
3	5.8861	17.6583	9	34.6461
4	5.9348	23.7392	16	35.2218
5	6.2915	31.4575	25	39.5829
6	6.2225	37.335	36	38.7195
7	6.1944	43.3608	49	38.3706
8	6.4738	51.7904	64	41.9100
Σt = 36	ΣlnY_t = 47.976	$\Sigma lnY_t \cdot t$ = 222.095	Σt^2 = 204	$\Sigma(lnY_t)^2$ = 288.83

Figure 15

where $\bar{t} = \Sigma t/N = 36/8 = 4.5$

$\overline{lnY} = \Sigma lnY/N = 47.976/8 = 5.997$

Example 5: (Method of Least Squares for Log-Linear Trends). Use the Method of Least Squares to determine the slope b and the y-intercept a for the best fit line for the log-linear trend for the Round Table Furniture Company past demand data given in Example 1.

Answer: Substituting the Log-Linear Regression Calculation Table data into the Method of Least Squares expressions for a and b given in Section 6.2 (where the term lnY_t has been substituted for the term Y_t) gives:

$b = \{N(\Sigma lnY_t \cdot t) - (\Sigma t) \cdot (\Sigma lnY_t)\} / \{N(\Sigma t^2) - (\Sigma t)^2\}$

$= \{(8)(222.095) - (36)(47.976)\} / \{(8)(204) - (36)^2\}$

$= 0.147$

$a = \overline{lnY} - b \cdot \bar{t}$

$= 5.997 - (0.147) \cdot (4.5)$

$$= 5.3355$$

indicating that the Least Squares best fit line for the log-linear trend for the past demand data is given by:

$$lnY_t^* = a + b \cdot t$$

$$= 5.3355 + 0.147 \cdot t$$

or, equivalently, the Least Squares best fit line for the exponential (log-linear) trend for past demand data is given by:

$$Y_t^* = e^{5.3355 + 0.147 \cdot t}$$

Example 6: (F-Statistic for Log-Linear Trends). Use the F-statistic to test the null hypothesis of no trend (b=0) versus the alternative hypothesis that the Round Table Furniture Company past demand data reflects a log-linear trend. Compare the results to those for the linear trend in Example 2.

Answer: To test how well the demand data reflects a log-linear trend, the F-statistic must be calculated:

$$F = [e/1] / [u/(N-2)]$$

where:

$$e = [\Sigma lnY_t \cdot t - (\Sigma t)(\Sigma lnY_t)/N]^2 / [\Sigma t^2 - (\Sigma t)^2/N] \ (Explained \ Deviation)$$

$$u = \Sigma(lnY_t)^2 - (\Sigma lnY_t)^2/N - e \qquad (Unexplained \ Deviation)$$

The term lnY_t has been substituted for term Y_t in the general expressions (see Section 6.3) for the explained and unexplained deviations. The F-statistic can be calculated directly from the Log-Linear Regression Calculation Table (see Figure 15):

$$e = [222.095 - (36)(47.976)/8]^2 / [204 - (36)^2/8]$$

$$= 0.916$$

$$u = 288.83 - (47.976)^2/8 - 0.916$$

$$= 0.202$$

The degrees of freedom associated with the explained deviation is 1 since the # independent variables in the regression is 1 (time). The degrees of freedom associated with the unexplained deviation is N - 2 = 6. The F-statistic for the log-linear trend of demand over time for tables from the Round Table Furniture Company is calculated as follows:

$$F = [0.916/1] / [0.202/6]$$

$$= 27.2$$

Note: The F-statistic for the linear trend (42.6) is larger than the F-statistic for the log-linear trend (27.2) though both are significant at the 99th percentile (since $F_{1,6,99} = 13.75$). The linear trend characterizes the past demand data better on a *statistical* basis.

Example 7: (R^2-Statistic for Log-Linear Trends). Use the R^2-statistic to determine the goodness of fit of the Least Squares line for the log-linear trend ($\ln Y_t^* = 5.3355 + 0.147 \cdot t$) for the Round Table Furniture Company past demand data and compare the results to those for the linear trend in Example 3.

Answer: To test the goodness of fit of the log-linear trend to the past demand data, the R^2-statistic must be calculated:

$$R = [N(\Sigma \ln Y_t \cdot t) - (\Sigma t)(\Sigma \ln Y_t)] \,/\, \{[N(\Sigma t^2) - (\Sigma t)^2][N\Sigma(\ln Y_t)^2 - (\Sigma \ln Y_t)^2]\}^{1/2}$$

where $\ln Y_t$ has been substituted for Y_t in the general equation for R (see Section 6.4). Using the data from the Log-Linear Regression Calculation Table (see Figure 15):

$$R = [8(222.095) - 36(47.976)] / \{[8(204) - (36)^2][8(288.83) - (47.976)^2]\}^{1/2}$$

$$= 0.905$$

The large positive coefficient of correlation R indicates that the demand for tables has a strong positive log-linear correlation with time (though not as strong as the linear correlation where R = 0.936). The value of the R^2-statistic is given as the square of R:

$$R^2 = 0.8184$$

The R^2-statistic for the Least Squares linear fit to the past demand data (0.877) is larger than the R^2-statistic for the Least Squares log-linear fit (0.8184). More of the total deviation from the sample mean $\bar{Y}$ in the past demand data is explained by the Least Squares line for the linear trend than for the Least Squares line for the log-linear (exponential) trend.

Due to the larger F-statistic and the larger R^2-statistic, a linear trend characterizes the past data better than the log-linear (exponential) trend on a *statistical* basis. However, the final decision whether a linear trend or an log-linear trend is a better characterization of the past demand for tables should be determined *logically* by the decision maker rather than statistically. The F-statistic and the R^2-statistic are only valuable inputs to the decision maker in making this determination.

Introduction to Multiple Regression

It has been noted that other *factors* besides time (e.g. growth in the local economy, population growth, advertising spending, etc.) may affect the demand for tables. These factors could be listed individually rather than aggregated into the surrogate factor time as was done for demand for Round Table Furniture Company tables in Example 1.

The regression of a dependent variable on two or more independent variables is called a _multiple regression_. Consider the following multiple regression for demand Y:

$$Y = a + b_1 X_1 + b_2 X_2 + \cdots + b_k X_k$$

where the demand Y is the dependent variable and X_1, X_2, $\cdots$, X_k are the independent variables (factors) thought to help explain past demand Y.

Multiple regression generalizes the theory used for simple regression (one independent variable). The Method of Least Squares is again used to determine the a and b_i's which provide the best fit line which minimizes the sum of the squares of the deviations between the actual and estimated demand Y.

The calculation of the parameters a, b_1, b_2, $\cdots$, b_k, the F-statistic, and the R^2-statistic for the multiple regression can be difficult to accomplish by hand. However, these calculations and other statistics required for multiple regression are straightforward using matrix algebra and are performed by many available regression software packages. The topic of multiple regression is presented formally in Chapter 8.

Generality of The Method of Least Squares

The Method of Least Squares can only be used to determine the best fit _line_ to past data. This is not as significant a limitation in applicability as it may seem since many non-linear functions may be transformed into the form $Y = a + b_1 X_1 + b_2 X_2 + \cdots + b_k X_k$ which is linear in the regression parameters a and the b_i's. Remember that the dependent variable Y and the independent variables $X_1, X_2, \cdots, X_k$ become known numerical quantities once the past demand data is collected.

The following are types of regression where the linear Method of Least Squares is applicable:

• **Simple Regression** $Y = a + b \cdot X$

   $Y = a + b \cdot X$
   Regression of Y on X

• **Multiple Regression** $Y = a + b_1 X_1 + b_2 X_2 + \cdots + b_k X_k$

   $Y = a + b_1 X_1 + b_2 X_2 + \cdots + b_k X_k$
   Regression of Y on $X_1, X_2, X_3, \cdots, X_k$

• **Polynomial Regression** $Y = a + b_1 X + b_2 X^2 + b_3 X^3 + \cdots + b_k X^k$

   $Y = a + b_1 X_1 (=X) + b_2 X_2 (=X^2) + b_3 X_3 (=X^3) + \cdots + b_k X_k (=X^k)$

Regression of Y on $X_1, X_2, X_3, \cdots, X_k$

• **Log-Linear (Exponential) Regression** $Y = e^{a + b \cdot X}$

> Note: $lnY = a + b \cdot X$
> $Y(=lnY) = a + b \cdot X$
> Regression of Y on X

• **Geometric Regression** $Y = a \cdot X^b$

> Note: $lnY = lna + b \cdot lnX$
> $Y(=ln \, Y) = a(=ln \, a) + b \cdot X(=ln \, X)$
> Regression of Y on X

• **Logarithmic Regression** $Y = a \cdot b^X$

> Note: $lnY = lna + X \cdot lnb$
> $Y(=lnY) = a(=lna) + b(=lnb) \cdot X$
> Regression of Y on X

• **Reciprocal Regression** $Y = a + \dfrac{b}{X}$

> $Y = a + b \cdot X(=\dfrac{1}{X})$
> Regression of Y on X

Notes:

• The Method of Least Squares can be applied to non-linear functions as long as the functions are linear in the parameters (a and the b_i's are to the first power) and the regression can be expressed in the form $Y = a + b_1 X_1 + b_2 X_2 + \cdots + b_k X_k$.

• If a non-linear function can not be transformed into a function that is linear in it parameters, the Method of Least Squares *can not be applied*.

6.7   Future Forecasts and Confidence Intervals

Making a Future Forecast

Example 8: (Future Forecasts). Provide a forecast for the demand for tables at the Round Table Furniture Company in year 10 based on the eight years of past demand data given in Example 1.

Answer: The Least Squares fit for the linear and log-linear trends as well as the null hypothesis of no trend for the demand for tables over the past 8 years are used to forecast the demand for tables in year 10:

- Linear Trend: $Y_t^* = a + b \cdot t$

$$Y_t^* = 176.2 + 55.95 \cdot t$$
$$Y_{10}^* = 176.2 + 55.95 \cdot (10) \approx 736 \; tables$$

- Log-Linear Trend: $Y_t^* = e^{a + b \cdot t}$

$$Y_t^* = e^{5.3355 + 0.147 \cdot t}$$
$$Y_{10}^* = e^{5.3355 + 0.147 \cdot (10)} \approx 902 \; tables$$

- No Trend: $\bar{Y} = 428$

$$Y_{10}^* = 428 \; tables$$

Notes:

• If neither the linear nor the log-linear trend (or any other logical trend hypothesized) provides a good fit to the demand data (that is, no logical trend is statistically significant), the best forecast for *any* year in the future would simply be the sample mean $\bar{Y} = 428$ (the most likely value for both the past and future demand).

• It is *not guaranteed* that models based on past behavior will provide a good estimate of future demand. The processes which produced the past demand may not be the same processes producing future demand (e.g. new competition may appear, the size of the surrounding community may dramatically increase or decrease, a new housing tract may be built nearby, etc.). Typically, it is usually safer only to forecast a *short time* into the future rather than assume a past trend will continue far into the future.

Confidence Intervals for Future Forecasts

Forecasting future demand based on past demand data is problematic for two reasons:

1. Processes which generated the past demand generally will not remain the same into the future. This has been examined previously.

2. Uncertainty in future demand forecasts increases *statistically* the further into the future the forecast is projected. This will be addressed in this section.

Using the linear trend, forecasted *(projected)* future demand for tables in year p (p = 9, 10, 11, ⋯) has an expected value Y_p^* given by:

$$Y_p^* = a + b \cdot t$$

$$= 176.2 + 55.95 \cdot p$$

and a <u>standard deviation of the regression</u> s_p given by:

$$s_p = \{[(\Sigma Y_t^2 - a\Sigma Y_t - b\Sigma Y \cdot t) / (N-2)] \cdot [(N+1/N) + ((p-\bar{t})^2/(\Sigma t^2 - N\bar{t}^2))]\}^{1/2}$$

where p is the forecasted time (e.g. here p = year 10) and $\bar{t}$ is the sample mean of the past time data (here $\bar{t}$ = year 4.5).

For a regression model with N past demand data points such that N-2 ≤ 30, the confidence interval around a forecast for demand is assumed to follow a Student t distribution with N-2 degrees of freedom (here N-2 = 6).

Example 9: (Confidence Interval for Future Forecasts). Find a 95% confidence interval around the expected value of the future demand forecast for demand for tables in year 10 using the linear trend in Example 8.

Answer: The expected value Y_p^* of the future demand in year p = 10 using the linear trend is given by:

$$Y_p^* = 176.2 + 55.95 \cdot p$$

$$Y_{10}^* = 176.2 + 55.95 \cdot (10)$$

$$= 735.7 \; tables$$

The standard deviation s_p is calculated using the data from the Linear Regression Calculation Table (see Figure 7):

$$s_p = \{[(\Sigma Y_t^2 - a\Sigma Y_t - b\Sigma Y \cdot t) / (N-2)] \cdot [(N+1/N) + ((p-\bar{t})^2/(\Sigma t^2 - N\bar{t}^2))]\}^{1/2}$$

$$s_{10} = \{[(1,615,480 - (176.2)(3424) - (55.95)(17758)) / (6)] \cdot$$
$$[(9/8) + ((10-4.5)^2/(204 - (8)(4.5)^2))]\}^{1/2}$$

$$= (5,723.65)^{1/2}$$

$$= 75.65 \; tables$$

A 95% confidence interval around the expected value Y_p^* for the forecast of demand at p = year 10 with N-2 = 6 degrees of freedom (df) is given by:

$$(Y_p^* + t_{df,.025} \cdot s_p \, , \, Y_p^* + t_{df,.975} \cdot s_p)$$

$$(Y_{10}^* + t_{6,.025} \cdot s_{10} \, , \, Y_{10}^* + t_{6,.975} \cdot s_{10})$$

where $t_{6,.975} = 2.45$ and $t_{6,.025} = -2.45$ since the Student t distribution (see Table 6) is symmetric and the 2.5th percentile has the same but opposite value as the 97.5th percentile.

Since in this example $Y_{10}^* = 735.7$ tables and $s_{10} = 75.65$ tables, the symmetric 95% confidence interval around the demand Y_{10}^* forecasted by the linear trend in year 10 is given by:

$$[735.7 - (2.45) \cdot (75.65) \, , \, 735.7 + (2.45) \cdot (75.65)]$$

$$[550.5 \; tables, 920.5 \; tables]$$

Notes:

• The 95% confidence interval for future demand for tables in year 10 is quite large even though the linear trend characterizes the past demand well (as evidenced by the large F-statistic and R^2-statistic). Inherently, there is significant *statistical* uncertainty in forecasting future demand even when the Least Squares line is a good fit to the past demand data.

• This statistical uncertainty occurs since the value of the standard deviation s_p *increases* the further the forecast time p (year) is projected into the future. This can be seen by the formula for s_p above which is a function of only time p when the Linear Regression Calculation Table (see Figure 7) used to give numerical values to all other terms:

$$s_p = \{ [(\Sigma Y_t^2 - a\Sigma Y_t - b\Sigma Y \cdot t) / (N-2)] \cdot [(N+1/N) + ((p - \bar{t})^2 / (\Sigma t^2 - N\bar{t}^2))] \}^{1/2}$$

$$s_p = \{ [3101.85] \cdot [1.125 + (p - 4.5)^2 / 42] \}^{1/2}$$

• The size of a given percentile confidence interval (e.g. 95%) around the forecasted demand becomes *larger* (see Figure 15) the *further* into the future that demand is forecasted (as p increases). This reflects the increasing statistical uncertainty as past demand is used to forecast further into the future.

• In general, the size of a given percentile confidence interval is smallest at the sample mean of the independent variable ($\bar{t} = 4.5$ years in this example) since this value minimizes s_p. The further the value of the estimate or the forecast deviates from the sample mean of the independent variable, the larger the size of the standard deviation s_p and the larger the size of a given percentile confidence interval (e. g. 95%) around the past estimate or future forecast.

Expanding 95% Confidence Interval For Future Forecasts

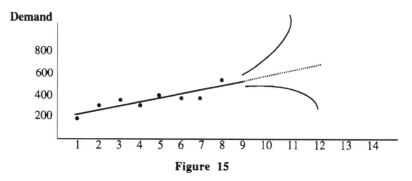

Figure 15

• Recall that the Student t distribution more closely approximates the Normal distribution as its number of degrees of freedom increases (see Section 1.13). For 30 or more degrees of freedom, the Student t distribution and the Normal distribution are virtually identical.

• When the number of degrees of freedom (N-2) exceeds 30, the Normal distribution percentiles can be used to determine confidence intervals around the forecasts for future demand. For example, when $N - 2 \geq 30$, the 95% confidence interval around the future forecast Y_p^* becomes:

$$(Y_p^* + Z_{.025} \cdot s_p , Y_p^* + Z_{.975} \cdot s_p)$$

$$(Y_p^* - 1.645 \cdot s_p , Y_p^* + 1.645 \cdot s_p).$$

Exercise 6

Simple Regression

Problem #1 (Method of Least Squares)

Agony Hospital's walk-in health clinic has measured daily demand (average # patients/day served) over the past five years to be as follows:

Time t (Year)	Demand D_t (Ave # Patients/Day)
1	20
2	22
3	25
4	29
5	30

a) Use the Method of Least Squares to fit the best line to the above demand data. Define "best".

b) Calculate the F-statistic to test the hypothesis of a linear trend in the past demand data.

c) Calculate the correlation coefficient R and the R^2-statistic to determine the goodness of fit of the Method of Least Squares line. The linear model should show a large correlation coefficient R. What does this demonstrate about the relationship between time (t) and the demand (D) in average # patients/day served? Interpret the R^2-statistic.

d) Plot the error terms (residuals). Do they appear to be random and of zero sample mean? (If not, check your calculations).

e) Use the Method of Least Squares for the linear model to predict the average # patients/day served in year 7, year 10, and year 100. Which prediction is the most reliable? Explain.

f) Find a 95% confidence interval for the forecast for year 7. How would the size of a 95% confidence interval around the year 7 forecast compare with a 95% confidence interval around the year 10 forecast?

Problem #2 (Linear Regression vs. Log-Linear Regression)

Fred Frail, staff to the Deputy Under Assistant Secretary of Agriculture, has been asked to forecast the nation's utilization of wheat. He has compiled the following data on wheat usage (in millions of tons) for the past nine years:

Year t	Wheat Usage Y
1	4.7
2	5.4
3	5.8
4	6.3
5	6.8
6	7.4
7	8.0
8	8.7
9	9.4

Mr. Frail has used a commercial software regression package (which uses the Method of Least Squares) to determine the best fit line to both a hypothesized linear trend and a log linear (exponential) trend in the above data: The results are as follows:

• Linear Trend and Analysis of Variance (ANOVA):

Wheat $(Y) = 4.0944 + 0.57 \cdot$ Year (t)

R^2-statistic $= 0.994479$

Linear Regression Analysis of Variance Table

Deviation Source	Deviation Sum of Squares	Degrees of Freedom	Variance
Explained	19.49	1	19.49
Unexplained	0.10822	7	0.01546
Total	19.602	8	F-Statistic = 1260.9

• Log-Linear Model and Analysis of Variance (ANOVA):

Ln Wheat $(Y) = 1.4976 + 0.083458 \cdot$ Year (t), or equivalently,

Wheat $(Y) = e^{1.4976 \ + \ 0.083458 \cdot \text{Year}(t)}$

R^2-statistic $= 0.995547$

Log-Linear Regression Analysis of Variance Table

Deviation Source	Deviation Sum of Squares	Degrees of Freedom	Variance
Explained	0.41791	1	0.41791
Unexplained	0.00187	7	0.000267
Total	0.41978	8	F-Statistic = 1565.1

a) Plot the Wheat Usage (Y) versus Year (t) data.

b) Is it expected that the correlation coefficient R is positive for the linear and log-linear (exponential) regression analysis based on the plot of the wheat usage data over time?

c) Explain the # degrees of freedom associated with the explained, unexplained, and total deviation.

d) Discuss whether the linear or log-linear trend provides the better characterization of the wheat usage over the past nine years.

e) How should Mr. Frail decide between using the linear and log-linear models to forecast future wheat usage for the nation?

Chapter 6

Simple Regression Additional Solved Problems

Problem #1 (Simple Regression)

The Federal Highway Administration (FHA) has received citizen complaints concerning the safety of Highway 95 through northern New Jersey. In response, FHA collected the following data over the last four years on the number of deaths per year on this stretch of Highway 95:

Year t	Deaths D
1	14
2	17
3	18
4	21

a) Use the Method of Least Squares to fit the best line to the above data collected by FHA.

Answer: The Method of Least Squares using a Linear Regression Calculation Table is used to determine the best fit line to the Federal Highway Administration (FHA) data:

FHA Linear Regression Calculation Table

Year t	Deaths D	D·t	t^2	D^2
1	14	14	1	196
2	17	34	4	289
3	18	54	9	324
4	21	84	16	441
$\Sigma t = 10$	$\Sigma D = 70$	$\Sigma D \cdot t = 186$	$\Sigma t^2 = 30$	$\Sigma D^2 = 1250$

$N = 4$
$\bar{t} = \Sigma t / N = 10 / 4 = 2.5$
$\bar{D} = \Sigma D / N = 70 / 4 = 17.5$

$$b = \frac{N \cdot \Sigma D \cdot t - (\Sigma D) \cdot (\Sigma t)}{N \cdot \Sigma t^2 - (\Sigma t)^2}$$
$$= \frac{(4)(186) - (70) \cdot (10)}{(4)(30) - (10)^2}$$
$$= 2.2$$

$\bar{D} = a + b \cdot \bar{t}$

$a = \bar{D} - b \cdot \bar{t}$
$\quad = 17.5 - 2.2 \cdot (2.5)$
$\quad = 12$

$$\boxed{D = 12 + 2.2 \cdot t}$$

b) A software regression package reports an R^2-statistic of 0.989. What does this R^2-statistic demonstrate about the statistical and causal relationship between time (t) and the number of deaths (D) per year?

Answer: A high R^2-statistic is indicative of high *statistical* linear correlation between the increase in time (years) and the increase in deaths on Highway 95 in northern New Jersey. That is, there is a strong tendency for the deaths to increase as time passes (increases).

A high R^2-statistic does not imply that the increase in the number of deaths is *caused* by the increase in time. Linear regression can establish a statistical relationship between independent and dependent variables. However, a causal relationship must be established by the model builder using logic and not by using statistics only.

Moreover, time is a *surrogate* for other variables (e.g. lighting conditions, road conditions, more cars on the road, change in the speed limit, etc.). The increase in deaths over time is not caused by time itself but other independent variable(s). Hypotheses must be made as to what are these independent variables. These hypothesized independent variables can be regressed versus the number of deaths and their statistical significance can be determined (using t-statistics). Logic would still have to be used to establish a causal relationship among the hypothesized variables which are statistically significant.

c) Find a 90% confidence interval for the forecast of the number of fatalities in year 6. A software regression package reports the standard deviation for the forecast for year 6, s_6, is 1.21 deaths.

Answer: A forecast for the number of deaths on this stretch of highway 95 is determined by substituting 6 for t in the Least Squares best fit line $D = 12 + 2.2 \cdot t$:

$$D_6^* = 12 + 2.2 \cdot (6)$$

$$D_6^* = 25.2 \; deaths$$

The standard deviation s_6 is given as 1.21. A symmetric 90% confidence interval is determined using the 5th and 95th percentiles of the Student t distribution (Note: the number of degrees of freedom in the data is less than 30) with $N-2=2$ degrees of freedom:

$$[25.2 + t_{2,.05} \cdot s_6 \; , \; 25.2 + t_{2,.95} \cdot s_6]$$
$$[25.2 - 2.92 \cdot (1.21) \; , \; 25.2 + 2.92 \cdot (1.21)]$$
$$[21.66 \; , \; 28.73]$$

Problem #2 (Simple Regression)

Barry Blarney has been a corporate officer at MegaMedical Corporation for the past four years. Barry's salary for the past four years is given in the following table:

Year t	Salary S
1	$1M
2	$3M
3	$4M
4	$6M

a) After examining the above the above data, identify the "most likely" point (t,S) though which the MegaMedical Salary (S) vs year (t) regression line must pass.

Answer: A regression line must pass through the mean value (the centroid) of the data. The MegaMedical regression line must pass though the most likely point $(\bar{t}, \bar{S}) = (2.5, \$3.5M)$..

b) After examining the above the above data, would you expect the correlation coefficient between Salary (S) and year (t) to be positive, negative, or zero? Explain.

Answer: The salary S is increasing as the year t increases (positive correlation). One would expect that the correlation coefficient between the salary S and year t data will be positive.

c) Apply the Method of Least Squares by hand to fit the best fit line to the above salary (S) data as a function of year (t).

Answer: The Method of Least Squares using a Linear Regression Calculation Table is used to determine the best fit line to the MegaMedical Salary (S) vs. year (t) data:

MegaMedical Linear Regression Calculation Table

Year t	Salary S	S·t	t^2	S^2
1	1	1	1	1
2	3	6	4	9
3	4	12	9	16
4	6	24	16	36
$\Sigma t = 10$	$\Sigma S = 14$	$\Sigma S{\cdot}t = 43$	$\Sigma t^2 = 30$	$\Sigma S^2 = 62$

$N = 4$
$\bar{t} = \Sigma t / N = 10 / 4 = 2.5$
$\bar{S} = \Sigma S / N = 14 / 4 = 3.5$

$$b = \frac{N \cdot \Sigma S{\cdot}t - (\Sigma S){\cdot}(\Sigma t)}{N \cdot \Sigma t^2 - (\Sigma t)^2}$$
$$= \frac{(4)(43) - (14){\cdot}(10)}{(4)(30) - (10)^2}$$
$$= 1.6$$

$$\hat{S} = a + b \cdot \bar{t}$$

$$a = \bar{S} - b \cdot \bar{t}$$
$$= 3.5 - 1.6 \cdot (2.5)$$
$$= -0.5$$

$$S = -0.5 + 1.6 \cdot t$$

d) Find an 80% confidence interval around the expected value of Barry's salary in year t=10 after joining MegaMedical Corporation. A software regression package reports the standard deviation for the forecast for year 10, s_{10}, is $1.25M.

Answer: The expected value S_{10}^{*} for Barry's salary in year 10 is determined by substituting 10 for t in the Least Squares best fit line S = -0.5 + 1.6·t:

$$S_{10}^{*} = -0.5 + 1.6 \cdot (10)$$

$$S_{10}^{*} = \$15.5M$$

The standard deviation s_{10} is reported to be $1.25M. A symmetric 80% confidence interval around the forecast is determined using the equal but opposite 10th and 90th percentiles (i.e. In Table 6, $t_{.90} = -t_{.10}$) of the Student t distribution with N-2=2 degrees of freedom. The Student t distribution is used since the number of degrees of freedom in the data is less than 30.

$$[15.5 + t_{2,.10} \cdot s_{10}, \ 15.5 + t_{2,.90} \cdot s_{10}]$$
$$[15.5 - 1.89 \cdot (1.25), \ 15.5 + 1.89 \cdot (1.25)]$$
$$[\$13.14M, \$17.86M]$$

"Science is the great antidote to the poison of enthusiasm and superstition."
Adam Smith, 1776

7 | Forecasting

7.1 Introduction

A time series is a set of observations of the dependent variable taken a specific times (usually at equal intervals). Time series models seek forecasts of the dependent variable as a function of time. It is understood that there are underlying factors *other than time* affecting the dependent variable. However, there are two reasons why time series models (rather than causal models) are often used to forecast a dependent variable:

1) The process (the correct set of independent variables) explaining the changes in the dependent variable may not be understood or may be extremely complicated.

2) The ability to forecast changes in the dependent variable may be more important than understanding the process (e.g. the 11 year cycle of changes in sunspot activity was determined by time series long before the underlying physical process was understood).

Notes: Some time-series models vs. causal models observations:

• Time series models are more useful for forecasting a dependent variable when the forecast itself is more urgent than determining the underlying factors (e. g. the increase in the nation's GNP over the next few years). Causal models which do attempt to characterize the underlying factors affecting the dependent variable are more useful for policy decisions (e.g. the effect of price and advertising budget on sales revenue over the next few years).

• Causal models require that the future values of the independent variables be *estimated* before a forecast can be made. For time series models, the future value of the independent variable need not be estimated since it is simply the time of the future forecast.

There is always concern when the past is used to forecast the future since things do not tend to remain the same over time. On the other hand, there is some tendency for history to repeat itself. The objective of time series methods is to determine the historical patterns in the past demand data and to forecast those patterns into the future.

The time series methods examined in this chapter assume that the deviations in the dependent variable are *random*. When the observations of the dependent variable are highly correlated (e.g. if the number of people in line at a point in time is high, it is likely to be high a short time later), other time series methods besides those in this chapter must be employed.

There are four types of patterns in the past demand data which can be distinguished by time series methods:

1) A horizontal pattern (H) or constant pattern (see Figure 1) in the past demand data.

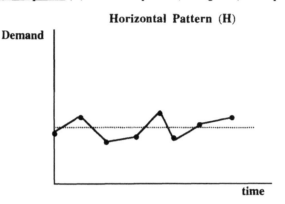

Figure 1

Demand data with a horizontal pattern fluctuates randomly around a value that is constant over time. A horizontal (constant) pattern is sometimes called a *stationary* pattern. The other three patterns which change over time are called *non-stationary* patterns.

2) A seasonal pattern (S) (see Figure 2) in the past demand data.

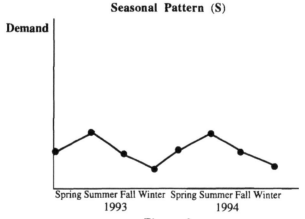

Figure 2

A seasonal pattern may occur when demand is influenced by seasonal factors. Typically, seasonal patterns in the past demand data are of constant length and recur at regular intervals of less than a year (e.g. quarterly, monthly, weekly, or daily).

3) A cyclical_ pattern (C) (see Figure 3) in the past demand data.

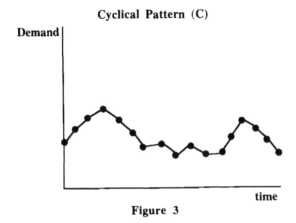

Cyclical Pattern (C)

Figure 3

A cyclical pattern may occur when demand is influenced by factors which occur at irregular intervals (e. g. economic cycles). Typically, patterns in the past demand data are not considered cyclical unless they recur over a time interval greater than a year.

4) A trend pattern (T) (see Figure 4) in the past demand data.

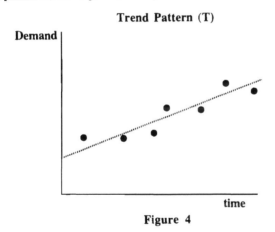

Trend Pattern (T)

Figure 4

A trend pattern is characterized by a *long term* increase or decrease in the demand data. Regression analysis (i.e. the Method of Least Squares) may be successfully applied to demand data with a trend pattern.

Note: Past demand data may include *more than one* of the above four patterns. Thus more than one of the time series methods presented in this chapter may be useful for (1) determining the patterns within a single set of demand data and (2) forecasting future

233

demand. Change in the demand data which can not be attributed to one of these four patterns (H, S, C, or T) is then considered unexplained or random deviation (ε):

Demand Data $= H + S + C + T + \varepsilon$

7.2 Method of Moving Averages

A disadvantage of using regression analysis to forecast demand at a time in the future is that a large amount of data (all the past demand vs. time data) must be stored and a fair amount of calculation complexity is involved (e.g. another application of the Method of Least Squares after each additional past demand value becomes known). Even if the demand data has a horizontal pattern and the sample mean is an appropriate forecast, all of the past data points must be stored to compute the new sample mean after each additional past demand value becomes known. If a business has many items for which it must forecast demand, the data storage and computational complexity required can become burdensome.

Regression analysis also has inherent difficulty reacting to a shift in trends in the demand data. When there is a shift in trends in demand over time, the Method of Least Squares uses the data which describes the old trend on an *equal* basis with the recent data which describes the new trend (each of the data points is given equal weight in the computations).

One approach often used to limit the amount of data that must be stored and the computational complexity is to keep the number of past data values used in forecasting future demand at a reasonably small number N. This approach has the additional advantage of being able to react to changes in a demand trend since only more recent past demand data values are used to forecast future demand.

A class of a time series methods which a reasonably small number N of past demand values to forecast demand is called moving average forecasts. The Method of Moving Averages is used most effectively to forecast demand data which varies slowly (i. e. a horizontal pattern or a slow-varying trend). Moving average forecasts are not typically useful when seasonal or cyclical effects are present in the demand data.

Moving average techniques for demand data involve computing the mean of the N *most recent* past values and using this mean as a forecast for the next period. For each period, the oldest past demand is dropped and the most recent past demand value is added in order to keep the number of demand values used in the computation at a constant N. The choice of how many data values N to use for the moving average is a function of the volatility of the random fluctuations in the demand data. Two types of moving averages will be presented: *simple* moving averages and *weighted* moving averages.

Simple Moving Averages

A simple moving average *equally* weights all N past demand data values used to make the forecast of future demand.

Example 1: (Simple Moving Averages). Consider the following table of past weekly demand for calculators in the electronics department of the Illinois Christian University (ICU) bookstore. Compute the three week *simple* moving average forecasts for future demand.

Week	Past Demand
1	43
2	42
3	45
4	42
5	44
6	40
7	43

Answer: The three week simple moving average forecasts use the actual past demand for the three most recent weeks (equally weighted by 1/3) to forecast demand for the next week. The first week a three week moving average forecast will be available is in week 4.

Forecast for week 4   = (Sum of actual demand for weeks 1, 2, and 3) / 3

                           = (43 + 42 + 45) / 3

                           = 43 1/3 calculators

Forecast for week 5   = (Sum of actual demand for weeks 2, 3, and 4) / 3

                           = (42 + 45 + 42) / 3

                           = 43 calculators

Forecast for week 6   = (Sum of actual demand for weeks 3, 4, and 5) / 3

                           = (45 + 42 + 44) / 3

                           = 43 2/3 calculators

Forecast for week 7   = (Sum of actual demand for weeks 4, 5, and 6) / 3

                           = (42 + 44 + 40) / 3

                           = 42 calculators

Forecast for week 8   = (Sum of actual demand for weeks 5, 6, and 7) / 3

                           = (44 + 40 + 43) / 3

                           = 42 1/3 calculators

A three week moving average forecast of 42 1/3 calculators has been made for week 8. The actual demand for calculators in week 8 is not known at this time (it is blank in Figure 5). When the actual demand in week 8 becomes known, it can be used with past demand in week 6 and week 7 to forecast demand in week 9.

The above three week simple moving average demand forecasts for calculators can now be added to the demand table provided in the description of Example 1 above (see Figure 5).

Three Week Simple Moving Average Forecasts

Week	Past Demand	3 Week Simple Moving Average
1	43	---
2	42	---
3	45	---
4	42	43 1/3
5	44	43
6	40	43 2/3
7	43	42
8	---	42 1/3

Figure 5

Desired Properties of Simple Moving Average Forecasts

Stability is defined as the ability of the forecast to remain close to the true level of demand despite random fluctuations in the demand data. Note the stability of the three week simple moving average forecast (see Figure 6) as compared with the random fluctuations of the actual weekly past demand data given in Example 1.

Three Week Simple Moving Average vs. Demand Data

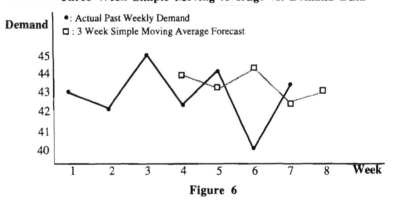

Figure 6

Note: The more periods that are included in the simple moving average, the more stable the simple moving average forecast will be. Random fluctuations in the demand data tend to average out making it more likely the simple moving average forecast will remain close to the true level of demand.

Responsiveness is the flexibility of the forecast to adjust to changes in the true level of demand. Consider the responsiveness of the simple moving average forecast to the shift in the level of demand given in Example 2.

Example 2: (Shift in Level of Demand). Consider the following weekly demand data for calculators at the ICU bookstore where (due to a reduction in the price to students) there is a shift in demand starting in week 4 from 40 to 44 calculators per week. Compute and plot the two week and the three week simple moving average forecasts. Compare the responsiveness of the two week vs. the three week simple moving average forecasts.

Week	Past Demand
1	40
2	40
3	40
4	44
5	44
6	44
7	44

Answer: The two week and three week simple moving average forecasts are calculated in the same manner as in Example 1. For example, the forecasts for demand in week 5 are computed as follows:

- Two week simple moving average:

Forecast in week 5   = (Sum of actual demand for weeks 3 and 4) / 2

= (40 + 44) / 2

= 42 calculators

- Three week simple moving average:

Forecast in week 5   = (Sum of actual demand for weeks 2, 3, and 4) / 3

= (40 + 40 + 44) / 3

= 41 2/3 calculators

The table associated with weekly calculator demand and its two and three week simple moving average forecasts is given in Figure 7. The two week simple moving average forecast is first available in week three while the three week simple moving average forecast is first available in week four.

Two and Three Week Simple Moving Average Forecasts

Week	Past Demand	2 Week Simple Moving Average	3 Week Simple Moving Average
1	40	---	---
2	40	---	---
3	40	40	---
4	44	40	40
5	44	42	41 1/3
6	44	44	42 2/3
7	44	44	44

Figure 7

The actual past weekly demand, the two week simple moving average forecasts, and the three week simple moving average forecasts are plotted in Figure 8.

Note: An N period simple moving average will take N periods to adjust *fully* to a shift in the true level of demand. Moving averages with a small number of periods will display more responsiveness to a shift in the true level of demand since old data is dropped from the simple moving average calculation more quickly.

Two and Three Week Simple Moving Averages vs. Shift in Demand

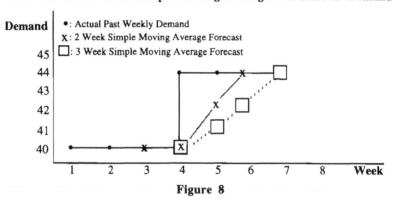

Figure 8

Notes: Stability and responsiveness for the forecast work in opposite directions:

• *Many* periods in a simple moving average $\Rightarrow$ Good stability in the forecast but a loss of responsiveness to adjust to shifts in the true level of demand.

    - Good Stability: Random fluctuations tend to average out due to the many demand periods used in the simple moving average.
    - Loss of Responsiveness: There is a longer lag time before values associated with the former level of demand have been dropped.

• *Few* periods in a simple moving average $\Rightarrow$ Good responsiveness in the forecast but a loss of stability to remain close to shifts in the true level of demand.

    - Good Responsiveness: There is a shorter lag time before all values associated with the former level of demand have been dropped.
    - Loss of Stability: Random fluctuations have a greater affect on the forecast due to the few demand periods used in the simple moving average.

• A general *qualitative* strategy in choosing the proper number N periods for a simple moving average forecast is as follows:

    - If demand data is relatively stable around the true level of demand $\Rightarrow$ Want the forecasts to be responsive to small changes in the demand data. Choose a smaller number N for the forecast.

- If demand data has large random fluctuations around the true level of demand $\Rightarrow$ Want the forecasts to be stable. Choose a larger number N for the forecast.

• The specific number of periods N for the simple moving average which provides the best forecasts is determined through *trial and error* on past demand data. As usual, "best" is defined as the number of periods N which provides the minimum squared error (see Section 7.5) between the actual demand data and the forecasted demand data.

Reaction to a Trend in Demand

Recall that a <u>trend</u> is a discernible long term increase or decrease in the demand data over time (see Section 6.1). Consider the effects of a trend in the demand data on simple moving average forecasts (see Example 3).

Example 3: (Trend in Demand). The electronics department at the ICU bookstore has been using calculators as a loss leader (trying to entice students once they arrive into buying computers which provide a high profit for the electronics department). Word of the low cost calculators has spread on campus and the demand for calculators (given below) has increased steadily since week 4 of the semester. Compute, plot, and interpret the responsiveness of the two week and the three week simple moving average forecasts.

Week	Past Demand
1	40
2	40
3	40
4	42
5	44
6	46
7	48
8	50

Answer: The two week and three week simple moving average forecasts are calculated in the same manner used in Examples 1 and 2. Figure 9 provides the actual past weekly calculator demand and the two and three week simple moving average forecasts.

Two and Three Week Simple Moving Averages

Week	Past Demand	2 Week Simple Moving Average	3 Week Simple Moving Average
1	40	---	---
2	40	---	---
3	40	40	---
4	42	40	40
5	44	41	40 2/3
6	46	43	42
7	48	45	44
8	50	47	46

Figure 9

The actual past demand data, the two week simple moving average forecasts, and the three week simple moving average forecasts are plotted in Figure 10.

Note: A simple moving average forecast *always lags* the change in demand due to a trend (see Figure 10) since old demand data is used to forecast future demand. The lag in the forecast is roughly equal to the number N periods chosen for the simple moving average.

Two and Three Week Simple Moving Averages vs. Upward Trend

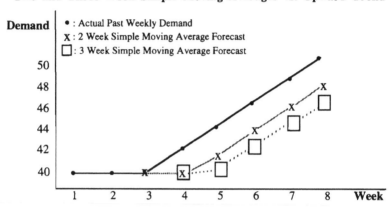

Figure 10

Weighted Moving Averages

A weighted moving average assigns *different weights* to some of the N past demand data values used to make the forecast of future demand than to others. Typically, more weight is given to the more *recent* values of past demand. In this way, if there is a trend in the demand data, the forecast will not lag as far behind. Again there is no analytical method to choose the appropriate weights (which must add to 1) to assign the past demand data.

Example 4: (Weighted Moving Averages). For the upward trend in calculator demand at the ICU bookstore described in Example 3 (the demand data is repeated below), compute the three week (0.2, 0.3, 0.5) weighted moving average forecasts of the demand for calculators.

Week	Past Demand
1	40
2	40
3	40
4	42
5	44
6	46
7	48
8	50

Answer: As with the simple moving average forecasts, the first three week (0.2, 0.3, 0.5) weighted moving average forecast is available in week 4. Three week weighted moving average forecasts (where 0.2 + 0.3 + 0.5 = 1.0) are calculated using the most recent three past weekly demands in the following manner:

Forecast for week 4 = (0.2)·(Actual demand week 1) + (0.3)·(Actual demand week 2) + (0.5)·(Actual demand week 3)

= (0.2)·(40) + (0.3)·(40) + (0.5)·(40)

= 40

Forecast for week 5 = (0.2)·(Actual demand week 2) + (0.3)·(Actual demand week 3) + (0.5)·(Actual demand week 4)

= (0.2)·(40) + (0.3)·(40) + (0.5)·(42)

= 41

The table associated with the actual past weekly calculator demand, the 3 week simple moving average forecasts, and the 3 week (0.2, 0.3, 0.5) weighted moving average forecasts for the upward trend is given in Figure 11.

Simple vs. Weighted Moving Average Forecasts For An Upward Trend

Week	Past Demand	3 Week Simple Moving Average	3 Week Weighted Moving Average
1	40	--	---
2	40	---	---
3	40	---	---
4	42	40	40
5	44	40 2/3	41
6	46	42	42.6
7	48	44	44.6
8	50	46	46.6

Figure 11

Note: Weighted moving average forecasts which place more weight on recent past demand values are more responsive to trends in the data than simple moving average forecasts. However, weighted moving average forecasts are less stable than simple moving average forecasts in that they respond more quickly to short lived changes in demand data (which could be just random fluctuations rather than a long term trend). Again, stability and responsiveness of the forecast work in opposite directions.

It would seem that weighted moving average forecasts which provide more responsiveness are always preferred to simple moving average forecasts when there is a trend in the demand data. However, consider when an overall trend exists in the past weekly demand data but there are large fluctuations in demand from week to week. One would like to limit the responsiveness of the forecast (follow the trend but not chase the large fluctuations around the trend). Such a demand scenario is displayed in Figure 12 and Figure 13 where demand fluctuates widely from week to week but follows a consistent upward trend.

Figure 12 displays the two week simple moving average forecasts which equally average the demand data and provide stability despite the large fluctuations in past weekly demand.

Simple Moving Average Forecast For Data Fluctuating Around A Trend

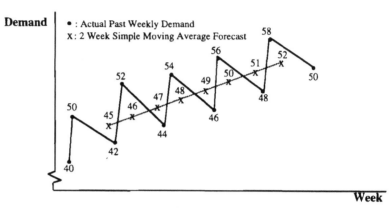

Figure 12

Figure 13 displays the two week (0.2, 0.8) weighted moving average forecasts which weight more heavily the most recent demand data. The two week weighted moving average forecasts are over-responsive. The forecasts are high when demand is low (and vice versa) overreacting to the large fluctuations in weekly demand for calculators.

Weighted Moving Average Forecast For Data Fluctuating Around A Trend

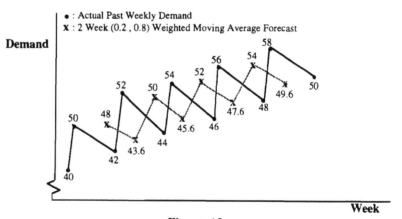

Figure 13

For such large fluctuations around a trend in the demand data, the more stable and less responsive simple moving average forecast is the more useful for forecasting future demand (i.e. minimum squared error between the actual and forecasted demand).

<u>Notes:</u> Final observations on the Method of Moving Averages (simple and weighted):

• The Method of Moving Averages can be used for data with a horizontal pattern (stationary) or a slow-varying trend pattern (non-stationary). The Method of Moving Averages will typically forecast demand data with a cyclical or a seasonal pattern poorly. Moving average forecasts always lag the repetitive cyclical or seasonal changes in demand.

• The advantages of the Method of Moving Averages for forecasting the dependent variable:

> - Limited data storage: Only N values of historical past demand data need be stored for an N period moving average. Unless N is quite large, this limits data storage requirements.

> - Computational simplicity: The computational complexity of computing moving averages is minimal.

> - Flexibility: The value of N can be changed to respond to volatility of the data.

• The disadvantage of the Method of Moving Averages for forecasting the dependent variable is as follows:

> - Limited Accuracy: The accuracy of the moving average forecasts is often low. A great deal of past demand data (greater than N periods old) is not used in the forecast in order to limit data storage and computational complexity. Thus, all the *information* in the data is not used.

• The Method of Exponential Smoothing (see Section 7.3) is used more often in practice since it has all the advantages of moving averages (limited data storage, computational simplicity, and flexibility) plus better forecasting accuracy. The surprising forecasting accuracy is a result of using *all* the past demand data in making the future forecast of demand. For an N period moving average forecast, the Method of Exponential Smoothing constant α (see Section 7.3) of $2/(N+1)$ gives equivalent smoothing results.

7.3 Method of Exponential Smoothing

The Method of Exponential Smoothing is used most effectively to forecast demand for demand data which exhibits a slow-varying (positive or negative) trend (see Figure 14). If there is a cyclical or seasonal pattern in the data, the forecast of future demand is best addressed by the Method of Seasonal Indices (see Section 7.4).

Slow-Varying Trend in the Demand Data

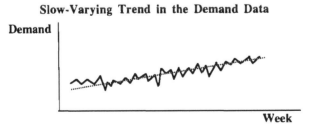

Figure 14

The <u>Method of Exponential Smoothing</u> takes advantage of the following logical concepts which apply to forecasting future demand:

- *Recent* past demand data is more relevant than older past demand data in making forecasts of future demand.

- When a forecast is made for the next period, the forecasted demand can be compared with the actual demand in that time period to compute an *error* in the forecast.

- The errors in the forecasts provide *information* about the demand data.

- This information can then be used to forecast demand in future periods *more accurately.*

Define the following notation:

Y_t = Actual demand for time period t
Y_t^* = Forecasted demand for time period t
a = Exponential smoothing constant (chosen between 0 and 1, inclusive)

Then the Method of Exponential Smoothing forecast Y_{t+1}^* for time period t+1 is calculated using the <u>exponential smoothing equation</u>:

$$Y_{t+1}^* = Y_t^* + \alpha \cdot (Y_t - Y_t^*)$$

In words, the forecast Y_{t+1}^* for period t+1 is the forecast Y_t^* for period t plus a fraction α of the error $(Y_t - Y_t^*)$ in the forecast for period t. The error is the difference between the actual demand and the forecasted demand in period t.

<u>Notes:</u> Properties of the Method of Exponential Smoothing (which are apparent when the exponential smoothing equation is examined) include the following:

• The *information* contained in the *error* $(Y_t - Y_t^*)$ between the actual demand Y_t and the forecasted demand Y_t^* in each period t is used to adjust the forecast for period t+1.

• Small values (e.g. 0.2 and below) for the exponential smoothing constant α will cause the sequence of future forecasts to react slowly to the errors $(Y_t - Y_t^*)$ in past forecasts (see α = 0.1 in Figure 15). In the extreme, if $\alpha = 0$:

$$Y_{t+1}^* = Y_t^*$$

$\Rightarrow$ The forecast for the first period (t = 1) is used for each succeeding period. Information contained in the error between forecasted demand and actual demand is not used at all to adjust future forecasts.

244

• Values over 0.5 for the exponential smoothing constant α will cause the sequence of future forecasts to react strongly to errors in past forecasts (see $\alpha = 0.9$ in Figure 15). In the extreme, if $\alpha = 1$:

$$Y_{t+1}^{*} = Y_{t}$$

$\Rightarrow$ Forecasts for the succeeding period are exactly the actual demand value in the preceding period. No *smoothing* of the random fluctuations in the data is performed.

Exponential Smoothing Constants $\alpha = 0.1, 0.9$

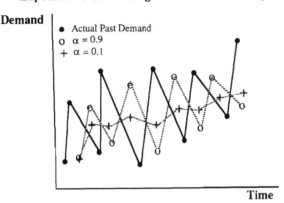

Figure 15

• A small exponential smoothing constant α has the same smoothing properties as a large number N periods for the moving average forecasts (i.e. both provide stability at the expense of responsiveness in the forecast). Conversely, a large α has the same lack of smoothing properties as a small number N periods for the moving average forecasts (i.e. both provide responsiveness at the expense of stability in the forecast).

• Typically, values for the exponential smoothing constant α are chosen between 0.1 and 0.2. However, the choice of α is dependent on the actual past demand data. *Trial and error* is typically used to find the value of α which provides the best future forecasts. Here, "best" is defined as the value of α which provides the minimum squared error (see Section 7.5) over representative sets of data.

• The Method of Exponential Smoothing provides its first forecast in period 2.

$$Y_{2}^{*} = Y_{1}^{*} + \alpha \cdot (Y_{1} - Y_{1}^{*})$$

A forecast Y_{1}^{*} for period 1 using some other forecasting method is required. If no forecast Y_{1}^{*} exists for the period 1, one can simply use the actual demand in period 1 (given by Y_{1}) as the forecast for period 1. From the exponential smoothing equation, the forecast for period 2 will then be Y_{1} itself since no error in the forecast will have occurred in period 1.

$$Y_{2}^{*} = Y_{1} + \alpha \cdot (Y_{1} - Y_{1}) = Y_{1}$$

Future forecasts for periods 3, 4, 5, etc. onwards proceed using the exponential smoothing equation applied to the actual past demand data Y_2, Y_3, Y_4, etc., respectively.

Example 5: (Method of Exponential Smoothing). Monthly demand for calculators has a slow-varying trend. An initial forecast for demand in period 1 (January) is 127 calculators (Y_1^*). The actual demand (Y_1) in January is 148 calculators. Assume that $\alpha = 0.2$ has been demonstrated to be a satisfactory exponential smoothing constant in past years. Use the Method of Exponential Smoothing to provide a forecast for February (Y_2^*).

Answer: The error between the actual demand for calculators in January and the forecasted demand is 148 - 127 = 21. The fraction $\alpha = 0.2$ will be applied to the information contained in this forecasting error in order to obtain the next month's (February) forecast for calculators:

$$Y_2^* = Y_1^* + \alpha \cdot (Y_1 - Y_1^*)$$
$$= 127 + 0.2 \cdot (148 - 121)$$
$$= 127 + 0.2 \cdot (21)$$
$$= 127 + 4.2$$
$$= 131.2 \text{ calculators}$$

The actual demand for January (148 calculators) was larger than the forecast for January (127 calculators). Consequently, the forecast for February (131.2 calculators) using the Method of Exponential Smoothing is larger than the forecast for January.

If the actual demand for calculators in February (Y_2) is 119, then the forecast for March (Y_3^*) using the Method of Exponential Smoothing is as follows:

$$Y_3^* = Y_2^* + \alpha \cdot (Y_2 - Y_2^*)$$
$$= 131.2 + 0.2 \cdot (119 - 131.2)$$
$$= 131.2 + 0.2 \cdot (- 12.2)$$
$$= 131.2 - 2.44$$
$$= 128.76 \text{ calculators}$$

The actual demand in February (119 calculators) was smaller than the forecast for February (131.2 calculators). Consequently, the forecast for March (128.76 calculators) using the Method of Exponential Smoothing is smaller than the forecast for February.

Notes:

• The exponential smoothing equation shows that when the actual demand in a period is smaller or larger than the forecast, the next period's forecast will *always* be adjusted to be smaller or larger than the current period's forecast, respectively:

Actual demand is smaller the forecasted demand: $(Y_t - Y_t^*) < 0$

$$Y_{t+1}^* = Y_t^* + \alpha \cdot (Y_t - Y_t^*)$$

$$Y_{t+1}^* = Y_t^* + \text{negative number}$$

$$Y_{t+1}^* < Y_t^*$$

$\Rightarrow$ Next period's forecast is smaller than the current period's forecast.

Actual demand is larger than the forecasted demand: $(Y_t - Y_t^*) > 0$

$$Y_{t+1}^* = Y_t^* + \alpha \cdot (Y_t - Y_t^*)$$

$$Y_{t+1}^* = Y_t^* + \text{positive number}$$

$$Y_{t+1}^* > Y_t^*$$

$\Rightarrow$ Next period's forecast is larger than current period's forecast.

• The forecasts using the Method of Exponential Smoothing will always lag a significant (not slow-varying) trend in the data because the most each forecast can do is make up a percentage α of the error in the forecast. The forecasts for demand generally falls *further behind* (see Figure 16) the actual past demand values over time for a significant trend.

Use of Method of Exponential Smoothing For a Significant Trend

Period t	1	2	3	4	5	6	7	8
Y_t	0	1	2	3	4	5	6	7
Y_t^*	0	0	0.2	0.56	1.05	1.64	2.31	3.04

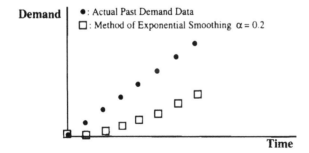

Demand

●: Actual Past Demand Data
□: Method of Exponential Smoothing $\alpha = 0.2$

Time

Figure 16

247

• If there is a slow-varying trend in the demand data (see Figure 17), the Method of Exponential Smoothing will continually adjust the forecast to correct for the trend in the data. Using continuous small corrections, eventually the forecast for demand will closely track the actual past demand values over time.

Use of Method of Exponential Smoothing For a Slow-Varying Trend

Period t	1	2	3	4	5	6	7	8	9	10	11	12	13	14
Y_t	0	1	0.50	0.75	1	0.75	1.25	0.75	1	1.25	1	0.75	1.50	1
Y_t^*	0	0	0.20	0.26	0.36	0.49	0.54	0.68	0.69	0.75	0.85	0.88	0.85	0.98

Demand | ●: Actual Past Demand Data
□: Method of Exponential Smoothing $\alpha = 0.2$

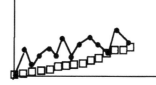

Time

Figure 17

• The exponential smoothing equation $Y_{t+1}^* = Y_t^* + \alpha \cdot (Y_t - Y_t^*)$ can be rewritten as follows:

$$Y_{t+1}^* = \alpha \cdot Y_t + (1 - \alpha) \cdot Y_t^*$$

This version of the smoothing equation indicates that when forecasting the next period's demand, α is the weight given to the most recent actual demand Y_t while $(1 - \alpha)$ is the weight given the most recent demand forecast Y_t^*. This form of the exponential smoothing equation again demonstrates for a large value of α, a time period's forecast will be strongly influenced by the most recent observed demand value.

• The alternate form of the exponential smoothing equation is able to demonstrate that *all* the past actual demand values are used in the forecast (i.e. all the *information* in the data is used) but that the more recent actual demand data values are given greater weight.

$$Y_{t+1}^* = \alpha \cdot Y_t + (1 - \alpha) \cdot Y_t^* \qquad \text{(1 period)}$$

$$Y_{t+1}^* = \alpha \cdot Y_t + (1 - \alpha) \cdot Y_t^*$$

$$= \alpha \cdot Y_t + (1 - \alpha) \cdot [\alpha \cdot Y_{t-1} + (1 - \alpha) \cdot Y_{t-1}^*]$$

$$= \alpha \cdot Y_t + \alpha \cdot (1 - \alpha) \cdot Y_{t-1} + (1 - \alpha)^2 \cdot Y_{t-1}^* \qquad \text{(2 periods)}$$

248

$$Y_{t+1}^* = \alpha \cdot Y_t + \alpha \cdot (1-\alpha) \cdot Y_{t-1} + (1-\alpha)^2 \cdot Y_{t-1}^*$$

$$= \alpha \cdot Y_t + \alpha \cdot (1-\alpha) \cdot Y_{t-1} + (1-\alpha)^2 \cdot [\alpha \cdot Y_{t-2} + (1-\alpha) \cdot Y_{t-2}^*]$$

$$= \alpha \cdot Y_t + \alpha \cdot (1-\alpha) \cdot Y_{t-1} + \alpha \cdot (1-\alpha)^2 Y_{t-2} + (1-\alpha)^3 \cdot Y_{t-2}^* \qquad \text{(3 periods)}$$

Actual demand N periods in the past receives only $(1-\alpha)^N < 1$ times as much weight as the most recent actual demand in making a forecast for the next period. This exponentially decreasing weighting $(1-\alpha)^N$ of past demand data (e.g. for $\alpha = 0.2$, values 1, 2, 3, 4 etc. periods in the past receive 0.8, 0.64, 0.512, 0.4096, etc. times less weight than the most recent demand, respectively) is the derivation of the name "exponential smoothing."

Notes:

• The Method of Exponential Smoothing (like the Method of Moving Averages) is used most effectively when the demand data has a slow-varying trend pattern. The Method of Exponential Smoothing will typically forecast significant trends poorly and will lag (often out of phase) the repetitive changes in cyclical or seasonal demand data.

• The advantages of the Method of Exponential Smoothing for forecasting demand data are as follows:

- Small data storage: Storage space is substantially reduced over even the Method of Moving Averages. The Method of Exponential Smoothing requires only the storage of the current forecast Y_t^* and the exponential smoothing constant α.

- Computation simplicity: Only two data points (the forecasted demand Y_t^* and the actual demand Y_t) along with a constant α are involved in computing the forecast for demand for the next period.

- Flexibility: The exponential smoothing constant α can be changed to respond to volatility in the data to provide the desired stability or responsiveness.

- All information contained in the past demand data is used: The relevant information in all the past demand data is included in the current forecast Y_t^*. Data N periods old receives a factor $(1-\alpha)^N$ less weight than the current demand value.

• The disadvantage of the Method of Exponential Smoothing for forecasting future demand data is as follows:

- Specification of the exponential smoothing constant α: There is no analytical method to choose α. The value of α must be chosen by trial and error using logical values and examining the minimum squared error of the forecast..

7.4 Method of Seasonal Indices

Consider the following plot of the past four years of monthly demand data for calculators at the ICU bookstore (see Figure 18).

249

Demand Data For Monthly Calculator Demand Over Four Years

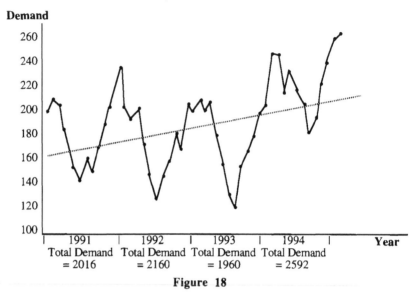

Figure 18

There seems to be two patterns within the change in the calculator demand over time:

- Positive trend pattern: An overall trend of increased yearly demand for calculators (dotted Least Squares line in Figure 18).

- Seasonal pattern: A repetitive pattern of increasing demand for calculators during the winter months (when the college is in session) and decreasing demand during the summer months (when the college is not in session).

Figure 19 provides the tabular monthly calculator demand data Y_t used to plot Figure 18.

Monthly Calculator Demand At ICU Bookstore Over Four Years

m t	1 (J)	2 (F)	3 (M)	4 (A)	5 (M)	6 (J)	7 (J)	8 (A)	9 (S)	10 (O)	11 (N)	12 (D)
1 (1991)	196	212	202	180	150	140	156	144	164	186	200	230
2 (1992)	200	188	192	164	140	122	132	144	176	168	196	194
3 (1993)	196	188	192	164	140	120	112	140	160	168	192	200
4 (1994)	242	240	196	220	200	192	176	184	204	228	250	260
Mean $\bar{Y}_m$	208.6	207.0	192.6	182.0	157.6	143.6	144.0	153.0	177.6	187.6	209.6	221.0

Figure 19

250

Define the following additional notation used in the table for the ICU Bookstore monthly calculator demand (see Figure 19):

m :   The index for the month of the demand data $\{m = 1$ (Jan), 2 (Feb), $\cdots$, 12 (Dec)$\}$.

t :   The index of the year of the demand data $\{t = 1$ (1991), 2 (1992), 3 (1993), and 4 (1994)$\}$.

$\bar{Y}_m$:   The monthly mean of the demand data over the 4 years for each specific month m (e.g. $\bar{Y}_1 = (196+200+196+242)/4 = 834/4 = 208.6$ for the month of January over the 4 years).

$\bar{\bar{Y}}$:   The grand mean for the means of the 12 monthly demands ($\bar{\bar{Y}} = \Sigma \bar{Y}_m /12 = 2082/12 = 181.84$).

The Method of Seasonal Indices forecasts demand associated with cyclical or seasonal (shorter) periods by applying percentages (called indices) to the average (mean) demand for an aggregate (longer) period forecast. Monthly indices (I_m) for the calculator demand data given in Figure 19 are as follows:

I_m =   Monthly Mean $\bar{Y}_m$ / Grand Mean $m = 1, 2, \cdots , 12$

    = $\bar{Y}_m / \bar{\bar{Y}}$

Example 6: (Computation of Monthly Indices). Determine the monthly indices for the past four years of monthly demand for calculators at the ICU bookstore given in Figure 19.

Answer: The monthly indices can be calculated by dividing the monthly means ($\bar{Y}_m$) over the four years of calculator demand from Figure 19 by the grand mean ($\bar{\bar{Y}}$):

I_1 (Jan) = 208.6 / 181.84 =   1.15

I_2 (Feb) = 207.0 / 181.84 =   1.14

I_3 (Mar) = 192.6 / 181.84 =   1.06

I_4 (Apr) = 182.0 / 181.84 =   1.00

I_5 (May) = 157.6 / 181.84 =   0.87

I_6 (Jun) = 143.6 / 181.84 =   0.79

I_7 (Jul) = 144.0 / 181.84 =   0.79

I_9 (Aug) = 153.0 / 181.84 = 0.84

I_9 (Sept) = 177.6 / 181.84 = 0.97

I_{10} (Oct) = 187.6 / 181.84 = 1.03

I_{11} (Nov) = 209.6 / 181.84 = 1.15

I_{12} (Dec) = 221.0 / 181.84 = <u>1.22</u>

$$12.00 \ (= \Sigma \, I_m)$$

Note: The mean (average value) for the monthly indices ($= \Sigma \, I_m / 12$) equals 1.

Example 7: (Method of Seasonal Indices). Assume that the demand data for calculators has been found to have a yearly trend pattern as well as a monthly seasonal pattern. The Method of Least Squares has been used to forecast demand in *year 5* to be 2400 calculators. Use the Method of Seasonal Indices to forecast demand for calculators for the *month* of July in year 5.

Answer: Given a yearly forecast (2400 calculators) in year 5, the forecast for an *average* month in year 5 can be determined by dividing the yearly forecast by 12. The Method of Seasonal Indices multiplies the *average* monthly demand in year 5 (200 calculators) by the appropriate monthly index (0.79 for July):

Method of Least Squares forecast for demand in year 5 = 2400 calculators

Average monthly demand in year 5 = 2400/12
= 200 calculators

Method of Seasonal Indices forecast for July demand in year 5   = (200)·(0.79)
= 158 calculators

Note: Since the average of the monthly indices is *always* 1, the sum of the monthly indices $\Sigma \, I_m$ is 12. If the appropriate index for each month had been applied to the average monthly demand of 200 calculators, the total sum of the monthly demand forecasts would equal 200·($\Sigma \, I_m$ = 12) = 2400 calculators, the Method of Least Squares forecast for the entire year 5.

The Method of Seasonal Indices is typically used to make forecasts for *shorter* periods of time (e.g. quarterly demands which show a seasonal pattern) when a forecast for a *longer* time period already exists (e.g. a yearly demand provided by the Method of Least Squares). When there is a cyclical or seasonal pattern in the demand data, the Method of Seasonal Indices is likely to improve the shorter period forecasts (i.e. a smaller mean squared error).

7.5 Measures of Forecasting Error

A forecaster will generally fit several models to the available past demand data. The choice of which model to use for future forecasts will naturally depend on which model provides the most accurate forecasts of past demand data. It is possible simply to visually *observe* the time series plots for each of the candidate forecast models and to choose the model which *seems* to match the past demand data most closely. However, this method is not acceptable to decision makers because it is both inaccurate and subjective.

Formal measurement of forecasting errors is required. Such measurements focus on the magnitude and sign of the *errors* between the actual and forecasted data in each time period. The objective of providing measures of forecasting error is to allow the decision maker to rank the candidate forecast models quantitatively.

Recall the forecasting error notation previously defined (see Section 7.3):

Y_t = Actual demand for time period t

Y_t^* = Forecasted demand for time period t

$(Y_t - Y_t^*)$ = Error (residual) of the forecast for time period t

Notes:

• There will typically be errors in the forecast whenever past data is used to forecast future demand. However, experience has demonstrated that the forecast errors using formal forecasting models (e.g. regression, Method of Moving Averages, etc.) tend to be *smaller* than forecasts using informal methods (e.g. visual projection from plotted time series data).

• The *magnitude* of the error of the forecast can provide information on the relative performance of different forecasting models. However, the magnitude of the error of the forecast does not provide information about the *source* of the error. The source of the error is not a mathematical proposition but is based on logic and intuition.

• A *bias error* (a consistent positive or negative error in the forecast) typically occurs when a relevant independent variable has not been included in the forecast model whereas *random error* (no pattern in the errors in the forecast) typically occurs when the error is unexplainable by the forecast model.

Mean Absolute Deviation

The mean absolute deviation (MAD) is the sum of the absolute values of the forecast errors divided by the number of periods N in the forecast. Hence, the MAD is the *average* absolute error between the actual and forecasted demand data:

$$MAD = \frac{\sum_{t=1}^{N} \left| Y_t - Y_t^* \right|}{N}$$

Notes:

• The mean absolute deviation (like the standard deviation σ of a data stream) is a measure of the *dispersion* of the actual demand values around the mean demand value.

• The mean absolute deviation is dynamic and is computed after each period t when another absolute error in the forecast ($\left| Y_t - Y_t^* \right|$) becomes available.

• If the errors of the forecast are assumed normally distributed (a typical assumption), then the standard deviation σ and the mean absolute deviation (MAD) are related as follows:

$$\sigma = \sqrt{\frac{\pi}{2}} \bullet MAD \approx 1.25 \bullet MAD$$

• If the MAD for a forecast is 20 units, the standard deviation σ for the actual demand data Y_t would be approximately 25 units. Under the assumption of normally distributed errors it should then be expected that approximately 95% of the actual demand data will fall into band plus or minus 2 standard deviations (50 units) around the mean demand value.

Mean Absolute Percentage Error

The mean absolute percentage error (MAPE) is the average of the absolute *percentage differences* between the actual and forecasted demand data:

$$MAPE = \frac{\sum_{i=1}^{N} \frac{\left| Y_t - Y_t^* \right|}{Y_t}}{N}$$

Notes:

• The mean absolute deviation (MAD) does not address differences in the relative magnitude of sets of demand data. A set of data with large values of demand is more likely to have a large value for MAD. This can be a problem for a forecaster comparing the accuracy of models over different sets of data.

• The mean absolute percentage error (MAPE) differs from the mean absolute deviation (MAD) in that the MAPE normalizes for the magnitude of the demand data.

• For example, say a demand data set with a mean value of 10,000 has a MAD of 100 while a second demand data set with a mean value of 50 has a MAD of 10. While the second demand data set has a smaller MAD, its MAD is 20% of the mean value of the data while the MAD for the first demand data set is only 10% of the mean value of the data.

Bias

The bias of a forecast is the average of the sum of the deviations:

$$Bias = \frac{\sum\limits_{t=1}^{N}\left(Y_t - Y_t^*\right)}{N}$$

Notes:

• The bias computation indicates whether a forecast on the whole tends to be higher or lower than the actual data. It is computed on a *one-time* basis for all the actual demand data and the forecasts to measure whether the forecast model tracks the actual demand data.

• The tracking signal (defined below) provides a *dynamic* running measure (computed at each time period) of whether the forecast tends to be higher or lower than the actual data.

Tracking Signal

The tracking signal (TS) is defined here (there are several definitions in the literature) as the average of the forecast errors (bias) divided by the mean absolute deviation (MAD):

$$TS = \frac{Bias}{MAD}$$
$$= \frac{\sum\limits_{t=1}^{N}\left(Y_t - Y_t^*\right)}{\sum\limits_{t=1}^{N}\left(\left|Y_t - Y_t^*\right|\right)}$$

Notes:

• The value of the tracking signal varies between -1 which occurs when all the errors in the forecast are negative (a consistent overestimating bias) and +1 which occurs when all the errors in the forecast are positive (a consistent underestimating bias).

• The dynamic tracking signal is computed after each period t when another error in the forecast ($Y_t - Y_t^*$) becomes available.

• A tracking signal which hovers near zero indicates there is little or no bias and that the errors are random in nature. The forecasted demand is tracking the actual demand.

• A tracking signal which hovers above zero indicates that the forecasted demand is consistently below the actual demand. Conversely, a tracking signal which hovers below zero indicates that the forecasted demand is consistently above the actual demand.

• A forecaster will often determine that the forecast model is running acceptably as long as the absolute value of the tracking signal is less than a preset limit. When the absolute value of the tracking signal exceeds the preset limit (e.g. 0.5), the forecaster then adjusts the model parameters in an appropriate manner to offset the errors.

Example 8: (MAD and Tracking Signal). Compute the MAD and the tracking signal for the following data:

(1) Period t	(2) Actual Demand Y_t	(3) Forecasted Demand Y_t^*
1	75	100
2	50	100
3	125	50
4	100	50
5	50	25

Answer: Column (4) in Figure 20 provides the deviation between the actual demand Y_t and the forecasted demand Y_t^* for each period t as well as the running sum of the deviations up to period t. Column (5) is the running sum of the absolute deviations up to period t. The mean absolute deviation (MAD) which appears in column (6) is computed a the sum of the absolute deviations given in column (5) divided by the number of period given in column (1).

The bias in the forecast which appears in Column (7) in Figure 20 is computed as the sum of the deviations given in column (4) divided by the number of periods. The tracking signal (TS) which appears in column (8) is computed as the bias given in column (7 divided by the MAD given in column (6).

Mean Absolute Deviation (MAD) and Tracking Signal (TS) Computations

(1) Period t	(2) Actual Demand Y_t	(3) Forecasted Demand Y_t^*	(4) Deviation/ Sum of Deviations $(Y_t-Y_t^*)$/ $\Sigma(Y_t-Y_t^*)$	(5) Sum of Absolute Deviations $\Sigma\|Y_t-Y_t^*\|$	(6) MAD $\dfrac{\Sigma\|Y_t-Y_t^*\|}{N}$	(7) Bias $\dfrac{\Sigma(Y_t-Y_t^*)}{N}$	(8) Tracking Signal TS $\dfrac{Bias}{MAD}$
1	75	100	-25/-25	25	25	-25	-1
2	50	100	-50/-75	75	37.50	-37.50	-1
3	125	75	50/-25	125	41.66	-8.33	-0.20
4	100	50	50/25	175	43.75	6.25	0.143
5	50	25	25/50	200	40	10	0.25

Figure 20

In Example 8, the forecasted demand over 5 periods deviated from the actual demand by the mean absolute deviation of 40 units. In a *unbiased* forecast, the sum of the deviations (see Column (4) in Example 8) should be close to zero indicating that the forecasted demand is generally neither above nor below the actual demand.

Note: The tracking signal in period 1 is always -1 or +1 since all the deviation is in one direction (negative or positive, respectively). The tracking signal of 0.25 after 5 periods is positive indicating that the forecasted demand was generally lower than the actual demand.

Mean Squared Error

The mean squared error (MSE) is the average of the squared errors between the actual and the forecasted demand data:

$$MSE = \frac{\sum_{t=1}^{N}\left(Y_t - Y_t^*\right)^2}{N}$$

Notes:

• Both the MAD and the MAPE treat large forecast errors in a *linear proportion* to small forecast errors. For example, a forecast error of 20 contributes 10 times as much to the magnitude of the MAD than a forecast error of 2 contributes.

• The MSE measure squares all the errors between the actual and forecasted data. For this measure, a outlier (say an error of 20) leads to a large addition to the value of the MSE (400). Such an outlier can dwarf the effects of several smaller errors (say errors of 2 or 3) which lead to relatively small additions to the value of the MSE (only 4 or 9, respectively).

• In some applications, large errors can have much larger than proportional effects. These effects can even lead to catastrophic results (e.g. when actual rainfall is consistently much greater than forecasted and this leads to a water dam overflowing onto a valley). Here the forecaster will probably prefer to use the MSE rather than the MAD (or the MAPE).

• It is not uncommon when assessing the relative accuracy of candidate forecast models to determine a *different* ranking when using the MAD (or the essentially equivalent MAPE) measure of forecasting error than when using the MSE measure of forecasting error. In such cases, the forecaster must use his or her judgment on which measure of forecasting error is most appropriate.

• Forecasters tend to prefer the MSE over the MAD and MAPE since many forecast models derive the values of parameters based on minimizing mean squared error. For purposes of comparing accuracy of forecast models, the mean squared error (MSE) measure of forecasting error will be used in the examples presented in this chapter unless otherwise stated.

7.6   Demand Times Series vs. Forecasting Methods (Notional)

Example 9: (Choosing Forecasting Methods). Consider the notional time series plot of monthly demand data in Figure 21. Determine an appropriate forecasting method(s).

Notional Times Series Plot of Monthly Demand Data

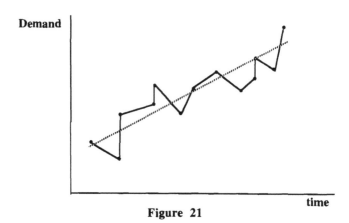

Figure 21

Answer: The monthly demand data seems to show a positive trend pattern over the long term. Regression using the Method of Least Squares can be used to determine how well the monthly data fits a linear trend. There doesn't *seem* to be a cyclical or seasonal pattern in the data. However, the Method of Seasonal Indices may be applied to determine if the forecasts provided by the Method of Least Squares can be improved. A Least Squares forecast can be made for each aggregate (e.g. year) time period. Monthly indices would then be applied to the yearly forecast in order to forecast monthly demand (which may have a cyclical or seasonal pattern) to determine if the monthly forecasts can be improved (i.e. smaller mean squared error).

Example 10: (Choosing Forecasting Methods). Consider the notional time series plot of monthly demand data in Figure 22. Determine an appropriate forecasting method(s).

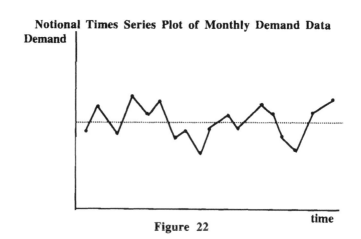

Notional Times Series Plot of Monthly Demand Data

Figure 22

258

Answer: The monthly data does not seem to show a trend pattern. The sample mean appears to be the best forecast for future monthly demand. When the Method of Least Squares is applied, the F-statistic can be used to verify whether the null hypothesis of no trend in the data (the sample mean provides the best forecast of future monthly demand). The data does seems to show a cyclical or seasonal pattern around a constant value. The Method of Seasonal Indices can be applied to the sample mean to determine if the monthly forecasts provided by the sample mean value can be improved (i. e. smaller mean squared error).

Example 11: (Choosing Forecasting Methods). Consider the notional time series plot of monthly demand data in Figure 23. Determine an appropriate forecasting method(s).

Notional Times Series Plot of Monthly Demand Data

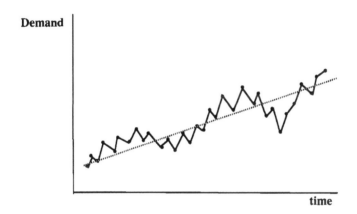

Figure 23

Answer: The monthly demand data seems to show a positive trend pattern over the long term but also a seasonal or cyclical pattern over shorter time periods. Regression using the Method of Least Squares can be used to determine how well the monthly data fits a linear trend. A Least Squares linear forecast can be made for each aggregate (e.g. year) time period. Monthly indices would then be applied to the yearly forecast in order to forecast monthly demand (which may have a seasonal or cyclical pattern) to determine if the monthly forecasts can be improved (i.e. smaller mean squared error).

Example 12: (Choosing Forecasting Methods). Consider the notional time series plot of monthly demand data in Figure 24. Determine an appropriate forecasting method(s).

Notional Times Series Plot of Monthly Demand Data

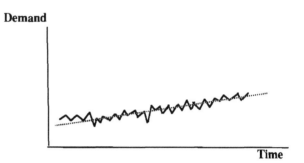

Figure 24

Answer: The monthly demand data seems to show a slow-varying trend increasing over time. The Method of Exponential Smoothing can be used to forecast future monthly demand based on the error in the previous month between the actual demand and the forecasted demand. There may also be a cyclical or seasonal pattern in the data. The Method of Seasonal Indices can be used to augment the Method of Exponential Smoothing to attempt to improve the monthly forecasts (i.e. smaller mean squared error).

Exercise 7

Forecasting

Problem #1 (Method of Moving Averages)

The Ann Guish Memorial Hospital has observed the following monthly demand for floral bouquets at its gift shop over the past year:

Monthly Demand for Floral Bouquets

Month	Past Demand	3 Month Simple Moving Average	5 Month Simple Moving Average
1 (Jan)	2000	---	---
2 (Feb)	1350	---	---
3 (Mar)	1950	---	---
4 (Apr)	1975		---
5 (May)	3100		---
6 (Jun)	1750		
7 (Jul)	1550		
8 (Aug)	1300		
9 (Sept)	2200		
10 (Oct)	2770		
11 (Nov)	2350		
12 (Dec)	2550		

1a) Use the Method of Moving Averages to provide the three month and five month simple moving average forecasts up to month 12.

1b) Plot the past demand, the three month simple moving average forecasts, and the five month simple moving average forecasts versus time together on a single graph.

1c) Which of the simple moving average forecasts provides the better stability? Which provides the better responsiveness? Explain.

Problem #2 (Method of Exponential Smoothing)

The Calamity Health Maintenance Organization has observed over the past year the following # Medicaid patients each month:

Month	# Patients Observed	α = 0.1	α = 0.5	α = 0.9
1 (Jan)	2000	2000	2000	2000
2 (Feb)	1350			
3 (Mar)	1950			
4 (Apr)	1975			
5 (May)	3100			
6 (Jun)	1750			
7 (Jul)	1550			
8 (Aug)	1300			
9 (Sept)	2200			
10 (Oct)	2770			
11 (Nov)	2350			
12 (Dec)	2550			

2a) Use the Method of Exponential Smoothing with α = 0.1, 0.5 and 0.9 to provide forecasts up to month 12. Since there is no forecast available for month 1, the actual (observed) # patients in month 1 is used as the forecast for month 1.

2b) Plot the observed (actual) demand and the Method of Exponential Smoothing forecasts for α = 0.1 and 0.9 versus time together on a single graph.

2c) Rank the Method of Exponential smoothing forecasts (α = 0.1, 0.5 and 0.9) for providing stability. Rank the Method of Exponential Smoothing forecasts (α = 0.1, 0.5 and 0.9) for providing responsiveness. Explain.

Problem #3 (Method of Seasonal Indices)

Amber Light, the administrator at the Payne Medical Center, must forecast future demand (# visits per quarter) for the Home Care Program based on quarterly demand for the past five years (given in the table below). The data is hypothesized to have a linear trend from year to year and a seasonal pattern over the quarters within the year.

Visits Y vs. Quarter (Q) Over Five Years

Quarter Q Year t	1 (Qtr 1)	2 (Qtr 2)	3 (Qtr 3)	4 (Qtr 4)	Year Total
1 (1989)	20	150	40	610	820
2 (1990)	630	710	990	1380	3710
3 (1991)	1250	1250	1640	2160	6300
4 (1992)	1860	1800	2280	2940	8880
5 (1993)	2480	2540	2930	3720	11670

a) Find the four quarterly (seasonal) indices. Note: These four indices should add to 4 (thus providing a mean of 1).

b) The Method of Least Squares has been applied to the above yearly data (total # visits per year in the last column above vs. year t) The Least Squares line forecasts a total of 15000 visits during the year 1995. Use the Method of Seasonal Indices to forecast the # visits in the second and fourth quarters of 1995.

262

Problem #4 (Times Series vs. Forecasting Methods)

Consider the following notional time series plots of monthly demand data and determine an appropriate forecasting method(s):

a)

Notional Times Series Plot of Monthly Demand Data

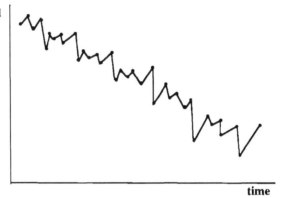

b)

Notional Times Series Plot of Monthly Demand Data

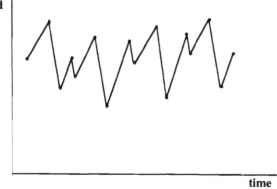

c)

Notional Times Series Plot of Monthly Demand Data

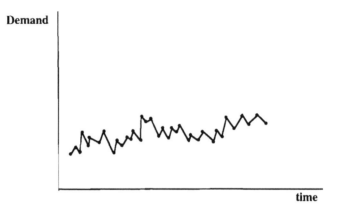

Chapter 7

Forecasting Additional Solved Problems

Problem #1 (Forecasing)

Consider the following two notional time series plots (demand data versus time). Suggest and justify appropriate demand forecasting method(s) for each time series.

a)

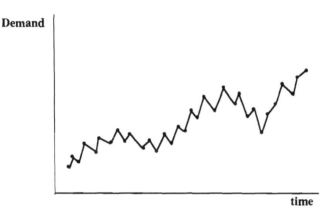

Answer: This time series data seems to show a linear trend pattern over the long term but also a cyclical pattern over shorter time periods. Regression using the Method of Least Squares can be used to determine how well the data fits a linear trend over the long term. Once the Least Squares linear forecast is made over the long term, the Method of Seasonal Indices can be applied to the forecasts in the shorter intervals (which may have a cyclical pattern) to determine if the linear trend forecasts can be improved.

b)

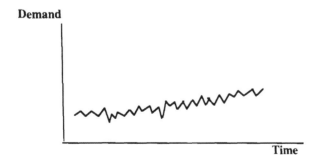

Answer: This time series data seems to show a stable, slow-varying pattern increasing over time. The Method of Exponential Smoothing can be used to forecast future demand based on the error in the previous time period between the actual demand and the forecasted demand. There may also be a cyclical pattern in the data. The Method of Seasonal Indices can be used to augment the Method of Exponential Smoothing to attempt to improve the forecasts.

"When I was younger, Statistics was the science of large numbers. Now, its seems to me rapidly to be becoming the science of no numbers at all."
Dr. Oswald George

8 | Multiple Regression

8.1 Introduction

The regression of a dependent variable on two or more independent variables is called <u>multiple regression</u>. The following general multiple regression equation is used:

$$Y = a + b_1 X_1 + b_2 X_2 + \cdots + b_k X_k$$

where $X_1, X_2, \cdots, X_k$ are the independent variables (factors) hypothesized to explain the change in the dependent variable Y (from its sample mean).

There are three principal uses of multiple regression:

1) Testing hypotheses: Does a change in a given independent variable have a significant effect on the dependent variable?

2) Determining coefficients of independent variables: How much does a unit change in a given independent variable (while all other independent variables remain constant) affect the dependent variable?

3) Forecasting: What is the expected value of the dependent variable given specific values for the independent variables?

Notes:

• It may not be possible to measure the values of independent variables with high accuracy. However, even without a highly accurate estimate of the actual value of a given independent variable, multiple regression can be used to determine whether a change in this independent variable produces a change in the dependent variable.

• A given independent variable is typically one of many affecting a dependent variable. If one could keep constant all the other independent variables and vary only this single independent variable, its effect on the dependent variable could be measured directly using

267

simple regression. Unfortunately, it is typically not possible to isolate the change in the dependent variable to only a single independent variable while all other independent variables remain constant. Rather, the independent variables naturally vary simultaneously.

• The utility of multiple regression in quantitative analysis is its ability to determine *statistically* the effect of varying one independent variable alone on the dependent variable (without actual data points where all other independent variables are held constant).

Multiple regression has the potential for greater accuracy than simple regression because more explanatory independent variables may be introduced into the regression model. Of course, introducing more independent variables increases the complexity of the regression model and the cost of collecting and storing the required past data. When developing a regression model for a real world system, one can take two extreme approaches:

  1) Develop a simple model which captures the overall system trend. This approach emphasizes simplicity at the expense of accuracy.

  2) Develop an intricate model which captures the details of the underlying system behavior. This approach emphasizes accuracy at the expense of simplicity.

The objective in the specification of a multiple regression model is to define a set of independent variables which provides an acceptable level of *accuracy* (defined by the proportion of the deviation explained by the regression) and *simplicity* (defined by the resources required to collect the data and to solve the proposed regression).

Due to the intricate nature of the real world, even complex models can not fully address every factor which affects the behavior of the system. An error term ε is added to the model to account for residual errors not accounted for in the model such as:

  1) Omission of independent variables which have small effects on the dependent variable.

  2) Incorrect specification of the model (e.g. using a linear model when the true functional form is exponential).

  3) Non-random (systematic) errors in *measuring* the data for the independent variables ($X_1, X_2, \cdots, X_k$) and the dependent variable Y used in the regression.

  4) Random effects (noise) in the model which can not be explained by the inclusion of independent variables.

When the unexplained error (deviation) in the model is accounted for, the multiple regression equation which denotes the linear trend becomes:

$$Y = a + b_1 X_1 + b_2 X_2 + \cdots + b_k X_k + \varepsilon.$$

The sum of the squares of the unexplained deviation (actual minus estimated data) $\varepsilon = (Y - Y^*)$ is again minimized in the Method of Least Squares to derive the coefficients a and b_i's. The sums of the squares of the deviations (explained, unexplained and total) are again used to determine the goodness of fit of the regression (i.e. the R^2-statistic) and to test the null hypothesis of no trend in the data (i.e. the F-statistic).

8.2 Method of Least Squares Applied to Multiple Regression

Example 1: (Multiple Regression). The Designer Pet Water Bottling Company sells meat, liver, and cheese flavored bottled water for dogs and cats belonging to discriminating pet owners who like to pamper their pets. Designer Pet has been experimenting with different monthly prices and monthly advertising budgets in order to determine their effects on monthly profit. Consider the following data collected by Designer Pet over the last 17 months on the profit (PROFIT) per month given in millions of dollars which is hypothesized to be a function of the average price (PRICE) per liter for the month given in dollars and the advertising budget (ADV) for the month given in millions of dollars.

Designer Pet Profit Data For Last 17 Months

Monthly Profit (PROFIT)	Price Per Liter (PRICE)	Advertising Budget (ADV)
$28.00 M	$0.39	$0.943 M
$28.15 M	$0.51	$1.036 M
$33.74 M	$0.67	$1.450 M
$40.42 M	$0.53	$1.575 M
$40.21M	$0.43	$1.678 M
$45.20 M	$0.61	$1.744 M
$43.17 M	$0.56	$1.977 M
$58.23 M	$0.79	$2.376 M
$59.66 M	$0.58	$3.161 M
$62.08 M	$0.61	$3.217 M
$51.36 M	$0.43	$3.509 M
$60.69 M	$0.56	$3.642 M
$62.90 M	$0.67	$3.658 M
$60.27 M	$0.55	$3.714 M
$65.67 M	$0.69	$4.130 M
$77.85 M	$0.76	$4.562 M
$87.76 M	$0.78	$4.738 M

For the hypothesized Designer Pet Water Bottling Company multiple regression model:

$$PROFIT = a + b_1 \cdot PRICE + b_2 \cdot ADV$$

determine and interpret the coefficients of the Least Squares best fit line.

Answer: A regression software package (using the Method of Least Squares which minimizes the sum of the squares of the errors between the actual and estimated monthly PROFIT) has been used to determine the following coefficient values:

$a\ (Constant) = 1.996$

$b_1\ (PRICE) = 35.565$

$b_2\ (ADV) = 10.858$

The Least Squares line for the Designer Pet multiple regression model is then given by:

$$PROFIT = 1.997 + 35.565 \cdot PRICE + 10.858 \cdot ADV$$

Notes:

• The magnitude of the coefficients estimated by the regression varies according to the *units* specified in the data. That is, if the advertising budget (ADV) data were given in billions of dollars (a decrease in the values of the advertising budget data by of a factor of 1000), the coefficient 10.858 for the independent variable ADV would and become 10858 (an increase in the coefficient for ADV by a factor of 1000).

• When an independent variable has a coefficient larger than that of another independent variable, this does not imply that the given independent variable has a larger effect on the dependent variable. A change in units can make any coefficient arbitrarily large or small.

• The correct interpretation of a coefficient is that it indicates how much the dependent variable will change with a unit change in that independent variable *alone* (i.e. a coefficient measures the marginal effect of that independent variable on the dependent variable):

  - An increase in PRICE (specified in dollars) of 1 dollar will increase expected monthly PROFIT (specified in Millions of dollars) by $35.565 M assuming all other independent variables remain constant. Hence, an increase in PRICE of 1 cent ($0.01) will increase expected monthly PROFIT by $0.36 M.

  - An increase in the advertising budget ADV (specified in $M) of $1 M will increase the expected monthly PROFIT (specified in Millions of dollars) by $10.858 M assuming all other independent variables remain constant.

• There is always danger in forecasting the effects of independent variables with values outside the range of the data collected for the regression. The price of the product can only be increased so much before profits start to decrease. Likewise, the advertising budget can only be increased so much before saturation takes effect and the increases in sales diminish to near zero (and profits decrease). But when forecasting close to the range of the data collected, an increase in price or advertising budget can be expected to increase profit.

The Correlation Matrix

A regression software package initially provides a correlation matrix (see Figure 1) for the regression. This matrix remains constant and need by computed only *once* prior to any regressions being performed (assuming no independent variables are added to the model).

Designer Pet Example Correlation Matrix

	PRICE	ADV	PROFIT
PRICE	1.0000	0.5316	0.6887
ADV	0.5316	1.0000	0.9481
PROFIT	0.6887	0.9481	1.0000

Figure 1

The *symmetric* correlation matrix indicates the statistical *linear* correlation among the independent variables and to the dependent variable. For example, there are positive correlations of 0.6887 and 0.9481 between the dependent variable PROFIT and the independent variables PRICE and ADV, respectively. The fact that change in each of these independent variables has a high correlation (in absolute value) to the change in the dependent variable provides indication that these variables are appropriate to consider in the specification of the Designer Pet monthly profit regression.

The correlation between the independent variables PRICE and ADV is reported as 0.5316. This is indicative of a moderate positive correlation between the two independent variables. A high value for the correlation coefficient (a rule of thumb is 0.70 and higher in absolute value) between two independent variables indicates that their is a strong linear relationship between the two independent variables. The regression may have difficulty determining the *individual* effects of these two independent variables on dependent variable PROFIT.

Notes:

• Multiple regression determines coefficients for each independent variable by taking advantage of the fact that independent variables rarely move in perfect unison with each other. Rather, as their name suggests, the independent variables tend to move in an independent fashion. It is this random variation in movement among the independent variables which allows multiple regression to determine the individual effect of changes in each independent variable alone on the dependent variable.

• Multicollinearity *for a data set* occurs when two or more independent variables have such a strong tendency to move together that the regression has difficulty in determining the effects each independent variable alone has on the change in the dependent variable.

• Consider an extreme case where two independent variables X1 and X2 move in perfect unison (called exact or *perfect multicollinearity*). That is, for every observation, each time X1 changes in a positive or negative manner, the change in X2 can be predicted exactly (see Figure 2 where X2 = X1 + 5). Then a regression would not be able to separate the individual effects of X1 and X2 on the change in the dependent variable Y.

• The more typical case is *near perfect multicollinearity* for two independent variables X1 and X2. When X1 changes in a given direction, X2 generally changes in a predictable direction (see Figure 3). If X1 and X2 are both included in a regression, it is difficult to isolate the effect on the dependent variable of changing one of these independent variables while the other correlated independent variable remains constant. A *rich* data set with large amounts of variation in the values of X1 and X2 is required to extract the individual effects.

• High correlation among independent variables causes uncertainty in the estimates of coefficients and difficulty in showing significance of coefficients of independent variables. High correlation between independent and dependent variables *is highly desirable* since it is hypothesized that the dependent variable is a function of the independent variables.

• There are two main indicators (symptoms) of possible multicollinearity for a data set:

1) When the value of the R^2-statistic is high and the F-statistic is significant but one or more of the independent variable coefficient t-statistics (described later in this section) is not significant. That is, while the overall regression is significant, one or more of the independent variables is not significant.

2) High correlation among two or more independent variables (≈ 0.70 or higher).

271

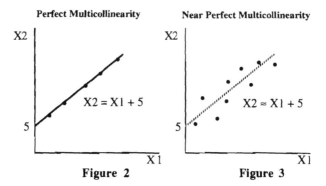

Figure 2   Figure 3

• The usual solution is to remove (or combine) variables so that only one of the highly correlated independent variables remains in the model. The likely independent variable to remove from the model is the one with the *lowest* (in absolute value) correlation coefficient with the dependent variable. This approach will cause the least decrease in the R^2-statistic (which is inevitable when an independent variable is dropped from a regression).

• Whenever an independent variable is removed, a *new regression* should be performed so that the coefficients of the remaining independent variables are adjusted. The coefficients for a regression will always change when an independent variable is dropped unless the independent variable is uncorrelated to all the other independent variables in the regression. Some of the power of the dropped independent variable to explain the change in the dependent variable will be attributed by the regression to other independent variables (thereby changing the remaining independent variable coefficient values).

• The situation where independent variables are highly correlated does not necessarily imply multicollinearity. As long as the independent variable coefficient t-statistics are significant, multicollinearity does not exist in the regression. In such a case, the data set is *rich* enough (contains enough variation in the values of the correlated independent variables) for the regression to determine the effect of each independent variable alone on the dependent variable to the desired level of significance.

• One technique for separating the effects of two highly correlated independent variables is to collect more data. A large sample is likely to include enough variation in the correlated independent variables for the regression to measure their individual effects. Of course, collection of more data may not be practical. Typically, one initially collects all the data possible subject to budget and time constraints. Adding more data may not be feasible.

• One should keep in mind the multicollinearity problem before beginning the expensive process of collecting data for regression analysis. If an independent variable is expected to have a high correlation with independent variables already in the model, this variable will have limited value in increasing the explanatory power of the regression.

The R^2-Statistic and the Adjusted R^2-Statistic

In simple regression, the R^2-statistic is used to measure the proportion of the total deviation explained by the single independent variable. In multiple regression, the R^2-statistic has the same interpretation except that it denotes the proportion of the total deviation explained by all the independent variables taken *together*. As expected, the R^2-statistic for multiple regression ranges from 0 to 1 (appropriate for a statistic which measures a proportion).

The regression software package reports an R^2-statistic of 0.9465 for the Designer Pet Water Bottling Company data in Example 1. This statistic is interpreted that 94.65% of the total deviation of the monthly PROFIT values from their sample mean is explained by the set of independent variables monthly PRICE and monthly ADV budget.

The addition of an independent variable (which is likely to have some correlation with the dependent variable) into the regression is likely to increase the R^2-statistic. The increase in the R^2-statistic occurs whether or not it makes sense logically to expect a change in the dependent variable due to a change in this independent variable. Hence, one method to increase the R^2-statistic to a satisfactory level (in the neighborhood of 0.90) is to include a large number of independent variables regardless of their logical explanatory value.

The adjusted R^2-statistic (sometimes called "R^2 adjusted for degrees of freedom") is designed to penalize this type of "fishing expedition" where independent variables are added to the regression without much consideration of their logical value. The adjusted R^2-statistic is a *better* measure of the goodness of fit of the regression. When additional independent variables are contemplated for addition to the regression, it recognizes the tradeoff between a higher R^2-statistic versus the loss of degrees of freedom in the data:

- The addition of an independent variable typically leads to a gain in the R^2-statistic since each independent variable will likely have some statistical power to explain the change in the dependent variable.

- The addition of an independent variable leads to a loss of one degree of freedom in the data since one extra coefficient must be estimated. A lower number of degrees of freedom makes it more difficult to show significance of individual regression coefficients (t-statistics) and the overall regression (F-statistic).

If there are N observations in the data and there are k independent variables being estimated, the adjusted R^2-statistic can be computed in terms of the R^2-statistic as follows:

$$Adjusted\ R^2 = 1 \ - \ \left[\frac{(N-1)}{(N-1)-k} \cdot (1-R^2) \right]$$

Notes: Properties of the adjusted R^2-statistic for both simple and multiple regression:

• The adjusted R^2-statistic is always *less* than the R^2-statistic unless $R^2 = 1$ (then both statistics equal 1).

• While the R^2-statistic is between 0 and 1, the adjusted R^2-statistic can be *negative*. A negative adjusted R^2-statistic is an indication of a regression whose explanatory power is largely the result of the number of independent variables included in the regression.

Example 2: (Adjusted R^2-statistic). Compute the value of the adjusted R^2-statistic for the Designer Per Bottling Company data given in Example 1.

Answer: The R^2-statistic was previously reported by a regression software package to be 0.9465. The number of observations $N = 17$ while the number of independent variables being estimated $k = 2$. The number of degrees of freedom in the data is $N - k - 1 = 14$.

The adjusted R^2-statistic is computed as follows:

$$Adjusted\ R^2 = 1 - \left[\frac{(N-1)}{(N-1)-k} \cdot (1 - R^2) \right]$$

$$= 1 - \left[\frac{(17-1)}{(17-1)-2} \cdot (1 - 0.9465) \right]$$

$$= 1 - \left[\frac{16}{14} \cdot (1 - 0.9465) \right]$$

$$= 0.9388\ .$$

A regression software package will report an adjusted R^2-statistic of 0.9388 (less than the R^2-statistic of 0.9465).

The Standard Error of The Estimate

The Standard Error of the Estimate is a goodness of fit statistic which provides a different measure than the R^2-statistic of how well the data fits the Least Squares best fit line. The Standard Error of the Estimate measures the average scatter of the dependent variable Y data around the Least Squares best fit line.

Note: The term "line" has been loosely used in this presentation to refer to the output of the Least Squares best fit to the data no matter how many independent variables in the regression. Mathematically, for one independent variable, the Least Squares best fit $Y^* = a + b \cdot X$ is indeed a line. For two independent variables, the Least Squares best fit $Y^* = a + b_1 \cdot X_1 + b_2 \cdot X_2$ is a plane. For three or more independent variables in the regression model, the Least Squares best fit $Y^* = a + b_1 \cdot X_1 + b_2 \cdot X_2 + \cdots + b_k \cdot X_k$ is a hyper plane. However, this presentation will refer to the Least Squares line even when it is understood that the multi-dimensional linear equation does not produce a two-dimensional line.

The Standard Error of the Estimate of the dependent variable is the square root of the sum of the squares of the unexplained deviations (the actual values of the dependent variable minus the values estimated by the regression) averaged on the basis of the number of degrees of freedom in the data $(N - k - 1)$:

$$Standard\ Error\ of\ the\ Estimate\ of\ the\ Dependent\ Variable = \left[\frac{\Sigma(Y - Y^*)^2}{N-k-1} \right]^{1/2}$$

A regression software package reports the Standard Error of the Estimate for the dependent variable PROFIT in Example 1 to be \$4.12 Million.

Notes:

• The properties of the Standard Error of the Estimate of the dependent variable Y include:

- A Standard Error of the Estimate of zero is indicative of a perfect fit of the data to the regression line (i.e. all the error terms $(Y - Y^*)$ equal zero).

- As the Standard Error of the Estimate increases, the scatter of the dependent variable data increases and the goodness of fit to the Least Squares line diminishes.

- A band two Standard Errors of the Estimate above and below (and parallel to) the best fit line will include roughly 95% of the data points (i.e. the Normal distribution is assumed for the scatter of the data points above and below the best fit line).

• The magnitude of the Standard Error of the Estimate is a function of the units in which the dependent variable is measured. If the dependent variable is measured in dollars instead of millions of dollars, the Standard Error of the Estimate will be a million times greater.

• A large Standard Error of the Estimate tells you do not understand all the change in the dependent variable (e.g. there is a large amount of unexplained deviation in the regression). But a large Standard Error of the Estimate is not cause for despair that you know nothing about what is occurring in the data. Even with a set of independent variables that is well specified and accurately estimated, random effects may cause unsettling deviations in the values of the dependent variable above and below the Least Squares best fit line.

• A method to determine how well the regression fits the data is to compare the Standard Error of the Estimate of the dependent variable Y versus the sample standard deviation (error) of the data for the dependent variable Y (see Section 1.10):

$$Sample\ Standard\ Deviation\ of\ the\ Dependent\ Variable = \left[\frac{\Sigma(Y - \bar{Y})^2}{N-1}\right]^{1/2}$$

• Without this regression, one would estimate the dependent variable Y by computing its sample mean value $\bar{Y}$. The error of this estimation scheme is the sample standard deviation of the dependent variable Y which a regression software package reports to be $16.67 M.

• With the regression, one is given the *additional information* that Y is related to the independent variables X_i's by the equation $Y = a + b_1X_1 + b_2X_2 + \cdots + b_kX_k$. Using this information, one is able to make a better estimate of Y than its sample mean $\bar{Y}$. In fact, the Standard Error of the Estimate for the dependent variable Y reported to be $4.12 Million as compared to the sample standard deviation for the dependent variable Y of $16.67 Million. The regression reduced the error in estimating Y by a factor of 4.

F-Statistic and t-Statistic

In multiple regression, there are two main statistics used to test significance:

1) The F-statistic is used to test the null hypothesis that the sample mean is the best estimate for the dependent variable (versus the alternative hypothesis of a linear trend in the data).

2) The t-statistic is used to test the null hypothesis that a coefficient is not statistically different from zero and the variable may be dropped from the regression (versus the alternative hypothesis that a coefficient is non-zero).

In simple regression with a single independent variable, the F-statistic is the ratio of the explained variance to the unexplained variance. In multiple regression, the F-statistic has a similar interpretation denoting the ratio of variance explained by the *all* the independent variables in the regression to the variance unexplained by the regression.

Example 3: (F-statistic). Determine the F-statistic for the Designer Pet Water Bottling Company data given in Example 1.

Answer: One method to compute the F-statistic is to determine the explained and unexplained deviations, to divide these deviations by the appropriate degrees of freedom to compute the explained and unexplained variances, and then to find the ratio of the explained variance to the unexplained variance (see Section 6.3).

An easier method computationally to compute the F-statistic makes use of the R^2-statistic which has been reported by a regression software package to be 0.9465. The F-statistic may then be computed as a function of the R^2-statistic, the number of observations $N = 17$, and the number of independent variables $k = 2$:

$$F = \frac{R^2/k}{(1-R^2)/(N-k-1)}$$

$$= \frac{0.9465/2}{(1-0.9465)/(17-2-1)}$$

$$= 123.84$$

Notes:

• The F-statistic can be compared to the 99th percentile of the F distribution with 2 (the number of independent variables) and 14 (N - # independent variables - 1) degrees of freedom which is found to be 6.51 (see $F_{2,14,.99}$ in Table 2a). One rejects with 99% confidence the null hypothesis of no trend ($b_1 = b_2 = 0$) where the data is characterized best by the sample mean and accepts the alternative hypothesis of a linear trend in the data.

• This formula for the computation of the F-statistic using the R^2-statistic is also valid for simple regression (k=1).

In simple regression, an Analysis of Variance (ANOVA) Table was used to concisely report the data used to compute the F-statistic. Regression software packages report an ANOVA Table for multiple regression as well. The ANOVA Table for the Designer Pet Water Bottling Company data in Example 1 is given in Figure 4.

Designer Pet Analysis of Variance (ANOVA) Table

Deviation Source	Deviation Sum of Squares	Degrees of Freedom	Variance
Explained	4207.01	2	2103.50
Unexplained	237.80	14	16.986
Total	4444.81	16	F-statistic = 123.84

Figure 4

Note: A regression software package typically reports a significance probability which is the probability that sample data from the dependent and independent variables in the regression that would fit a linear trend and provide this value for the F-statistic could have been produced purely by *chance*. The significance probability for the Designer Pet Water Bottling Company data in Example 1 is 0.0000001 (close to zero probability).

The second significance test applies to the individual coefficients of the regression (a, b_1, and b_2 in Example 1). The t-statistic is used to test the null hypothesis that an individual coefficient is not statistically different from zero (and the variable can be dropped from the regression) versus the alternative hypothesis that the coefficient is non-zero (and a change in the independent variable has a measurable effect on the dependent variable).

The Method of Least Squares provides an estimate of the mean of each coefficient (used in the Least Squares best fit line) and the standard deviation of the estimate of each coefficient. This standard deviation of the Least Squares estimate for each regression coefficient is called the standard error of the coefficient σ. The t-statistic for a coefficient is the ratio of the Least Squares estimate to its standard error:

$$t\text{–}statistic = \frac{Estimated\ Coefficient}{Standard\ Error\ of\ the\ Coefficient}$$

Notes:

• The t-statistic for a coefficient indicates whether the value zero is within a chosen (e.g. 95%) confidence interval around the Least Squares coefficient. If zero is not included in the appropriate confidence interval for the coefficient, the coefficient is statistically different from zero and is said to be significant.

• Mathematically, a significant t-statistic is one where:

$$0 \notin (Coefficient \pm t_{N-k-1,\ 95}\cdot\sigma)$$

• At higher significance levels, the confidence interval becomes wider and the coefficient must be larger in order that the confidence interval not contain the value zero.

A regression software package will report t-statistics for the Designer Pet Water Bottling Company Data in Example 1 (see Figure 5).

Designer Pet t-Statistics

Variable	Least Squares Estimated Coefficient	Standard Error of Coefficient	t-Statistic
Constant	1.99635	5.2046	0.38357
PRICE	35.5646	10.084	3.52675
ADV	10.8577	0.9768	11.1162

Figure 5

Notes:

• Neither the magnitude of the Least Squares estimated values of the means of the coefficients nor the Least Squares estimated values of the standard errors of the coefficients *individually* provide information whether a coefficient is significant (non-zero). The magnitude of these values both vary as a function of the units chosen.

• It is the *ratio* of the magnitude of the mean of a coefficient to the magnitude of the standard error of a coefficient (i.e. the *dimensionless* t-statistic) that provides the statistical information whether an individual Least Squares coefficient is significant (non-zero).

• The Standard Error of the Coefficient can be used to make statistical statements about the true value of a regression coefficient. For example, the Standard Error of the Coefficient will become smaller as the sample size increases. That is, as the sample size N increases, the true value of the coefficients a and the b_i's will be in smaller neighborhoods around the (mean) values of the coefficients estimated by the Method of Least Squares.

Uncertainty (given by the Standard Error of the Coefficient σ) in the value of any of the coefficients a and the b_i's derived by the Method of Least Squares has an effect on the intercept and slope of the Least Squares line. For a simple regression, the effect on the Least Squares line due to uncertainty in the coefficients a (y-intercept) and b (slope) is given in Figure 6 and Figure 7, respectively, for a Standard Error of the Coefficient σ.

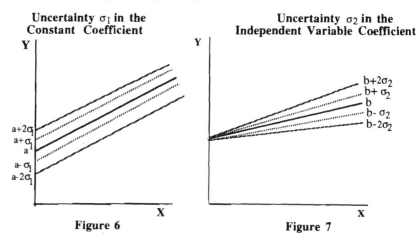

Uncertainty σ_1 in the Constant Coefficient

Uncertainty σ_2 in the Independent Variable Coefficient

Figure 6 **Figure 7**

Example 4: (t-Statistics). Use the t-statistics given in Figure 5 to determine whether the coefficients of the Least Squares regression of the Designer Pet Water Bottling Company data performed in Example 1 are significant (statistically non- zero) with 95% confidence.

Answer: The absolute value of the t-statistics will be compared vs. the 95th percentile of the Student t distribution with the number of degrees of freedom equal to N minus the number of independent variables minus 1 (i. e. 17 - 2 - 1) or 14. Since the 95th percentile of the Student t distribution with 14 degrees of freedom is 1.761 (see Table 6):

Constant: 1.99635/5.20456 = 0.038357 < 1.761

PRICE: 35.5646/10.084 = 3.52675 > 1.761

ADV: 10.8577/0.9768 = 11.1162 > 1.761

The hypothesis that the constant term is not statistically different from zero is not rejected but the hypotheses that the coefficients of the independent variables PRICE and ADV are not statistically different from zero are rejected. The fact that the constant term is not significant indicates that there is no more reason to specify the constant as 1.997 as there is to specify the constant to be zero. Consequently, it is acceptable statistically to drop the constant term from the *final* equation for the Designer Pet example which then becomes:

$$PROFIT = 35.5646 \cdot PRICE + 10.8577 \cdot ADV$$

Notes:

• It is recommended that the constant term always be retained unless there is a strong reason not to do so. The constant term captures effects of omitted variables and other random and non-random effects. Also, not including a constant term forces the regression to go through the origin which can result in biased estimates of the coefficients (slopes) for the independent variables.

• When an independent variable has a non-significant coefficient and is dropped from the regression, the correct procedure is to perform the regression (Method of Least Squares) again. This is required since the independent variables are not typically totally independent (some amount of correlation will exist). Dropping an independent variables will typically cause a change to the values of the coefficients of the variables remaining in the regression.

• If the absolute value of the t-statistic is less than 1, dropping this independent variable will increase the adjusted R^2-statistic for the regression. The decrease in the explained deviation is more than compensated by the increase in the number of degrees of freedom. Likewise, if the t-statistic is greater than 1, dropping this independent variable will decrease the adjusted R^2-statistic for the regression. Hence it is good practice when the absolute value of the t-statistic is at least 1 to retain the non-significant independent variable if there is a logical foundation for this independent variable to affect the dependent variable.

• Significance levels of 95% and 99% are generally used to accept the alternative hypothesis that a coefficient of an independent variable is significant (non-zero). This may seem like a high standard to impose. However, even with large samples it is possible to obtain quite different estimates of a coefficient depending on the sample data outcomes (governed by randomness). Thus it is wise to insure that there is high confidence that the coefficient is non-zero before rejecting the null hypothesis of a zero coefficient.

• When a t-statistic indicates non-significance for an independent variable, it is incorrect to say that this independent variable is not an important factor in explaining the change in the dependent variable. The correct interpretation is that *in the data set provided* the regression was not able to determine for the independent variable a significant effect on the dependent variable.

• Consider an extreme case where the data does not show *any* variability for an independent variable X (see Figure 8). Regression will not be able to determine the effect on the dependent variable Y due to a change in this independent variable X (since there is no change in X). A regression software package will eliminate this variable due to zero sample standard deviation for X. However, a richer set of sample data with variation in the values of X could indicate that the independent variable X is significant in explaining the change in the dependent variable Y.

No Variability in the Independent Variable X

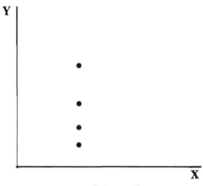

Figure 8

• A t-statistic significance of 95% for a coefficient does not mean there is only a 0.05 probability that the true coefficient value equals zero. In fact, the probability of obtaining the value zero from the continuous Student t distribution (or the Normal distribution applicable when the number of degrees of freedom is large) is zero. The correct interpretation is that if the true coefficient value *were* zero (the null hypothesis), there is less than a 0.05 probability of obtaining data with such a large t-statistic by chance.

• Recall that the F-statistic is also a measure of the significance of the regression. For multiple regression, the F-statistic provides a different significance measure than a t-statistic. The F-statistic provides a measure of the confidence that the data set follows a linear trend. A t-statistic provides a measure of the confidence that the coefficient of an individual independent variable is non-zero. That is, the F-statistic could indicate significance for a trend in the data (versus using the sample mean value as the best estimate of the dependent variable) despite the t-statistics indicating non-significance of some of the individual independent variables.

• The F-statistic tests whether all the overall regression (all the independent variables taken together) is significant. The t-statistic tests the significance of each independent variable individually. In a simple regression (where there is only one independent variable), the F-statistic and the t-statistic provide equivalent measures of significance. In a simple regression, the t-statistic is computed as the square root of the F-statistic.

Forecasting

As in simple regression, once the Least Squares line has been determined to fit the data to an acceptable level, the derived equation can be used to forecast the dependent variable. For the Designer Pet Water Bottling Company, values for the average monthly PRICE per liter and the monthly advertising budget ADV (the independent variables) may be used to forecast the dependent variable PROFIT.

Example 5: (Forecasting). Next month, the Designer Pet Water Bottling Company plans to set the average price of a liter of bottled water at $0.80 and plans a $5 Million dollar advertising campaign to promote the therapeutic value of the minerals and vitamins included in its flavored bottled water. Forecast the expected value of next month's PROFIT.

Answer: The values of $0.80 for PRICE and $5 Million for ADV are substituted into the Least Squares linear equation to forecast the expected value of next month's PROFIT:

$$PROFIT = 1.99635 + 35.5646 \cdot PRICE + 10.8577 \cdot ADV$$

$$= 1.99635 + 35.5646 \cdot (\$0.80) + 10.8577 \cdot (\$5\ M)$$

$$= \$84.74\ M$$

The fact that a price of $0.80 and an advertising budget of $5M are near the range of the data used to determine the Least Squares coefficients provides confidence in this forecast.

8.3 Assumptions of Multiple Regression Analysis

A basic assumption of a multiple regression analysis which determines the best fit line to a set of data is the *linearity* of the data analyzed. It has been demonstrated that many non-linear functions can be transformed into linear functions (see Section 6.6). When data that is non-linear (e.g. exponential) can not be transformed mathematically into data that is linear, regression analysis can not be applied successfully.

Another basic assumption of multiple regression is the *robustness* of the data. There must be at least k + 1 observations when estimating a model with k independent variables (e. g. there must be at least two observations to estimate a line which includes a constant and one independent variable). As the number of observations N exceeds the minimum number required, the number of degrees of freedom in the data (given by N - k - 1) increases and the data set for estimating coefficients of the regression tends to be richer (contain more variation in the values of the independent variables). Also, as the degrees of freedom increase, the corresponding F distribution and Student t distributions contract. The F-statistic and t-statistics indicate significance at higher levels of confidence.

Other key assumptions of multiple regression have to do with the error terms (also called the residuals) for the observations. It is important to examine these assumptions closely. They provide insight into the most appropriate functional form of the regression to specify and the most appropriate variables to include or exclude from the regression. The assumptions for the error terms ε_i of a well-specified multiple regression are as follows:

1) Independence: Error terms are independent of each other and are independent of (have no non-random effect on) the dependent variable Y.

2) Zero Mean: The error terms have an expected value of zero.

3) Constant Variance: The variance (i.e. the average squared size of the error terms) does not change in a non-random fashion over the entire range of the observations.

4) Normally Distributed: Error terms are generally small with few large errors. The distribution of the error terms is in the shape of a "bell curve".

The statistical consequences to the accuracy of the regression results become more serious as the violation of the above four error term assumptions becomes greater. These assumptions for the N error terms of the regression are written mathematically as follows:

$\varepsilon_i \sim N(0,\sigma^2)$ $\qquad\qquad$ $i = 1, 2, \cdots, N$

$\varepsilon_i, \varepsilon_j$ *independent* $\qquad\qquad$ *for all observations i and j*

Independence of the Error Terms

When the error terms are independent, a plot of the error terms $\varepsilon_i = (Y_i - Y_i^*)$ over the range of the N observations will show no systematic pattern either in the magnitude or the sign of the error terms (see Figure 9). A series of large or small error terms generally grouped together (see Figure 10) is indicative of a *positive* dependence among successive error terms. A series of positive and negative error terms that are in a general flip-flop pattern (see Figure 11) is indicative of *negative* dependence among successive error terms.

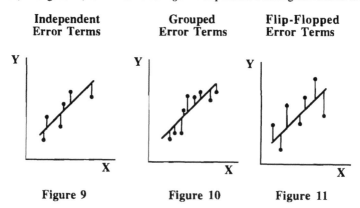

Independent Error Terms	Grouped Error Terms	Flip-Flopped Error Terms
Figure 9	Figure 10	Figure 11

Correlation among the error terms occurs when two or more observations are dependent. Dependence among any of the error terms in a regression model is called autocorrelation. Most theoretical treatments of autocorrelation examine the special case of autocorrelation called serial autocorrelation which is the dependence between *successive* error terms.

Serial autocorrelation is a common problem among successive error terms in time-series models. A random factor in one time period (e. g. an earthquake this year) typically will have an effect on the dependent variable over several successive periods (e. g. construction activity over the next few years). In fact, serial autocorrelation is so common in time-series models that one should always suspect serial autocorrelation in time-series data and test for it (e.g. using the Durbin-Watson statistic described below).

Lack of independence among the error terms will affect the Standard Error of the Coefficient σ for Least Squares coefficients (and thus the t-statistics) in an *unpredictable* manner. Regression coefficients with t-statistics which demonstrate significance may not actually be significant and those with non-significant t-statistics may actually be significant.

Significant autocorrelation will also render the calculation of the explained and unexplained variances invalid. The R^2-statistic and the F-statistic can be larger or smaller than their true value. A regression whose F-statistic indicates significance may not be significant and a regression whose F-statistic indicates non-significance may be significant.

Notes: Common examples of the effects of serial autocorrelation are as follows:

• **The serial autocorrelation is positive** (typically the case for time series models): Successive observations tend to be in groups above or below the regression line (see Figure 10). The unexplained variance for the regression and the standard errors of the individual coefficients are underestimated (smaller than their true value) resulting in a larger F-statistic and larger t-statistics, respectively, than the true values of these statistics.

• **The serial autocorrelation is negative**: Successive observations tend to flip-flop above and below the regression line (see Figure 11). The unexplained variance for the regression and the standard errors of the individual coefficients are overestimated (larger than their true value) resulting in a smaller F-statistic and smaller t-statistics, respectively, than the true values of these statistics.

Autocorrelation has three main sources:

1) An incorrect functional model chosen for the regression.

2) A significant independent variable omitted from the regression.

3) A random event occurring in one period propagates over several periods (causing serial autocorrelation).

The first two sources of autocorrelation are regression specification errors which can be addressed by correctly specifying the regression. These are discussed briefly. The third source which results in serial autocorrelation is discussed more fully.

The fact that the specification of an incorrect functional model for the regression can cause autocorrelation is evident. Consider where the data is linear (see Figure 12) but an exponential model is specified (see Figure 13). The resulting pattern for the error terms of the exponential model is no longer random. The exponential model at first typically underestimates the data (positive error terms) and later typically overestimates the data (negative error terms).

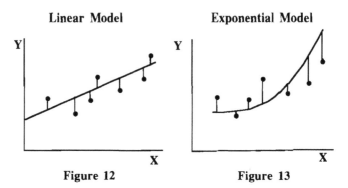

Linear Model   **Exponential Model**

Figure 12   Figure 13

Next, consider the case where X_1 and X_2 are the two significant independent variables explaining the variation in the dependent variable Y:

$$Y = a + b_1 \cdot X_1 + b_2 \cdot X_2 + \varepsilon$$

When the independent variable X_2 is omitted from the regression, its effects will be attributed to the error term:

$$Y = a + b_1 \cdot X_1 + (\varepsilon + b_2 \cdot X_2)$$

$$Y = a + b_1 \cdot X_1 + \mathcal{E}$$

The new error term $\mathcal{E}$ is then no longer has zero mean and will now be correlated to the trend in the dependent variable Y since X_2 is correlated to the dependent variable Y. That is, the error terms which generally vary along with the values of the omitted significant independent variable X_2 are now correlated to the dependent variable Y. Since the error terms are correlated to (generally follow the trend in) the dependent variable Y, the error terms are no longer random.

Testing for Serial Autocorrelation

When successive error terms are positively or negatively correlated, the error terms are in a grouped pattern (see Figure 10) and in a flip-flop pattern (see Figure 11), respectively, and the difference between the values of successive error terms is generally smaller and larger, respectively, than when there is no correlation between successive error terms (see Figure 9). The <u>Durbin-Watson</u> statistic d, defined as the ratio of sum of the squared *differences* between successive error terms to the sum of the squared error terms themselves, is used to test for *serial autocorrelation* in the error terms:

$$d = \frac{\sum_{i=2}^{N}(\varepsilon_i - \varepsilon_{i-1})^2}{\sum_{i=1}^{N}\varepsilon_i^2}$$

The Durbin-Watson statistic is *not* a valid test for general autocorrelaton among error terms but may only be used to test for serial autocorrelation in the data. The null hypothesis is that there is no serial autocorrelation in the data. The value of the Durbin-Watson statistic is compared to the 95% Durbin-Watson significance values (see Table 11) to determine whether the data set indicates no serial autocorrelation, positive serial autocorrelation, negative serial autocorrelation, or whether the data is *inconclusive* (neither can the null hypothesis of no serial autocorrelation be rejected nor can the alternative hypothesis of serial autocorrelation be rejected).

The Durbin-Watson 95% significance table (see Table 11) provides two significance values d_L and d_U as a function of the number of observations N and the number of independent variables k. The two values supplied by the Durbin-Watson significance table are used to specify 5 confidence regions for the Durbin-Watson distribution. The Durbin Watson distribution (see Figure 14) is symmetric around the value 2 (its mean value) and ranges from 0 to 4.

For the Designer Pet Water Bottling Company data in Example 1, there are N=17 observations and k=2 independent variables. From Table 11, the 95% Durbin-Watson values d_L and d_U supplied are 1.02 and 1.54. This leads to the 5 confidence regions symmetric around the mean value 2 indicated (see Figure 14).

Durbin-Watson Regions For Designer Pet Example

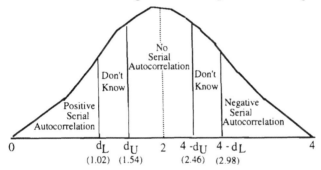

Figure 14

The 5 Durbin-Wastson confidence regions for the Designer Pet Water Bottling Company data from Example 1 are as follows:

1) Less than d_L (< 1.02) $\Rightarrow$ Positive autocorrelation
2) Between d_L and d_U (1.02, 1.54) $\Rightarrow$ Test not conclusive ("Don't Know")
3) Between d_U and 4 - d_U (1.54, 2.46) $\Rightarrow$ No autocorrelation
4) Between 4 - d_U and 4 - d_L (2.46, 2.98) $\Rightarrow$ Test not conclusive ("Don't Know")
5) Greater than 4 - d_L (> 2.98) $\Rightarrow$ Negative autocorrelation

A regression software package reports a Durbin-Watson statistic of 1.44 which is in the inconclusive region (d_L, d_U) on the positive serial correlation side of the mean value of 2. Neither the null hypothesis of no serial autocorrelation in the data is rejected nor the alternative hypothesis of serial correlation (most likely, positive) is rejected.

Notes:

• The Durbin-Watson statistic d can be used to *estimate* the serial correlation coefficient ρ between successive error terms (residuals) using the following approximation:

$$d \approx 2(1 - \rho)$$

Since the Durbin-Watson statistic ranges from 0 to 4, the serial correlation coefficient ρ ranges between 1 (positive serial correlation) and -1 (negative serial autocorrelation), respectively. A Durbin-Watson statistic at the mean value of 2 results in a serial correlation coefficient ρ = 0 (which indicates no serial autocorrelation).

• The weakness of the Durbin-Watson statistic is its two inconclusive regions. In these regions, no hypothesis on serial autocorrelation in the data can be validated statistically.

• For time series data, given the likelihood of serial autocorrelation, it is common practice to reject the null hypothesis of no serial autocorrelation even when the Durbin-Watson statistic is in one of the two inconclusive regions.

• Values of the Durbin-Watson statistic between 0 and 2 demonstrates at least some positive serial autocorrelation among the error terms while values between 2 and 4 demonstrates at least some negative serial autocorrelation among the error terms.

• Negative serial autocorrelation (where the values of successive error terms switch signs often) is more commonly caused by a model specification error rather than a random event in one period propagating over multiple periods.

• A simple technique which may help correct for serial autocorrelation is to increase the length of the unit of time. A random event which extends to future periods when day, week, or even month time periods are specified for the regression is less likely to propagate into longer periods such as years.

• If none of the simple techniques (specification of the proper functional model, addition of significant variables omitted, or lengthening the time periods) mitigate the serial autocorrelation problem, the method of first differences is used to try to subtract out the correlation between successive error terms (i.e. $\mathcal{E}_i = (\varepsilon_i - \rho \cdot \varepsilon_{i-1})$). This technique to restore independence among error terms is beyond the scope of this presentation.

Zero Mean for the Error Terms

The error terms should have a zero sample mean over the N observations. If not, the change in the dependent variable Y is not only dependent on the independent variables (as expected) but also has a non-zero dependence on the error terms. The assumption of zero mean of the error terms may be violated when relevant independent variables have been left out of the regression model (and thus become part of the error term). The assumption of zero mean may also be violated if there are systematic (non-random) measurement errors in determining the values of the data for the regression.

When the error terms ε_i have zero mean and are independent of both each other and the dependent variable, the Least Squares best fit line has the following desirable properties:

- The estimates of the a's and b_i's will be correct on the average (the Least Squares estimates of the a's and b_i's are their true expected values) for all sample sizes N. That is, Least Squares estimation is <u>unbiased</u>.

- The Standard Errors of the Coefficient for the a's and b_i's decrease uniformly as the sample size N increases. That is, Least Squares estimation is <u>consistent</u>.

- The Standard Errors of the Coefficient for the a's and b_i's have their smallest values for a given sample size N. That is, Least Squares estimation is <u>efficient</u>.

Error terms which have zero sample mean should be randomly scattered above and below the regression line when plotted over the N observations. When the error terms have a sample mean $\Sigma(Y_i - Y_i^*) / N$ which is not quite small, *bias* (where the Least Squares estimates of the coefficients are not their true expected values) should be suspected.

The Method of Least Squares deals with bias (defined mathematically as the difference between the estimated and true value of the coefficients) in the regression by adding the equal but opposite amount of the bias to the *constant term* in the regression. This is the simplest way of forcing the error terms back to a sample mean that is quite small. This approach has a number of implications for the constant term in a regression:

- One should not interpret the value of the constant term a as providing any physical meaning. It is often thought of as a "garbage can" term which is adjusted to provide an unbiased estimation of the b_i's. However, the b_i's do have the physical meaning of providing the marginal change in the dependent variable per unit change in that independent variable alone.

- One should always include a constant term a in specifying a regression. When a constant term is omitted, the regression is forced to go through the origin.

-- If the error terms have a non-zero mean (e. g. a significant variable has been omitted from the regression) and there is a constant term, then the estimate of the constant term rather than the coefficients for the independent variables will be biased. This is not consequential in a statistical sense.

-- If the error terms have a non-zero mean and there is no constant term, then the estimate of the independent variable coefficient b_i's will be biased. This is consequential since the b_i's define the marginal change in the dependent variable.

- One may drop the constant term a in the *final* regression model if the constant has been demonstrated to be non-significant. However, it is recommended that a constant term (whether it is significant or not) be carried in all intermediate regression steps.

Constant Variance For Error Terms Over Observations

When the error terms have the same likelihood of being large or small over the entire range of the observations, the assumption of constant variance for the error terms (called <u>homoskedasticity</u>) holds. When some observations have a higher likelihood to have large error terms than other observations, the data is said to exhibit <u>heteroskedasticity</u>. Figure 15 and Figure 16 provide notional examples of homoskedasticity and heteroskedasticity, respectively.

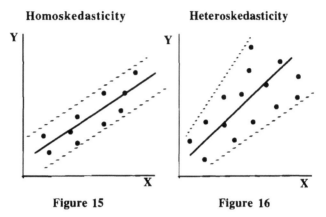

Figure 15   **Figure 16**

Consider a causal regression on family food expenditures versus family income. Low income families must spend a minimum amount in order to survive but do not generally have the option of spending a large amount of money splurging on expensive items. High income families have the option of spending in a wide range (from the minimum amount to buy the basics to a large amount splurging on expensive items). Hence the variation in expenditures on food is expected to increase significantly as the family income level rises. One would suspect heteroskedasticity in family food expenditures over the set of family income observations.

Unlike serial autocorrelation, heteroskedasticity occurs less often in times series models (which would typically focus on the food expenditures of a single family over time) than in causal models (which would typically examine the expenditures of many families perhaps having a wide range of income).

One check for heteroskedasticity is to plot the data (see Figures 15 and 16) and visually check for constant variation over the set of observations. Another check for heteroskedasticity is performed by tabulating the error terms (actual minus predicted values of the dependent variable). If the absolute value of the magnitudes (in terms of percentage of the dependent variable) of the error terms remains generally constant over the range of the observations, then one should not suspect heteroskedasticity. If the absolute values of the magnitudes of the error terms vary over the range of the observations, one should suspect heteroskedasticity. There are formal statistical tests for heteroskedasticity but these are beyond the scope of this presentation.

Notes:

• Heteroskedasticity (like autocorrelation) may be caused by specification of the incorrect functional form for the regression (e.g. the error terms become larger over time when a linear model is specified if the true model is exponential). Heteroskedasticity (like autocorrelation) may also be caused by the omission of a significant independent variable from the regression (e.g. the error terms become correlated to the dependent variable Y and increase or decrease as the value of the dependent variable Y increases).

• If some observations are *known* to have larger variances than others, then a regression which estimates the best fit line should compensate for this by giving these observations less weight relative to observations with smaller variances. However, regression using the Method of Least Squares weights all observations equally (the squared deviations are summed without weights attached to the observations) and thus the outliers in the data (with large squared deviations) have a disproportionate effect on the derived best fit line.

• A technique similar to the Method of Least Squares where weights (inversely proportional to individual error term variances) are attached to observations is called Generalized Least Squares. A treatment of the Method of Generalized Least Squares is beyond the scope of this presentation.

Normal Distribution For Error Terms

Since the error terms for a well-specified regression are typically the result of a large number of small sized random events, the error terms should follow a Normal or "bell-shaped" distribution (this occurs due to the Central Limit Theorem). It is difficult to construct examples where the assumption of a Normal distribution would be questioned without questioning one or more of the other assumptions for the errors terms. Hence, while there are statistical tests for departure from the Normal distribution, such tests are not frequently applied to the error terms.

One of the main advantages of the error terms following the Normal distribution is that statistical statements (e.g. using the Standard Error of the Coefficient σ) can be made on the deviation of each Least Squares estimated coefficient from its true value as a function of the number of observations N.

Notes:

• Small violations of the above four error term assumptions will have small effects on the validity of the Method of Least Squares. In practice, the Method of Least Squares is abandoned only when large violations of the error term assumptions occur. This is the case because the attributes of Method of Least Squares coefficient estimates (i. e. unbiased, consistent, and efficient) are so highly desirable.

• Even when the above error term assumptions are not strictly in effect, coefficient estimates using the Method of Least Squares are often more *robust* (i.e. small changes in the data do not lead to large changes in the statistics of the regression) than those of the complex alternative estimation techniques required to handle violations of these error term assumptions.

8.4 Specification of the Multiple Regression Model

It is critical to specify correctly which variable among the set of candidate variables included in the regression model is the dependent variable.

Example 6: (Dependent Variable Specification). Consider the following data for four randomly chosen dairy farms in McDonald County where Y represents the number of cows on the farm and X represents the number of acres for the farm.

Y (Cows)	X (Acres)
5	9
10	16
12	12
15	20

Determine the Least Squares best fit lines when Y and then X is specified as the dependent variable.

Answer: A regression software package is used to determine the Least Squares best fit lines using:

(1) Y (Cows) as the dependent variable and X (Acres) as the independent variable.

(2) X (Acres) as the dependent variable and Y (Cows) as the independent variable.

The results of these two simple regressions are reported as follows:

(1) $Y = 0.03 + 0.73 \cdot X$

(2) $X = 4.25 + 0.95 \cdot Y$

The regression (2) must be solved for Y as a function of X to compare directly with regression (1). All terms in regression (2) are divided by 0.95 and are rearranged to derive the equivalent regression (2'). The two regressions (1) and (2') may now be compared directly as follows:

(1) $Y = 0.03 + 0.73 \cdot X$

(2') $Y = -4.46 + 1.05 \cdot X$

Depending on which variable is designated as the dependent variable, the best fit line derived by the Method of Least Squares is different. This situation occurs since different error terms (see Figures 17 and 18) are minimized by regressions (1) and (2).

Figure 17 displays the deviations $(Y_i - Y_i^*)$ minimized by the Least Squares best fit line $Y = 0.03 + 0.73 \cdot X$ when Y is the dependent variable. Figure 18 displays the deviations $(X_i - X_i^*)$ minimized by the Least Squares best fit line $Y = -4.46 + 1.05 \cdot X$ when X is the dependent variable.

Y as Dependent Variable   **X as Dependent Variable**

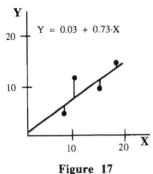

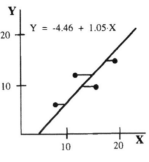

Figure 17   **Figure 18**

<u>Note:</u> Forecasts of the value of Y based on the value of X will be different depending on which variable is chosen as the dependent variable. For example, if $X = 10$, then the forecast Y^* using the regressions (1) and (2') are as follows:

(1) $Y^* = 0.03 + 0.73 \cdot (X = 10) = 7.33$

(2') $Y^* = -4.46 + 1.05 \cdot (X = 10) = 6.04$

Inclusion and Exclusion of Independent Variables

Since the desirable properties of the Least Squares regression (i.e. unbiased, consistent, efficient) occur most strongly when all the significant independent variables affecting the dependent variable are included in the regression model, it is important to define a strategy to ensure that all such independent variables are likely to be included in the model. The best place to start is to determine the set of independent variables which have some likelihood of being important in predicting the change in the dependent variable.

One should perform this search for independent variables affecting the dependent variable *logically* rather than by including a large number of variables in the regression, exercising a regression software package, and allowing the software package to choose those variables to remain in the regression. Such an approach can lead to misleading or incorrect results since correlation does not imply causation. The resulting regression may have a pleasing R^2-statistic yet may not be credible in explaining the changes in the dependent variable.

Remember that if two independent variables are highly correlated, multicollinearity may occur and the regression will then drop one of the two independent variables from the regression. The software package will always drop the independent variable with the lower correlation with the dependent variable (causing the smaller decrease in the R^2-statistic). But this independent variable may the one which logically explains the change in the dependent variable. Hence, independent variables which are highly correlated with the dependent variable *statistically* may be chosen at the expense of variables which have the capability to explain changes in the dependent variable logically.

<u>Notes</u>: The R^2-statistic should not be used as a score for the credibility of a regression:

• It is quite conceivable that even when all independent variables significantly affecting the dependent variable are included in the model, a large amount of the deviation in the data may still be unexplained by the regression. The appropriate response is to accept this situation rather than try to manufacture explanatory power for the regression statistically when it does not exist logically.

• A large amount of unexplained deviation in the final regression does not automatically imply that the regression is poorly specified; it may simply mean that the change in this dependent variable is largely explained by random effects.

• The fact that the error terms are generally independent with zero mean (as required for a well specified model) does not mean that the effect of randomness is small. Remember that a random variable with zero mean can have large standard deviation about its mean.

Use of a Software Package to Specify the Final Regression

Computers have the potential to substitute for logical reasoning in specifying the regression model. Regression software packages typically offer two options to assist the user to specify the final regression: <u>forward step-wise regression</u> and <u>backward step-wise regression</u>.

The forward step-wise regression option generally proceeds in the following manner:

- The software determines the independent variable (among the candidates specified by the user) with the highest correlation to the dependent variable. This independent variable is designated for inclusion in the final regression model. A regression is performed using this independent variable. The error terms (the difference between the actual and estimated values of the dependent variable) are calculated.

- At each succeeding step, the software searches for the independent variable (among those remaining) which is *most* highly correlated with the error terms of the current regression and includes this variable in the regression model. The inclusion of this independent variable has the effect of providing the greatest step-wise increase in the explained deviation (and thus the greatest increase in the R^2-statistic) for the regression.

- The software proceeds in this manner adding the independent variable most highly correlated with the error terms until the increase in the explained deviation by adding the next independent variable to the subsequent regression falls below a minimum threshold set by the software (or possibly the user). Remember that each independent variable added to the regression decreases the degrees of freedom in the data.

The backwards step-wise regression option (sometimes called <u>maximal regression</u>) generally proceeds in the following manner:

- The software performs a regression using *all* the candidate independent variables specified by the user. Highly collinear independent variables (based on the correlation matrix) are identified as candidates for leaving the regression.

- At each succeeding step, the software searches for the independent variable (among those identified as being highly collinear) which is the *least* highly correlated with the error terms of the current regression and excludes this variable from the next regression performed. The exclusion of this variable has the effect of providing the least step-wise decrease in the explained deviation (and thus the smallest decrease in the R^2-statistic) for the regression.

- The software proceeds in this manner excluding the independent variable least correlated to the error terms until the decrease in the explained deviation by excluding the next independent variable from the subsequent regression performed rises above a maximum threshold set by the software (or possibly the user).

Notes: The practical problems with having computers complete a stepwise regression (either forward or backward) on its own without being guided each step by the user are two-fold:

• The software can not reason logically which among the candidate independent variables identified by the user should be included or excluded from the regression. In a forward step-wise regression, once an independent variable is included, all highly correlated independent variables are not likely to be included. Likewise, in a backward step-wise regression, only independent variables which are highly collinear are candidates for exclusion from the regression.

• The forward step-wise regression technique has a tendency to omit logically significant independent variables from the regression. An independent variable which belongs in the final model but which is correlated to an independent variable chosen earlier by the computer for inclusion (due to its higher correlation with the error terms) will tend not to be included in the final model. Thus the set of independent variables included in the final regression model is influenced by the *order* in which independent variables are added to the regression.

• The backward step-wise regression has a tendency to include non-significant independent variables in the regression. When the regression reaches the step at which excluding another collinear independent variable causes the R^2-statistic to decrease by more than a threshold amount, the regression stops excluding independent variables. All non-significant independent variables will remain in the regression.

Omitting significant independent variables from the regression and including non-significant independent variables in the regression have different statistical consequences:

- Omitting a significant variable from the regression has serious statistical consequences. The exclusion of significant independent variables (and throwing over their effects of this variable to the error terms) results in a lack of independence among the error terms. This may lead to biased estimates (too large or too small) for the Standard Error of the Coefficients and the explained variance. If the bias in these estimates is large, the tests of significance for the regression (i.e. the F-statistic and t-statistic) become invalid.

- Inclusion of non-significant variables has less serious statistical consequences. The assumption of zero mean and independence for the error terms is not seriously violated. The estimates of the standard error of the coefficients and the explained variance are unbiased and the tests of significance are valid. A degree of freedom in the data is lost determining each additional coefficient. Each loss of a degree of freedom makes it more difficult to demonstrate significance (i.e. the F-statistic and t-statistics must be larger to demonstrate significance).

Notes:

• In general, it is better to start with an over-specified model and build down (backward step-wise regression) rather than starting with a null set of variables and build up (forward stepwise regression). Neither is recommended to be performed to completion by a regression software package *unguided by the user.*

• The most prudent approach to specifying a regression is to start with a basic logical model of the system and add or subtract independent variables based on *logical hypotheses* on those independent variables affecting the underlying behavior of the dependent variable.

• Regression results which are the result of a large amount of mechanical applications of a regression software package to achieve a high R^2-statistic should be considered suspect. One should seek a reasoned approach to determine what change in the dependent variable is logically explainable and what change must be attributed to random effects.

Multiple Regression Analysis Process

Example 7 (East of Eden Multiple GNP Regression). The GNP of the small nation of East of Eden has been measured over the last 38 years by its prestigious Congressional Budget Office (CBO). The CBO hypothesizes that the yearly GNP is based on 8 factors:

X1: Population
X2: Education Level
X3: Government Investment
X4: Weather
X5: Foreign Aid
X6: Consumer Confidence
X7: # of World Cup Soccer Games Won During Year
X8: # Border Skirmishes

Data has been collected for these 8 independent variables and also the dependent variable Y:GNP over the last 38 years. The correlation matrix, the Least Squares best fit line, the t-statistics, and other statistics for the East of Eden GNP regression are given below.

Correlation Matrix for East of Eden GNP Regression

Variable	X1	X2	X3	X4	X5	X6	X7	X8	Y
X1	1.00	-0.69	0.55	0.16	0.13	0.20	0.90	-0.20	0.74
X2	-0.69	1.00	0.03	0.01	-0.15	-0.12	0.05	-0.15	0.01
X3	0.55	0.03	1.00	0.44	-0.06	0.25	0.63	-0.18	0.29
X4	0.16	0.01	0.44	1.00	0.22	0.10	0.36	-0.13	0.03
X5	0.13	-0.15	-0.06	0.22	1.00	0.28	0.23	-0.63	0.41
X6	0.20	-0.12	0.25	0.10	0.28	1.00	0.13	-0.20	0.53
X7	0.90	-0.05	0.63	0.36	0.23	0.13	1.00	-0.02	0.67
X8	-0.20	-0.15	-0.18	-0.13	-0.63	-0.20	0.02	1.00	-0.18
Y	0.74	0.01	0.29	0.03	0.41	0.53	0.67	-0.18	1.00

Least Squares Best Fit Line

$$Y = 292.6 + 0.38 \cdot X1 + 0.51 \cdot X2 - 1.71 \cdot X3 - 1.03 \cdot X4$$
$$+ 0.15 \cdot X5 + 0.81 \cdot X6 + 0.39 \cdot X7 - 0.54 \cdot X8$$

t-Statistics for East of Eden GNP Regression

Variable	Least Squares Estimated Coefficient	Standard Error of the Coefficient	t-statistic
Constant	292.6	61.24	4.7779
X1	0.381	0.153	2.4902
X2	0.506	0.314	1.6115
X3	-1.713	0.800	-2.1413
X4	-1.026	0.627	-1.6364
X5	0.152	0.075	2.0267
X6	0.805	0.178	4.5225
X7	0.386	0.270	1.4286
X8	-0.054	0.038	-1.4211

R^2-statistic = 0.912
Adjusted R^2-statistic = 0.888
Standard Error of the Estimate = 24.3
F-statistic = 1144
Durbin-Watson statistic = 2.39

Determine the next steps in the multiple regression analysis based on the above results of the CBO GNP regression and further results as they become available later in this Example.

Answer: The table of t-statistics indicates that independent variables X2, X4, X7, and X8 are below the 95% significance level with N - k -1 = 38 -8 - 1 = 29 degrees of freedom in the data since the absolute value of their t-statistics is below 1.699 (see Table 6). However, all these independent variables have t-statistics above 1 (i.e. they provide an increase in the adjusted R^2-statistic when included) and should be given consideration for inclusion in the final regression.

The R^2-statistic and the F-statistic for the overall regression are quite large. A large F-statistic along with small t-statistics for some independent variables implies either: (1) multicollinearity (high linear correlation) exists among the set of independent variables; or (2) some or all the independent variables X2, X4, X7, and X8 have only small linear correlation to the dependent variable Y:GNP.

The correlation matrix is examined for multicollinearity among the independent variables. The correlation matrix indicates that the independent variables X1:Population and X7:# of World Cup Soccer Games Won During Year are highly correlated at 0.90 and hence one of them is a candidate for exclusion from the regression model (if appropriate logically).

For an indication of whether it is more appropriate *statistically* to exclude X1 or X7, the correlation between each of these independent variables and the dependent variable Y:GNP is examined. Since X1 and X7 have correlation coefficient with the dependent variable Y of 0.74 and 0.67, respectively, it seems X7 is the collinear independent variable to exclude from a statistical point of view. Logically, it also seems that the growth in the GNP will be affected more by the growth in the population than by the # World Cup games won.

Note: Remember that if X1 were not included in the regression model, X7 might have a significant t-statistic (i.e. some of the deviation in the dependent variable attributed to X1 might be attributed by the regression to X7). Thus X7 which does not have a significant t-statistic is not automatically the independent variable between X1 and X7 to exclude. The additional criteria of examining the correlation with the dependent variable Y is used to help decide which of the collinear independent variables to exclude.

Of all the independent variables with non-significant t-statistics, X2 and X4 have very small correlation coefficients with the dependent variable Y (0.01 and 0.03, respectively). These variables provide very little explanatory power (explained deviation) to the regression and may be excluded. The correlation coefficient -0.18 for the independent variable X8 is low but not low enough to conclude that X8 should be dropped from the regression. It is decided that X8 will remain in the regression for now.

The actions in the determination of the final regression model taken up to now have been:

- Exclude independent variable X7 (highly correlated with X1 and has a smaller correlation coefficient than X1 with the dependent variable Y)

- Exclude independent variables X2 and X4 (each has a non-significant t-statistic and a small correlation coefficient with the dependent variable Y)

A new multiple regression for the East of Eden GNP is performed using the remaining independent variables X1, X3, X5, X6, and X8 and a constant. There is no reason to recompute the correlation matrix. The correlation matrix will not change unless an independent variable is added to regression. However, a new regression should always be performed when correlated variables are excluded from the regression. The resulting intermediate Least Squares best fit line and the t-statistics are given in Figure 19.

Intermediate Least Squares Best Fit Line

$$Y = 364.6 + 0.42 \cdot X1 - 1.62 \cdot X3 + 0.15 \cdot X5 + 0.78 \cdot X6 - 0.62 \cdot X8$$

Intermediate t-Statistics for East of Eden GNP Regression

Variable	Least Squares Estimated Coefficient	Standard Error of the Coefficient	t-statistic
Constant	351.7	41.96	8.3818
X1	0.583	0.074	7.8784
X3	-1.709	0.696	-2.4555
X5	0.150	0.070	2.1429
X6	0.720	0.176	4.0909
X8	-0.576	0.389	-1.4807

Figure 19

R^2-statistic = 0.795
Adjusted R^2-statistic = 0.738
Standard Error of the Estimate = 25.53
F-statistic = 621.2
Durbin-Watson Statistic = 2.37

The table of t-statistics indicates that independent variable X8 is below the 95% significance level with 38 -1 - 5 = 32 degrees of freedom in the data since the absolute value of its t-statistic is below ≈ 1.697 (see Table 6).

Notes:

• It now is appropriate to exclude the independent variable X8. There was always a possibility that another independent variable was masking the effect of X8 on the dependent variable Y. But the new regression excluding the independent variables X2, X4, and X7 indicates that X8 is still non-significant and may be excluded.

• The values of the R^2-statistic and the F-statistic have decreased. This is expected since three independent variables have been excluded from the regression and the explained deviation has decreased. The standard error of the estimate (the unexplained deviation normalized by the degrees of freedom in the data) has increased slightly indicating a slightly worse scatter of the data points around the derived line.

A final regression for the East of Eden GNP is performed using the remaining independent variables X1, X3, X5, X6, and a constant. The final Least Squares best fit line and t-statistics for the East of Eden GNP are given in Figure 20.

Final Least Squares Best Fit Line

$$Y = 327.7 + 0.57 \cdot X1 - 1.52 \cdot X3 + 0.16 \cdot X5 + 0.76 \cdot X6$$

Final t-Statistics for East of Eden GNP Regression

Variable	Least Squares Estimated Coefficient	Standard Error of the Coefficient	t-statistic
Constant	327.6	39.37	8.3211
X1	0.570	0.074	7.7027
X3	-1.518	0.700	-2.1686
X5	0.155	0.070	2.2143
X6	0.757	0.178	4.2528

Figure 20

R^2-statistic = 0.781
Adjusted R^2-statistic = 0.672
Standard Error of the Estimate = 26.0
F-statistic = 471.2
Durbin- Watson Statistic = 2.25

The table of t-statistics indicates that all the independent variables are above the 95% significance level with 38 -1 - 4 = 33 degrees of freedom which is again found to be ≈ 1.697 (see Table 6).

Notes:

• For the final East of Eden GNP multiple regression, the independent variables X1:Population, X3:Government Investment, X5:Foreign Aid, and X6:Consumer Confidence explain 78.1% of the deviation in the dependent variable Y:GNP from its sample mean.

• The F-statistic of 471.2 (vs. $F_{4,33,0.99} \approx 13.8$) indicates high confidence in rejecting the null hypothesis that the change in the data represents only random fluctuations around the sample mean of the dependent variable Y:GNP. The alternative hypothesis of a linear trend in the data is accepted.

• Since there are over 30 degrees of freedom in the data, one can use the Normal distribution rather than the Student-t distribution) as an *approximation* for the 95% significance level. Remember (see Section 1.13) that the Student-t distribution converges to the Normal distribution when the number of degrees of freedom becomes large. Using the Normal distribution (see Table 1b) would result in a 95% significance level of 1.645 (as opposed to ≈ 1.697 indicated in Table 6 for the Student-t distribution).

• The values supplied by the Durbin-Watson significance table (see Table 11) for N=38 observations and k=4 independent variables estimated are $d_L = 1.26$ and $d_U = 1.72$. The Durbin-Watson statistic of 2.25 is in the region $(d_U, 4 - d_U) = (1.72, 2.28)$ where no serial autocorrelation among the error terms is indicated.

• A plot by a regression software package of the 38 error terms (residuals) would indicate that the error terms have no systematic pattern (i.e. grouped or flip-flopped) over the range of observations. This is indicative of no serial autocorrelation among the error terms.

• A table of the 38 error terms $(Y_i - Y_i^{\bullet})$ would indicate that the size of the absolute magnitude of the error terms (in percentage of the dependent variable) is generally constant over the range of observations. This is indicative of homoskedasticity among the error terms.

8.5 Conclusions

Recall that there are three principal uses of multiple regression (see Section 8.1):

1) Testing hypotheses: Does a change in a given independent variable have a significant effect on the dependent variable?

2) Determining coefficients of independent variables: How much does a unit change in a given independent variable (while all other independent variables remain constant) affect the dependent variable?

3) Forecasting: What is the expected value of the dependent variable given specific values for the independent variables?

Testing hypotheses (e.g. a coefficient is non-zero) depends on measuring the significance (i.e. t-statistic) of the effect of a change in the independent variable on the change in the dependent variable. Regression applications that depend on measuring the effect of a change in the independent variable (as opposed to measuring the expected value of its coefficient) are the most likely to be successful.

Determination of the marginal effect of independent variables on the dependent variable requires that the coefficients of each independent variable be accurately estimated. Regression applications that depend on the accurate estimation of the coefficients of independent variables have a good likelihood of being successful in a well-specified model.

Forecasting the dependent variable requires that the model accurately estimates values of the dependent variable inside and outside the range of the data. Regression applications that depend on the accurate forecast of the dependent variable outside of the range of the data collected become increasingly less likely to be successful as the forecast deviates from the sample mean value of the independent variables. The accuracy of the forecast also depends strongly on the correct regression model (e.g. linear, log-linear, geometric, etc.) being chosen. The exact model to use for a regression is often subjective and an unverifiable assumption. Also, for causal models, the independent variables may have to be estimated prior to forecasting the dependent variable.

The complexity vs. accuracy tradeoff for a regression model posed at the start of this chapter is without a well defined answer. In the final analysis, it is difficult to decide the proper balance between the inclusion and exclusion of independent variables in the regression model since:

- Inclusion of a minimum number of independent variables limits complexity but may lead to a less than pleasing R^2-statistic.

- Inclusion of a full complement of independent variables will lead to a higher R^2-statistic. However, a greater number of independent variables may lead to less significance for the coefficients of the regression due to:

-- More coefficients to estimate (less degrees of freedom).

-- Multicollinearity (larger Standard Error of the Coefficients) leading to smaller t-statistics.

The only "solution" to the dilemma of choosing which independent variables to include in a regression is the intuition which comes through practice with applying both logic and the techniques of multiple regression.

Exercise 8

Multiple Regression

Problem #1 (Method of Least Squares)

Monthly Profit (PROFIT)	Price Per Liter (PRICE)	Advertising Budget (ADV)
$28.00 M	$0.39	$0.943 M
$28.15 M	$0.51	$1.036 M
$33.74 M	$0.67	$1.450 M
$40.42 M	$0.53	$1.575 M
$40.21M	$0.43	$1.678 M
$45.20 M	$0.61	$1.744 M
$43.17 M	$0.56	$1.977 M
$58.23 M	$0.79	$2.376 M
$59.66 M	$0.58	$3.161 M
$62.08 M	$0.61	$3.217 M
$51.36 M	$0.43	$3.509 M
$60.69 M	$0.56	$3.642 M
$62.90 M	$0.67	$3.658 M
$60.27 M	$0.55	$3.714 M
$65.67 M	$0.69	$4.130 M
$77.85 M	$0.76	$4.562 M
$87.76 M	$0.78	$4.738 M

Use a regression software package to answer the following questions:

a) For the above data provided for the Designer Pet Bottling Company in Example 1, use the Method of Least Squares to determine the coefficients for the regression (1):

(1) $PROFIT = a + b_1 \cdot PRICE + b_2 \cdot ADV$

b) Assume that the above data for the advertising budget (ADV) is given instead in units of thousands of dollars (e. g. $943K, $1036K, etc.) instead of millions of dollars. Use the Method of Least Squares to determine the coefficients regression (2):

(2) $PROFIT = c + d_1 \cdot PRICE + d_2 \cdot ADV$

c) How do the coefficients for regression (1) compare to those of regression (2)? Is it correct to say that ADV has a larger effect on PROFIT in regression (1) than in regression (2)? What is the correct interpretation of the coefficients of the independent variable in regression (1) and (2)?

d) Determine the following goodness of fit statistics for the Designer Pet regression (1):

- R^2-statistic
- Adjusted R^2-statistic
- Standard Error of the Estimate
- F-statistic

How do you expect the values of the above statistics to change for regression (2)?

e) Determine the correlation matrix for the Designer Pet data (for the variables PROFIT, PRICE, and AGE). Does a regression have to performed in order to determine the correlation matrix for the data?

Problem #2 (Method of Least Squares)

Add a third independent variable to the Designer Pet regression. The third independent variable called MONTH ranges from 1 to 17 corresponding to the 17 months of Designer Pet Water Bottling Company data.

Monthly Profit (PROFIT)	Price Per Liter (PRICE)	Advertising Budget (ADV)	Month (MONTH)
$28.00 M	$0.39	S0.943 M	1
$28.15 M	$0.51	S1.036 M	2
$33.74 M	$0.67	$1.450 M	3
$40.42 M	$0.53	S1.575 M	4
$40.21M	$0.43	S1.678 M	5
$45.20 M	$0.61	$1.744 M	6
$43.17 M	$0.56	$1.977 M	7
$58.23 M	$0.79	$2.376 M	8
$59.66 M	$0.58	$3.161 M	9
$62.08 M	$0.61	$3.217 M	10
$51.36 M	$0.43	$3.509 M	11
$60.69 M	$0.56	$3.642 M	12
$62.90 M	$0.67	$3.658 M	13
$60.27 M	$0.55	$3.714 M	14
$65.67 M	$0.69	$4.130 M	15
$77.85 M	$0.76	$4.562 M	16
$87.76 M	$0.78	$4.738 M	17

Use a regression software package to answer the following questions:

a) Use the Method of Least Squares to determine the coefficients of the regression (3):

(3) $PROFIT = a + b_1 \cdot PRICE + b_2 \cdot ADV + b_3 \cdot MONTH$

b) For regression (3), determine the following overall goodness of fit statistics:

- R^2-statistic
- Adjusted R^2-statistic
- Standard Error of the Estimate
- F-statistic

c) Compare the goodness of fit statistics for regressions (1) and (3). Interpret the similarities and differences.

d) Compare the sample standard deviation (from its sample mean) of the dependent variable PROFIT in the Designer Pet data versus the Standard Error of the Estimate of the dependent variable PROFIT for the regression (3). Does the value of Standard Error of the Estimate for the dependent variable PROFIT indicate explanatory power for regression (3)?

e) Determine the correlation matrix for regression (3). The independent variables ADV and MONTH should show a large correlation coefficient. If at least one of these independent variables had a non-significant t-statistic, which variable should be dropped from the regression model based on the statistical information provided by the correlation matrix?

Problem #3 (Calculation of the Adjusted R^2-Statistic)

Verify *by hand calculation* the adjusted R^2-statistic for regression (3) using the R^2-statistic reported by the regression software package.

Problem #4 (Calculation of the F-Statistic)

In Example 3 in Chapter 6, the simple regression using the Method of Least Squares performed on the Round Table Furniture Company demand vs. time data indicated that the R^2-statistic for the Least Squares best fit line is 0.876541. Verify *by hand calculation* the value of the F-statistic of 42.6 computed in Chapter 6 (using the ratio of the explained variance to the unexplained variance) can also be computed by the general formula given in Example 3 of Chapter 8 (using the value of the R^2-statistic).

Problem #5 (Calculation of the t-Statistic in Simple Regression)

Apply the Method of Least Squares to the demand vs. time data given in Example 1 of Chapter 6. This data is repeated below:

t (YEAR)	Y_t (DEMAND)
1	180
2	324
3	360
4	378
5	540
6	504
7	490
8	648

a) Determine the F-statistic for the simple regression of the dependent variable DEMAND on the single independent variable YEAR.

b) Determine the t-statistic for the independent variable YEAR.

c) Verify that the value of the t-statistic for the single independent variable is the square root of the F-statistic in a simple regression.

Problem #6 (Method of Least Squares)

For the Designer Pet regression (3) from Problem #2 (including PROFIT, PRICE, ADV, and MONTH), use a regression software package to answer the following questions:

a) What are the t-statistics for each of the coefficients of the regression (3)?

b) Compute the number of degrees of freedom in the data for regression (3)? Is the Normal distribution or the Student t distribution applicable to setting levels of significance for the t-statistic?

c) To what value must the absolute value of the t-statistic be compared when testing for significance to the 95th percentile? Which of the coefficients for the independent variables of the Designer Pet regression (3) are significant (non-zero)?

d) The coefficient of the independent variable MONTH should not demonstrate significance to the 95th level of significance. And yet the F-statistic demonstrates a high level of significance (> 99th percentile) while the R^2-statistic is also high. What should one suspect about the independent variables?

e) The constant term is not significant. If this were not the final regression (but an intermediate step), should the constant term be dropped in this regression?

f) In the month 18, the PRICE of flavored water is estimated to be $0.75 and the budget allocation for advertising flavored water is estimated to be $4.5 Million. What is the forecast for profit for month 18? How much statistical confidence should be placed in this forecast?

Problem #7 (Durbin-Watson Statistic)

Consider the following 37 years of time series data (1948-1983) from the *Economic Report to the President*, 1987, Table B-95, p. 354 on the percentage of the population of the United States which lives on a farm:

Year (TIME)	Percentage (PERCENT)	Year (TIME)	Percentage (PERCENT)	Year (TIME)	Percentage (PERCENT)
1	17.9	14	8.7	27	4.5
2	16.6	15	8.1	28	4.3
3	16.2	16	7.7	29	4.1
4	15.2	17	7.1	30	3.8
5	14.2	18	6.7	31	3.6
6	13.9	19	6.4	32	3.6
7	12.5	20	5.9	33	3.4
8	11.7	21	5.5	34	3.2
9	11.7	22	5.2	35	3.0
10	11.5	23	5.1	36	3.0
11	10.3	24	4.7	37	3.0
12	9.8	25	4.5		
13	9.3	26	4.6		

Using a regression software package, determine the answers to the following questions:

a) Perform a simple regression of percentage of the U. S. population which lives on a farm (PERCENT as the dependent variable) on TIME (as the independent variable).

b) How many observations are included in the data? How independent variables are included in the regression? Provide the appropriate Durbin-Watson values d_L and d_U at the 95% significance level?

c) Determine the Durbin-Watson statistic for the above regression. Interpret whether the Durbin-Watson statistic indicates positive autocorrelation, negative correlation, or is inconclusive.

d) The Durbin-Watson statistic is used to indicate the correlation between successive error terms (residuals). If there is a positive error for a given (PERCENT, TIME) data point and the data point is thus above the derived regression line, what should we expect about the error term (residual) of other data points in vicinity of this data point?

Problem #8 (Dependent vs. Independent Variable)

Consider the data provided in Example 6 which provides data on the # cows on four farms of given acreage in McDonald County:

Y (Cows)	X (Acres)
5	9
10	16
12	12
15	20

Using a regression software package, determine the answers to the following questions:

a) Perform a simple regression using Y (Cows) as the dependent variable and X (Acres) as the independent variable.

b) Perform a simple regression using Y (Cows) as the independent variable and X (Acres) as the dependent variable.

c) Contrast these two regressions in terms of the Least Squares best fit lines derived, the R^2-statistics, t-statistics, etc.

Chapter 8

Multiple Regression Additional Solved Problems

Problem #1 (Multiple Regression)

Consider a multiple regression model for the dependent variable Y and the independent variables X_1 and X_2:

$$Y = a + b_1 \cdot X_1 + b_2 \cdot X_2 + \varepsilon$$

a) Define multicollinearity for a general data set.

Answer: Multicollinearity is the situation where two or more independent variables have such a strong tendency to move together that the regression has difficulty in determining the effects of a change in the individual independent variables on the dependent variable.

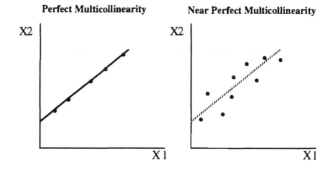

The correlation matrix has been determined by a regression software package as follows:

	X_1	X_2	Y
X_1	1.00	0.75	0.91
X_2	0.75	1.00	0.85
Y	0.91	0.85	1.00

b) The independent variables X_1 and X_2 have a high correlation coefficient. Does this necessarily indicate multicollinearity? How would one demonstrate that the data set does or does not have a multicollinearity problem?

Answer: There are two main indicators of multicollinearity:

  1) When the value of the R^2-statistic is high and the F-statistic is significant but one or more of the independent variable coefficient t-statistics is not significant. That is,

while the overall regression is significant, one or more of the independent variables is not significant.

2) High correlation among two or more independent variables (≈ 0.70 or higher).

A high correlation between independent variables does not necessarily imply multicollinearity. If the data set is rich enough (there exists enough variation in the values of the independent variables), the data set will not have a multicollinearity problem.

If the t-statistic for each of the independent variables is significant, a multicollinearity problem does not exist.

c) The adjusted R^2-statistic is reported by the regression software package to be 0.854. Why is the adjusted R^2-statistic a better measure of the goodness of fit of a regression than the R^2-statistic?

Answer: The R^2-statistic can be made higher by adding independent variables (which are likely to have some correlation with the dependent variable) to the regression. Hence, one can increase the number of independent variables until one obtains a pleasing R^2-statistic. The adjusted R^2-statistic is adjusted for the number of independent variables included in the regression. A high adjusted R^2-statistic indicates that the explanatory power is due to the quality and not just the number of independent variables in the regression.

d) The Standard Error of the Estimate of the Dependent Variable Y is reported by a regression software package to be $12.4M. What does the Standard Error of the Estimate a measure for a data set?

Answer: The Standard Error of the Estimate measures the average scatter of the points in the data set around the Least Squares best fit line. Ninety-five percent of the data points should be within two Standard Error of the Estimate values above and below the regression line determined by the Method of Least Squares.

e) What should the Standard Error of the Estimate be compared with to provide a true measure of the explanatory power of the regression?

Answer: The magnitude of the Standard Error of the Estimate is a function of the units of the dependent variable and can be made large or small by changing the units of the dependent variable Y. One should compare the Standard Error of the Estimate with the sample standard deviation of the dependent variable (both are measured in the same units).

If the Standard Error of the Estimate is significantly smaller than the sample standard deviation of the dependent variable, the regression has significantly more explanatory power in estimating the value of the dependent variable than does using the sample mean $\bar{Y}$ of the dependent variable data to estimate the value of the dependent variable Y.

f) The F-statistic and the t-statistics both provide measures of the significance of the regression. Explain the difference in the information provided by these two types of statistics.

Answer: The F-statistic provides a measure of the confidence that the overall data follows a linear trend. The t-statistics provide a measure of the confidence that the coefficients of the individual independent variables are non-zero. Thus, the F-statistic could indicate significance for a trend in the data (versus using the sample mean as the best estimate of the dependent variable) despite the t-statistics indicating non-significance of individual independent variables.

The t-statistics for the constant a and the independent variables X_1 and X_2 for N = 20 data points are reported by a regression software package as follows:

Variable	t-Statistic
Constant	2.54
X_1	1.25
X_2	3.89

g) Why should a constant be included in all intermediate regressions leading up to the final regression?

Answer: The constant term is sometimes called the "garbage can" term in that it captures the effects of omitted variables and other random and non-random effects. A biased estimate of the constant does not have statistical consequences since the constant in a regression is given no physical meaning. However, dropping the constant term from a regression forces the line to pass through the origin. In general, this causes the values of the coefficients (slopes) of the independent variables to be biased. This has statistical consequences since the slopes have the physical meaning of the marginal effect of the independent variables on the dependent variable.

h) Determine whether the above t-statistics are significant to the 95th percentile.

Answer: The t-statistics for the constant and the independent variables X_1 and X_2 must be compared with the 95th percentile of the Student t-distribution with N - k - 1 = 20 - 2 - 1 = 17 degrees of freedom. Since the value of the 95th percentile of the Student t-distribution for 17 degrees of freedom is 1.740 (see Table 6), the constant a and the independent variable X_2 are significant with 95% confidence. The independent variable X_1 is not significant at the 95th percentile.

i) Define autocorrelation for a regression model.

Answer: Autocorrelation is the situation where the error terms are not independent. A plot of the error terms $\varepsilon = (Y - Y^*)$ over the values of the dependent variable Y will show a systematic pattern either in the magnitude or the sign of the error terms. A series of large or small error terms generally grouped together is indicative of a positive dependence among successive error terms. A series of positive and negative error terms that are in a general flip-flop pattern is indicative of negative dependence among successive error terms.

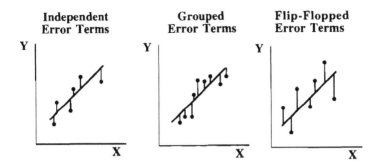

Independent Error Terms   **Grouped Error Terms**   **Flip-Flopped Error Terms**

The assumption of independent error terms is violated when the error terms for two or more observations are correlated. This correlation among error terms in a regression model is called autocorrelation.

Autocorrelation has three main sources:

    1) An incorrect functional form chosen for the regression.

    2) An independent variable relevant to the regression omitted.

    3) Serial autocorrelation caused by the effect of a random event in one period propagating over several periods.

When successive error terms are positively or negatively correlated, the error terms are in a grouped pattern and in a flip-flop pattern respectively, and the difference between the values of successive error terms is generally smaller and larger, respectively, than when there is no correlation between successive error terms. The Durbin-Watson statistic d (which examines the ratio of sum of the squared *differences* between successive error terms to the sum of the squared error terms themselves) is used to test for serial autocorrelation in the error terms. The Durbin-Watson statistic is used to test the hypothesis of *serial autocorrelation only* in the data. The null hypothesis is that there is no autocorrelation in the data. The value of the Durbin-Watson statistic is compared to a Durbin-Watson significance table to determine whether the data indicates no autocorrelation, positive autocorrelation, negative autocorrelation, or whether the data is inconclusive.

j) For the 20 data points collected, the software regression package reported a Durbin-Watson statistic of 0.98. What does this tell you concerning serial autocorrelation in the regression model?

Answer: For $N = 20$ and the number of independent variables $= 2$, the upper and lower values given in the 95th percentile Durbin-Watson Table (see Table 11) are $d_L = 1.10$ and $d_U = 1.54$. Since the Durbin-Watson statistic of 0.98 is in the region $(0, d_L)$, the null hypothesis of no serial autocorrelation in the regression model is rejected and the hypothesis of positive serial autocorrelation is accepted at the 95th percentile.

k) Define homoskedasticity and heteroskedasticity for a regression model. How would one check the data set for heteroskedasticity?

Answer: Heteroskedasticity is the situation where the error terms do not have the same likelihood of being large or small over the entire range of the observations of the dependent variable. Some observations have a higher likelihood to have large random effects than other observations.

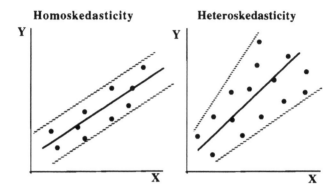

Heteroskedasticity (like autocorrelation) may be caused by specification of the incorrect functional form for the regression (e.g. the error terms become larger over time when a linear model is specified but the true model is exponential). Heteroskedasticity (like autocorrelation) may also be caused by the omission of a relevant independent variable from the regression (e.g. the error terms become correlated to the dependent variable Y and increase or decrease along with an increase in the value of Y).

The most common test for heteroskedasticity is to plot the data and visually check for constant deviation over the observations. Another common test is performed by tabulating the error terms (actual minus predicted values of the dependent variable). If the absolute value of the magnitudes of the error terms remain generally constant over the range of the observations, then one should not suspect heteroskedasticity. If the absolute values of the magnitudes of the error terms vary significantly over the range of the observations, one should suspect heteroskedasticity.

l) In a time series model, is serial autocorrelation or heteroskedasticity more likely to occur? Explain.

Answer: Serial autocorrelation is very common in time-series data (e.g. a random event in one period has a non-random effect in future periods). One should always suspect and test for serial autocorrelation in time-series data. Since time-series data typically focuses on the change in the characteristics of a single entity over time (as opposed to causal models which focus on the change in the characteristics of several entities), heteroskedasticity is not likely to be a problem for time-series data.

"There is no more common error than to assume that, because prolonged and accurate mathematical calculations have been made, the application of the result to some fact of nature is absolutely certain."

Alfred North Whitehead

9 | Linear Programming (LP)

9.1 Introduction

The total number of possible actions for decisions with even a moderate number of decision variables and a moderate number of alternative actions per decision variable can quickly become quite large.

Example 1: The California Lottery Pick 6 consists of choosing six *different* numbers from the set 1 to 54. What is the total number of possible Pick 6 choices (possible actions) for a person buying a lottery ticket (the decision)?

Answer: The first number chosen can be any of the 54 numbers. Since no number may be picked more than once, the second number chosen may be any of the 53 remaining numbers, and so on. The total number of possible actions for the Pick 6 is computed as follows:

$$\frac{54}{1^{st}\#} \bullet \frac{53}{2^{nd}\#} \bullet \frac{52}{3^{rd}\#} \bullet \frac{51}{4^{th}\#} \bullet \frac{50}{5^{th}\#} \bullet \frac{49}{6^{th}\#}$$

The total number of different choices (possible actions) for a Pick 6 entry (the decision) is $54 \bullet 53 \bullet 52 \bullet 51 \bullet 50 \bullet 49 = 18.6$ Billion!

Many quantitative methods become difficult to apply to decisions where there are many decision variables and/or many possible actions per decision variable. Determination of the best action through brute force enumeration of the possible actions even using high speed computers often is too expensive and/or time-consuming to be used on a regular basis to solve large complex problems.

A quantitative method that lends itself to the derivation of the optimal decision even among an *infinite* set of possible alternative actions is <u>linear programming</u>.

Notes: Characteristics of linear programming are as follows:

• A linear program (LP) is expressed as an objective function to be optimized subject to resource constraints and consists of decision variables and constants. For example:

$$\begin{array}{ll} \text{Objective Function} & Max\ 2 \cdot x_1\ +\ 6 \cdot x_2 \\ \text{Resource Constraints} & 3 \cdot x_1\ -\ 5 \cdot x_2\ \le\ 12 \\ & 4 \cdot x_1\ +\ 7 \cdot x_2\ \le\ 9 \\ \text{Decision Variables} & x_1 \ge 0,\ x_2 \ge 0 \end{array}$$

• Linear programming is applied to *deterministic* decision analyses. In deterministic decision analyses, *constants* (e.g. mean values or most likely values) are used rather than Random Variables to specify coefficients of decision variables. The accuracy of the LP solution will diminish to the extent that a deterministic decision analysis is not a reasonable approximation.

• The expressions for the objective function to be optimized and the constraints on the decision are restricted to be *linear expressions*. Linear expressions are of the form $c_1 \cdot x_1 + c_2 \cdot x_2 + \cdots + c_n \cdot x_n = d$. The $x_i's$ are decision variables raised to the first power while the $c_i's$ and d are constants.

• The decision variables in LP linear expressions are *independent* of each other and stand *alone* in the objective function and the resource constraints. LP linear expressions have no multiplication of decision variables (e.g. $5 \cdot x_1 x_2$). Multiplication rather than addition of the LP decision variables (e.g. $5 \cdot x_1 + 5 \cdot x_2$) indicates an interaction (a non-linearity) among the decision variables.

Examples of linear and non-linear expressions are provided in Figure 1.

Linear vs. Non-Linear Expressions

Expression	Linear/Non-Linear
$x_1 + 5 \cdot x_2 = 12$	Linear; all decision variables raised to the first power
$x_1 + \sqrt{2} \cdot x_2 + 7 \cdot x_3 = 5$	Linear; coefficients must be constants but need not be integers or fractions
$x_1^2 + 3 \cdot x_2 = 8$	Non-linear; expression is quadratic (x_1 raised to the second power)
$x_1 + 2 \cdot x_1 x_2 + x_2 = 3$	Non-linear; $2 \cdot x_1 x_2$ indicates an interaction between the x_1 and x_2 decision variables

Figure 1

9.2 Linear Program Formulation

Consider a typical production process (see Figure 2). *Inputs* (e.g. resources such as labor, raw materials, equipment, etc.) are transformed by the production process into an *output* (e.g. a finished product for sale).

The Production Process

Figure 2

This presentation of linear programming examines the two standard types of linear program (LP) formulations involving the optimization of the production process:

- **Type 1 LP:** Maximize production output given a *fixed* level of input resources.

- **Type 2 LP:** Minimize input resources to achieve a *fixed* level of production output.

Notes:

• Another type of LP formulation (called goal programming) optimizes two or more linear objective functions or *goals* simultaneously (e.g. maximize the production output *and* minimize the facility floor space used given a fixed level of manpower and equipment). Solving this formulation using linear programming requires that a single linear objective function be determined as a weighted average of the goals (e.g. 0.80 applied to maximizing production output and 0.20 applied to minimizing facility floor space) in accordance with the relative importance of the goals. The weighting of the goals in the single linear objective function is typically *subjective* which dilutes the usefulness of the LP solution.

• Linear programming is most useful (successful) when a *single* objective function is optimized. Quantitative methods for decision making in general are most useful when a single objective function is optimized.

Type 1 LP: Maximize Production Output Given a Fixed Level of Input Resources

Example 2: (GeeWhiz Computer Company LP). The demand for the high end special-purpose workstations produced at the GeeWhiz Computer Company plant has been decreasing over time. Rather than scale back plant operation and the number of employees, management has decided to use the available excess workstation resources including labor to make IBM-compatible laptop (L) computers and desktop (D) computers. GeeWhiz wishes to maximize the yearly profit generated when these laptops (L) and desktops (D) are produced.

Management has determined that an extra 36500 workdays, 9500 units of computer components, and 9000 hours of quality inspection are currently available per year. Flat screens required for laptop computers are not available from excess workstation resources and GeeWhiz must buy these from outside vendors. It is forecasted that GeeWhiz is limited to procurement of only 3500 flat screens per year. Each laptop (L) and desktop (D) computer produced uses the following amount of the four main input resources:

	Laptop	Desktop
Workdays	6	10
Computer components	1	3
Inspection hours	1.1	2.1
Flat screens	1	0

It is forecasted that each laptop will generate a profit of $1000 and each desktop will generate a profit of $2000. What number of laptops and desktops should be produced per year by the GeeWhiz Computer Company using the available workstation resources in order to maximize the yearly profit?

Answer: The *decision variables* are L (the number of laptops produced per year) and D (the number of desktops produced per year). Then the *objective function* (maximize the yearly profit) is given by:

$$Maximize \ \$1000 \cdot L + \$2000 \cdot D$$

The number of laptop and desktop computers that can be produced per year is limited by the following four *input constraints* on available resources:

1) The number of workdays available per year is 36500. Since each laptop and desktop requires 6 and 10 workdays, respectively, a workday (W) constraint holds:

$$6 \cdot L + 10 \cdot D \leq 36500 \qquad (W)$$

2) The number of computer components available per year is 9500. Since each laptop and desktop requires 1 and 3 computer components, respectively, a material (M) constraint holds:

$$1 \cdot L + 3 \cdot D \leq 9500 \qquad (M)$$

3) The number of product quality inspection hours available per year is 9000. Since each laptop and desktop requires 1.1 and 2.1 inspection hours, respectively, an inspection (I) constraint holds:

$$1.1 \cdot L + 2.1 \cdot D \leq 9000 \qquad (I)$$

4) Finally, only 3500 flat screens are available from vendors each year. Since each laptop and desktop requires 1 and 0 flat screens, respectively, a flat screen (F) constraint holds:

$$1 \cdot L + 0 \cdot D \leq 3500 \qquad (F)$$

The decision variables L and D can take on only *non-negative values* (it is not possible to make negative amounts of laptops or desktops). The following additional constraints hold:

$$L \geq 0 \, , \, D \geq 0$$

Then the GeeWhiz Computer Company LP (Type 1) formulation is given by:

Decision Variables	L (# *laptops*) , D (# *desktops*)
Objective Function	*Max* $1000 \cdot L + \$2000 \cdot D$
Resource Constraints	$6 \cdot L + 10 \cdot D \leq 36500 \qquad (W)$
	$1 \cdot L + 3 \cdot D \leq 9500 \qquad (M)$
	$1.1 \cdot L + 2.1 \cdot D \leq 9000 \qquad (I)$
	$1 \cdot L + 0 \cdot D \leq 3500 \qquad (F)$
Non-Negative Decision Variables	$L \geq 0 \, , \, D \geq 0$

Type 2 LP: Minimize Input Resources to Achieve a Fixed Level of Production Output

Example 3: (Roots N' Berries Trail Mix LP). The Roots N' Berries Health Food Company wishes to produce packages of trail mix for vending machines. The nutritionist for Roots N' Berries has set the following nutrition standards for each package of trail mix:

1) No more than 350 calories (cal.)
2) At least 125 grams of protein
3) No more than 50 grams of fat
4) At least 20 milligrams (mg.) of iron

Roots N' Berries plans to include 8 trail mix items in each package. The cost and nutritional values of each trail mix item per pound (lb.) are given as follows:

Trail Mix Item	Cost/lb.	Cal./lb.	Protein/lb.	Fat/lb.	Iron/lb.
1. Raisins	$0.49	350	20 grams	25 grams	6 mg.
2. Peanuts	$1.25	690	85 grams	100 grams	10 mg.
3. Almonds	$2.50	860	72 grams	90 grams	12 mg.
4. Sunflower seeds	$0.35	290	35 grams	35 grams	8 mg.
5. Coconut flakes	$1.89	780	40 grams	80 grams	5 mg.
6. Dried bananas	$1.59	550	65 grams	95 grams	18 mg.
7. Dried apricots	$2.35	370	30 grams	20 grams	14 mg.
8. Dried figs	$1.60	420	55 grams	30 grams	16 mg.

What amount (in lbs.) of the eight trail mix items per package will meet the nutrition standard at the lowest total cost to the Roots N' Berries Health Food Company?

Answer: The *decision variables* are X_1, X_2, $\cdots$, X_8 where X_i is the amount (in lbs.) of the ith trail mix item in each package. The *objective function*, to minimize the total cost to Roots N' Berries of a package of trail mix, is given by:

$$Minimize \ \$0.49 \cdot X_1 + \$1.25 \cdot X_2 + \$2.50 \cdot X_3 + \cdots + \$1.60 \cdot X_8$$

The fixed level of nutrition (output) supplied by a package of Roots N' Berries trail mix is set by the four nutrition level *output constraints*:

1) A package of trail mix must supply no more than 350 calories. Since the number of calories in a lb. of raisins, peanuts, almonds, $\cdots$, and dried figs is 350, 690, 860, $\cdots$, and 420, respectively, a calorie (C) output constraint holds:

$$350 \cdot X_1 + 690 \cdot X_2 + 860 \cdot X_3 + \cdots + 420 \cdot X_8 \leq 350 \qquad (C)$$

2) A package of trail mix must supply at least 125 grams of protein. Since the number of grams of protein in a lb. of raisins, peanuts, almonds, $\cdots$, and dried figs is 20, 85, 72, $\cdots$, and 55, respectively, a protein (P) output constraint holds:

$$20 \cdot X_1 + 85 \cdot X_2 + 72 \cdot X_3 + \cdots + 55 \cdot X_8 \geq 125 \qquad (P)$$

3) A package of trail mix must supply no more than 50 grams of fat. Since the number of grams of fat in a lb. of raisins, peanuts, almonds, $\cdots$, and dried figs is 25, 100, 90, $\cdots$, and 30, respectively, a fat (F) output constraint holds:

$$25 \cdot X_1 + 100 \cdot X_2 + 90 \cdot X_3 + \cdots + 30 \cdot X_8 \leq 50 \qquad (F)$$

4) Finally, a package of trail mix must supply at least 20 milligrams of iron. Since the number of milligrams of iron in a lb. of raisins, peanuts, almonds, $\cdots$, and dried figs is 6, 10, 12, $\cdots$, and 16, respectively, an iron (I) output constraint holds:

$$6 \cdot X_1 + 10 \cdot X_2 + 12 \cdot X_3 + \cdots + 16 \cdot X_8 \geq 20 \qquad (I)$$

The decision variables (lbs. of each trail mix item in a package) can take on only *non-negative values* and hence the following additional constraints hold:

$$X_1 \geq 0 \ , \ X_2 \geq 0 \ , \ X_3 \geq 0 \ , \ \cdots \ , \ X_8 \geq 0$$

Then the Roots N' Berries Trail Mix LP (Type 2) formulation is given by:

Decision Variables	$X_1, X_2, X_3, \cdots, X_8$ *(lbs. of each trail mix item in a package)*
Objective Function	*Minimize* $\$0.49 \cdot X_1 + \$1.25 \cdot X_2 + \$2.50 \cdot X_3 + \cdots + \$1.60 \cdot X_8$
Output Constraints	$350 \cdot X_1 + 690 \cdot X_2 + 860 \cdot X_3 + \cdots + 420 \cdot X_8 \leq 350 \qquad (C)$
	$20 \cdot X_1 + 85 \cdot X_2 + 72 \cdot X_3 + \cdots + 55 \cdot X_8 \geq 125 \qquad (P)$
	$25 \cdot X_1 + 100 \cdot X_2 + 90 \cdot X_3 + \cdots + 30 \cdot X_8 \leq 50 \qquad (F)$

$$6 \cdot X_1 + 10 \cdot X_2 + 12 \cdot X_3 + \cdots + 16 \cdot X_8 \geq 20 \qquad (I)$$

Non-Negative
Decision Variables
$$X_1 \geq 0 \ , \ X_2 \geq 0 \ , \ X_3 \geq 0 \ , \ \cdots \ , \ X_8 \geq 0$$

Notes:

• In the GeeWhiz Computer Company Computer Company LP (Type 1), the input (available resources) of the production process was fixed. Output (yearly profit) was maximized. The decision variables (number of laptops, number of desktops) were chosen to describe the *output* of the production process.

• In the Roots N' Berries Trail Mix LP (Type 2), the output (nutrition standard) of the production process was fixed. Input (total cost of a trail mix package to Roots N' Berries) was minimized. The decision variables (amount in lbs. of each trail mix item in a package) were chosen to describe the *input* of the production process.

• In general, the decision variables of an LP are chosen to describe mathematically the quantity being optimized (i.e. output maximized or input minimized).

LP Formulation Assumptions

The decision analysis formulations given in Example 2 (Type 1 LP) and Example 3 (Type 2 LP) make important assumptions which allow their solution by linear programming:

- The objective functions and resource constraints are specified as *linear expressions*. This implies no economies of scale (e.g. the cost of making 10 laptop computers is exactly 10 times the cost of making a single laptop computer) and no interactions between decision variables (e.g. the cost of buying almonds is the same even if both almonds and peanuts are purchased from the same supplier). If linearity is not a reasonable approximation, non-linear programming is required.

- The coefficients for decision variables in the objective functions and resource constraints are specified using *constants* (i.e. average or most likely values rather than Random Variables). If a deterministic model is not a reasonable approximation and Random Variables are required, stochastic programming is required.

– The decision variables (e.g. number of laptops, lbs. of raisins per package) are continuous and may take on *fractional values* rather than only integer values (this is not strictly true when making units of laptop and desktop computers). If fractional values are not a reasonable approximation (e.g. if the optimal number of laptops can not be successfully rounded off to the nearest integer value) and only integer values for decision variables are appropriate, integer programming is required.

Notes:

• Many "real world" decisions involve non-linearities and/or stochastic elements. However, only quite *small* non-linear programs or stochastic programs can be solved and then only for a few special and limited formulations.

• Often the essence of a non-linear or stochastic decision can be captured using a linear program (LP) formulation. When this approach is reasonable, linear programming is a very powerful analytical tool because large problems (hundreds of decision variables and thousands of resource constraints) can be solved quickly on a computer using an efficient LP solution algorithm called the Simplex Method (see Sections 9.4 and 9.5).

9.3 Solving the LP (Graphical Method)

The Graphical Method is an instructional technique for solving a linear program (LP). Initially, the Graphical Method requires the feasible region to be determined. The feasible region is determined by first plotting the lines associated with each linear constraint on the non-negative decision variables and then identifying the area (region) of intersection.

Example 4: Determine the feasible region for the GeeWhiz Computer Company LP described in Example 2.

Answer: The non-negative decision variables $L \geq 0$ and $D \geq 0$ limit the feasible LP solutions to the upper right quadrant of a two dimensional (L,D) coordinate system.

The workday constraint (W) given by $6 \cdot L + 10 \cdot D \leq 36500$ is plotted next (the order of plotting the linear constraints is not important). The feasible region considering the non-negative decision variables L and D and only the workday constraint (W) is provided in Figure 3.

GeeWhiz LP Feasible Region Including Only The Constraint (W)

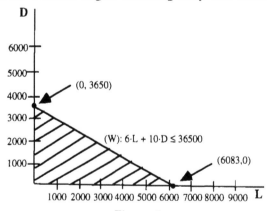

Figure 3

The workday constraint (W) intersects the L-axis near the point (L,D) = (6083,0). This point is a feasible solution if only laptops are made with the available workday resources. It is not feasible to make more than 6083 laptops since the 36500 available workdays will be exceeded (i.e. 6084 laptops x 6 workdays/laptop = 36504 workdays).

The workday constraint (W) intersects the D-axis at the point (L,D) = (0,3650). This point is a feasible solution if only desktops are made with the available workday resources. It is not feasible to make more than 3650 desktops since the 36500 available workdays will be exceeded (i.e. 3651 desktops x 10 workdays/desktop = 36510 workdays).

Note: Every point on a constraint line will use *exactly* the available resource (e.g. exactly 36500 workdays). Every point above or to the right of a constraint line will use more than the available resource. Every point below or to the left of a constraint line will use less than the available resource.

The workday constraint (W) and the non-negative decision variables (L $\geq$ 0 , D $\geq$ 0) describe a feasible region indicated by the hatched region in Figure 3. The other three linear constraints (M, I, and F) are graphed in the same fashion labeling the points where the constraint lines intersect. The intersection point for any two lines is determined by solving the set of simultaneous equations (two linear equations and two unknowns).

The feasible region and its intersection points described by the four linear constraints and the non-negative decision variables (L,D) are indicated by the hatched region in Figure 4.

GeeWhiz LP Feasible Region Using All Constraints (W, M, I, and F)

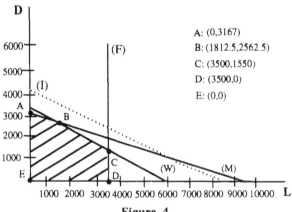

Figure 4

A: (0,3167)
B: (1812.5,2562.5)
C: (3500,1550)
D: (3500,0)
E: (0,0)

Notes:

• The quality inspection hours constraint (I) displayed as a dotted line in Figure 4 can be seen to have no effect on the final size of the feasible region defined by the hatched polygon ABCDE (since the line associated with constraint (I) does not touch the feasible region). Constraint (I) is sometimes said to be a <u>redundant</u> <u>constraint</u> to the other three constraints. That is, as more laptops and/or desktops are produced (one moves away from the origin in any direction), one will *always* run out of one of other three resources before the upper limit on the product quality inspection hours resource is reached. Constraints (W), (M), and (F) do affect the final size of the feasible region.

• The optimal solution for an LP must first be feasible. Thus the optimal solution must come from *within* the feasible region. It has been determined that the final size of the feasible region is not affected by the inspection constraint (I). Hence, this constraint may be dropped from consideration without affecting the size of the feasible region for the LP, the optimal solution (number of laptops and desktops produced) for the LP chosen from within the feasible region, or the optimal value (maximum yearly profit) for the LP.

Example 5: Use the Graphical Method to find the optimal solution to the GeeWhiz Computer Company LP described in Example 2.

Answer: The optimal solution for the GeeWhiz LP is that point in the feasible region (see Figure 4) which maximizes the yearly profit objective function $1000·L + $2000·D.

To find this optimal point, the Graphical Method uses the concept of iso-objective lines. The general iso-objective line for the GeeWhiz Computer Company LP is determined by setting the objective function equal to a constant:

$$\$1000 \cdot L + \$2000 \cdot D = constant$$

This equation is linear and graphs onto the feasible region as a family of lines as the value "constant" is changed (see Figure 5). Every point (L,D) on a single iso-objective line will provide a yearly profit equal to its value for "constant". That is, a decision maker is *indifferent* to all points (mix of laptops and desktops produced) *on* a single iso-objective line since all such points provide the same yearly profit equal to "constant".

As the value of "constant" changes, another distinct and parallel iso-objective line is generated. A decision maker is *not* indifferent between distinct iso-objective lines since each distinct line provides a different value for the yearly profit generated. At the origin (L,D) = (0,0), no laptops or desktops are produced so that $0 yearly profit is generated. As the values of "constant" increase, the parallel iso-objective lines (which have the same slope but different y-intercepts) move away from the origin and greater yearly profit is generated. The optimal solution is determined by increasing the value "constant" (yearly profit) of the iso-objective line to the maximum extent possible where the line still intersects *some portion* of the feasible region (see Figure 5).

GeeWhiz LP Iso-Objective Yearly Profit Lines

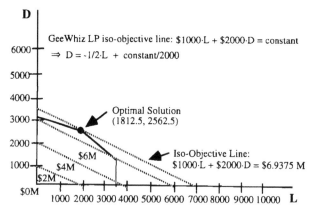

Figure 5

320

The maximum value "constant" can be increased and still have the iso-objective line $1000·L + $2000·D = "constant" touch the feasible region is $6,937,500 (see Figure 5). For this value of yearly profit, the iso-objective line will touch the feasible region for the GeeWhiz Computer Company LP only at the intersection point of the material (M) and workday (W) constraints (see Figure 6) given by the point (L,D) = (1812.5, 2562.5).

Note: Intersection points of the feasible region described by the linear constraints and the non-negative decision variables are referred to as corner points. The corner points for the GeeWhiz Computer Company LP are given in Figure 6. It is *the fundamental property* of linear programming that the optimal solution will be found at a corner point of the feasible region.

GeeWhiz LP Feasible Region with Corner Points

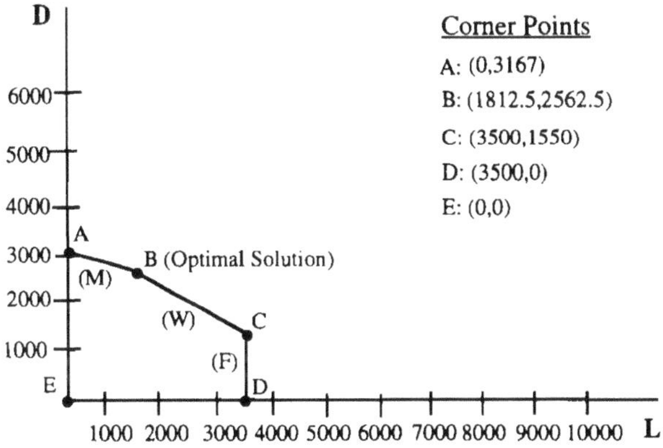

Corner Points

A: (0,3167)

B: (1812.5,2562.5)

C: (3500,1550)

D: (3500,0)

E: (0,0)

Figure 6

The Graphical Method featuring iso-objective lines for solving an LP is instructive for understanding of the LP optimization process. However, there are simpler and more general LP solution techniques (see Sections 9.4 and 9.5). LP solution techniques take advantage of the fundamental LP property that the optimal solution *must* occur at a corner point of the feasible region.

As a brute force technique, one could simply evaluate the objective function (= $1000·L + $2000·D for the GeeWhiz Computer Company LP) at *all* corner points of the feasible region. An evaluation of the yearly profit objective function at the corner points of the feasible region for the GeeWhiz Computer Company LP is provided in Figure 7.

GeeWhiz LP Corner Point Evaluation

	L	D	Yearly Profit ($1000·L + $2000·D)
A:	0	3167	$6,334,000
B:	1812.5	2562.5	$6,937,500 (Optimal)
C:	3500	1550	$6,600,000
D:	3500	0	$3,500,000
E:	0	0	$0

Figure 7

321

Notes: Properties of a linear program (LP) solution include:

• The optimal solution for an LP must occur at a corner point of the feasible region formed by the linear constraints and the non-negative decision variables.

• It is possible that *more than one* corner point of the feasible region is optimal for an LP.

• If two corner points are optimal (e.g. generate the same maximum yearly profit), then *every* point along the line segment joining the two corner points is also optimal. Graphically, the maximum or minimum iso-objective line intersects an *entire linear edge* of the feasible region rather than just a corner point.

• The Graphical Method is not readily applicable to the solving LP's with *more than two* decision variables since plotting in three or more dimensions is difficult or impossible. However, the properties of the feasible region (its edges are linear) and the optimal solution (occurs at a corner point) presented for an LP of two decision variables hold true for *any* number of decision variables.

• These two main features of an LP allow any size LP to be solved by an efficient mathematical algorithm called the Simplex Method (see Sections 9.4 and 9.5).

• In short, the Simplex Method starts at an initial feasible corner point (often the origin). It moves from a corner point of the feasible region to the *adjacent* corner point with the most *improved* objective function value. The Simplex Method keeps moving from corner point to adjacent corner point (improving the value of the objective function) until it arrives at a corner point where no adjacent corner point provides an improved objective function value. This corner point and its objective function value are optimal.

• An LP with hundreds of decision variables and thousands of constraints can have a feasible region with millions of corner points. Brute force enumeration and evaluation of the corner points is not practical. However, starting at an initial corner point, the efficient Simplex Method will arrive at the optimal corner point solution in typically fewer than 20 movements to adjacent corner points. For smaller LP's, the Simplex Method typically will arrive at the optimal corner point in much fewer movements.

• The optimal solution (L,D) = (1812.5, 2562.5) occurs at the intersection (see Figure 6) of the material (M) and workday (W) constraints (i.e. the optimal solution lies exactly on the (M) and (W) constraints). This denotes that the (M) and (W) constraints are binding constraints on the optimal solution (i.e. all of these resources are used up). (I) and (F) are non-binding constraints on the optimal solution (i.e. not all of the available inspection hours and available flat screen resources are used up). The optimal solution lies below the inspection hour (I) and to the left of the flat screen (F) constraints (see Figure 4).

• The fact that the constraints (M) and (W) are binding (all of these available resources are used up at the optimal solution) indicates that if one could increase the amount of workdays *or* computer component materials available, the optimal yearly profit value would *increase*.

• The fact that (I) and (F) are non-binding (not all of these available resources are used up at the optimal solution) indicates that even if more inspection hours and flat screens were made available, the optimal yearly profit would *not increase*. The yearly profit at the GeeWhiz Computer Company is not limited by the inspection hours or flat screens available but by the lack of workdays and computer components available.

322

• The change in the optimal value of the objective function due to a unit increase in the amount of a binding resource (e.g. an increase in the 36500 available workdays described by the (W) constraint 6 L + 10 M ≤ 36500) is called the <u>shadow price</u> (or marginal utility) for that resource.

- If the amount of workdays is increased by one unit from 36500 to 36501, the optimal objective function (the maximum yearly profit) value will increase by $125 (see Section 9.5). The $125 increase in the maximum yearly profit for each additional workday available is the shadow price for the workday resource (W).

- If the amount of computer components is increased by one unit from 9500 to 9501, the optimal objective function (maximum yearly profit) value will increase by $250. The $250 increase in the maximum yearly profit for each additional computer component available is the shadow price for the material resource (M).

• The shadow prices (marginal utilities) for the non-binding inspection hours (I) and flat screen (F) resources are *zero* since adding more of these resources (which are already in excess at the optimal solution) will not increase the maximum yearly profit.

• The shadow price (marginal utility) for a resource assumes an increase in that one resource only while all other resources and the objective function remain constant.

9.4 Solving the LP (Quick Overview of Simplex Method)

Recall the formulation of the GeeWhiz Computer Company LP provided in Example 2:

Decision Variables	L (# *laptops*) , D (# *desktops*)
Objective Function	*Max* $1000 \cdot L$ + $2000 \cdot D$
Resource Constraints	$6 \cdot L + 10 \cdot D \leq 36500$ (W)
	$1 \cdot L + 3 \cdot D \leq 9500$ (M)
	$1.1 \cdot L + 2.1 \cdot D \leq 9000$ (I)
	$1 \cdot L + 0 \cdot D \leq 3500$ (F)
Non-Negative Decision Variables	$L \geq 0$, $D \geq 0$

Though the Simplex Method is a *mathematical* algorithm, its progress towards finding the optimal solution can be traced *graphically* for an LP with two decision variables. Examine the feasible region with corner points and objective function (yearly profit) values labeled in Figure 8.

GeeWhiz LP Feasible Region with Corner Points and Yearly Profits

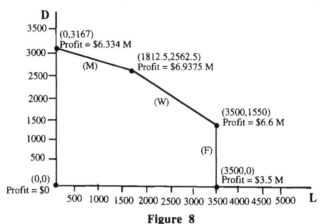

Figure 8

The Simplex Method will typically start its movement through the corner points of the GeeWhiz LP feasible region at the origin corner point (L, D) = (0, 0) where no laptops or desktops are made (the "do nothing" corner point). The value of the objective function (yearly profit) for the origin ("do nothing") corner point is $0 per year.

Notes:

• The Simplex Method may be given an initial feasible corner point solution other than the origin ("do nothing") corner point. Suppose a previous LP for the GeeWhiz Computer Company had been solved with the same constraints but a different objective function (i.e. a different forecasted profit per laptop and desktop produced). Since the feasible region has not changed (the constraints are the same), the optimal solution for the previous LP will be at least a *feasible corner point* for the current GeeWhiz LP.

• Depending on how close this previous optimal corner point is to the new optimal corner point, the number of movements to adjacent corner points before the optimal corner point is found could be significantly reduced. Typically, however, the origin ("do nothing") corner point is used as the initial feasible corner point solution for the Simplex Method.

Graphically, the Simplex Method starts at the origin (L,D) = (0,0) and moves to an adjacent corner point. Starting at the origin of the GeeWhiz LP feasible region, there are two adjacent corner point options available (see Figure 9):

    - Move along the D-axis. Go from corner point (0,0) to corner point (0,3167). Yearly profit ($1000·L + $2000·D) is improved from $0 to $6,334,000.

    - Move along the L-axis. Go from corner point (0,0) to corner point (3500,0). Yearly profit ($1000·L + $2000·D) is improved from $0 to $3,500,000.

324

The Simplex Method decides to move from corner point (0,0) to corner point (0,3167). This adjacent corner point is chosen because it provides the *most improved* value of the yearly profit objective function (see Figure 9).

Movement From Corner Point (0,0) to Corner Point (0,3167)

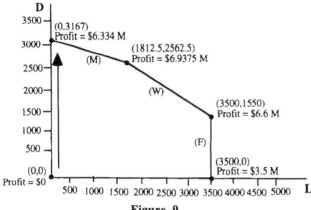

Figure 9

The Simplex Method will attempt to move from corner point (0,3167) to an adjacent corner point which provides the most improvement in the value of the objective function (yearly profit). Only one adjacent corner point improves the value of the yearly profit. The Simplex Method will move from corner point (0,3167) to corner point (1812.5,2562.5) with the yearly profit improving from $6,334,000 to $6,937,500 (see Figure 10).

Movement From Corner Point (0,3167) to Corner Point (1812.5,2562.5)

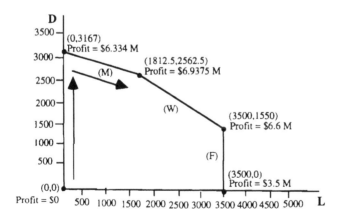

Figure 10

325

This corner point is *optimal* because it is no longer possible to move to an adjacent corner point with an improved value of the objective function (yearly profit). The optimal solution to the GeeWhiz LP is (L,D) = (1812.5,2562.5) with the maximum yearly profit of $6,937,500.

The optimal corner point (1812.5, 2562.5) is at the intersection of the material (M) and workday (W) constraints (see Figure 10). These resource constraints are binding and have positive shadow prices. That is, additional yearly profit is generated if the amount of available workdays or computer components can be increased.

The inspection hours (I) and flat screen (F) resource constraints are non-binding on the optimal solution and have shadow prices of zero. That is, no additional yearly profit is generated if the amount of inspection hours or flat screens (which are already in excess at the optimal solution) is increased.

9.5 Mechanics of the Simplex Method

At any corner point of the LP feasible region, the values of the variables will be non-negative (positive or equal to zero). Those variables which have positive values are called basic variables. Those variables which have zero values are called non-basic variables.

Example 6: At each of the five corner points (L,D) of the feasible region of the GeeWhiz Computer Company LP described in Example 2, determine whether the decision variables L and D are basic or non-basic.

Answer: The basic and non-basic decision variables for the five Gee Whiz LP corner points (see Figure 6) are identified in Figure 11.

Basic vs. Non-Basic Decision Variables

	L	D	Basic/Non-Basic
A:	0	3167	L non-basic, D basic
B:	1812.5	2562.5	L basic, D basic
C:	3500	1550	L basic, D basic
D:	3500	0	L basic, D non-basic
E:	0	0	L non-basic, D non-basic

Figure 11

Example 7: Solve the GeeWhiz Computer Company LP described in Example 2 using the Simplex Method.

Answer: Recall the formulation of the GeeWhiz Computer Company LP in Example 2:

Decision Variables	L (*# laptops*) , D (*# desktops*)	
Objective Function	*Max* $\$1000 \cdot L + \$2000 \cdot D$	
Resource Constraints	$6 \cdot L + 10 \cdot D \leq 36500$	(*W*)
	$1 \cdot L + 3 \cdot D \leq 9500$	(*M*)
	$1.1 \cdot L + 2.1 \cdot D \leq 9000$	(*I*)
	$1 \cdot L + 0 \cdot D \leq 3500$	(*F*)
Non-Negative Decision Variables	$L \geq 0$, $D \geq 0$	

Note: Whether or not the inspection hours constraint (I) is included in the GeeWhiz LP formulation does not affect the size of the feasible region (see Figure 4). Since the size of the feasible region is not affected by dropping the inspection hours (I) constraint, the optimal corner point and the optimal objective function value will not be affected. The constraint (I) which is redundant to the other three constraints and will be dropped from the Simplex Method solution of this LP.

The Simplex Method first converts the resource constraint inequalities to equalities by using a slack variable for each constraint (see Figure 12). The slack variables denote *unused resources* when any feasible corner point (L,D) of the feasible region is produced. The slack variables for the three resources (workdays, computer components, and flat screens) will be denoted by W, M, and F, respectively. The slack variables have coefficients of *zero* in the objective function since no laptops or desktops can be produced from resources which are unused and thus zero yearly profit will be generated.

Slack Variables W, M, F Added to the GeeWhiz LP

Max $1000 \cdot L + 2000 \cdot D + 0 \cdot W + 0 \cdot M + 0 \cdot F$

$$6 \cdot L + 10 \cdot D + 1 \cdot W \qquad = 36500 \, (W)$$

$$1 \cdot L + 3 \cdot D + \qquad + 1 \cdot M \qquad = 9500 \, (M)$$

$$1 \cdot L + 0 \cdot D + \qquad \qquad + 1 \cdot F = 3500 \, (F)$$

Figure 12

Figure 12 provides an equivalent formulation (using equalities rather than inequalities) to the GeeWhiz Computer Company LP. This is the formulation is required to begin solving the LP using the Simplex Method algorithm.

<u>Note:</u> At the initial corner point (L,D) = (0,0), both decision variables L and D are non-basic (have zero values) indicating that no laptop or desktop computers are produced (and all of the available resources are unused). The slack variables (W, M, and F) are basic (non-zero) with values equal to the total available resources (W = 36500, M = 9500, and F = 3500) as given on the right hand side of the three resource constraints (see Figure 12). This is the starting point for the Simplex Method.

The Simplex Method *mathematically* (rather than graphically) moves from the current corner point to the adjacent corner point which provides the most improvement in the objective function. This is done by converting one current non-basic (zero) variable (either a decision variable or a slack variable) into a basic (positive) variable while converting one current basic (positive) variable into a non-basic (zero) variable.

<u>Note:</u> The number of basic variables is *equal to the number of constraints* for an LP and remains the *same* after each matrix manipulation (movement from the current corner point to an adjacent corner point) called a *pivot*. In the GeeWhiz LP formulation, the number of basic (positive) variables remains at three. For example, at the origin corner point (0,0), the three slack variables W, M, and F have positive values 36500, 9500, and 3500, respectively, and represent the three basic variables.

The Simplex Method matrix, called a <u>tableau</u>, for the GeeWhiz LP (after the inclusion of slack variables) for the initial feasible solution (the 'do nothing" corner point) is given in Figure 13.

Initial Simplex Method Tableau For Corner Point (0,0)

L	D	W	M	F	RHS	
1000	2000	0	0	0	$0	Profit
6	10	1	0	0	36500	(W)
1	3	0	1	0	9500	(M)
1	0	0	0	1	3500	(F)

Figure 13

Initial Feasible Corner Point Solution (0,0)

<u>Notes:</u> The initial feasible corner point solution (0,0) for the GeeWhiz LP (see Figure 14) has the following properties:

• The basic variables correspond to those variable columns which have all zeros except for a single 1 (called identity matrix columns). The basic variables (positive) = W, M, F = 36500, 9500, 3500 (all excess resource is unused)

• The non-basic variables (those variables whose columns are not identity columns) L,D = 0,0 (the "do nothing" or origin corner point).

• The profit generated = $1000·L + $2000·D = $1000·(0) + $2000·(0) = $0 per year which is found in the upper right hand corner of the tableau.

Initial GeeWhiz LP Corner Point (0,0)

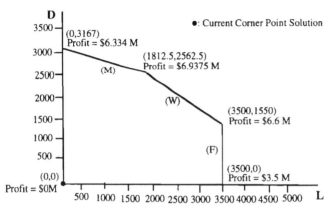

Figure 14

Next Simplex Method Corner Point (Two Choices)

<u>Notes</u>: Choice #1 (0,3167) for the adjacent corner point (see Figure 15) has the following properties:

• The Simplex Method will move along the D-axis going from corner point (L,D) = (0,0) to corner point (L,D) = (0,3167). The former non-basic (zero) D becomes positive (3167 desktops) and converts into a basic variable. The slack variable M becomes 0 since the corner point (0,3167) is on the (M) constraint denoting that all of the computer component resource (M) is used up. The slack variable M converts into a non-basic (zero) variable.

- The basic variables (positive) = W, D, F = 4830, 3167, 3500.

- The non-basic variables (zero) = L, M = 0, 0.

- The values for the slack variables W, M, and F are determined by substituting the current corner point (L,D) = (0,3167) into the (W), (M), and (F) constraints.

-- (W): $6 \cdot (L=0) + 10 \cdot (D=3167) + W = 35000 \Rightarrow W = 35000 - 31670 = 4830$ unused workdays.

-- (M): $1 \cdot (L=0) + 3 \cdot (D=3167) + M = 9500 \Rightarrow L = 9500 - 9500 = 0$ unused computer components.

-- (F): $1 \cdot (L=0) + 0 \cdot (D=3167) + F = 3500 \Rightarrow F = 3500 - 0 = 3500$ unused flat screens.

• The profit generated = $1000·L + $2000·D = $1000·(0) + $2000·(3167) = $6,334,000 per year.

329

Choice #1 (0,3167) for the Adjacent Corner Point

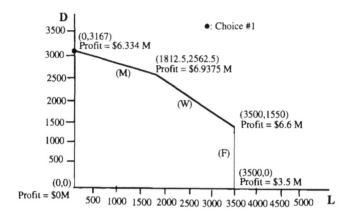

Figure 15

<u>Notes</u>: Choice #2 (3500, 0) for the adjacent corner (see Figure 16) has the following properties:

• The Simplex Method will move along the L-axis going from corner point $(L,D) = (0, 0)$ to corner point $(L,D) = (3500,0)$. The former non-basic (zero) L becomes positive (3500 laptops) and converts into a basic variable. The slack variable F becomes 0 (since the corner point (3500,0) is on the (F) constraint denoting that all of the flat screen resource (F) is used up) and converts into a non-basic (zero) variable.

- The basic variables (positive) = W, M, L = 14000, 6000, 3500.

- The non-basic variables (zero) = F, D = 0, 0.

- The values for the slack variables W, M and F are determined by substituting the current corner point $(L,D) = (3500,0)$ into the (W), (M), and (F) constraints.

-- (W): $6 \cdot (L=3500) + 10 \cdot (D=0) + W = 3500 \Rightarrow W = 35000 - 21000 = 14000$ unused workdays.

-- (M): $1 \cdot (L=3500) + 3 \cdot (D=0) + M = 9500 \Rightarrow L = 9500 - 3500 = 6000$ unused computer components.

-- (F): $1 \cdot (L=3500) + 0 \cdot (D=0) + F = 3500 \Rightarrow F = 3500 - 3500 = 0$ unused flat screens.

• The profit generated = $1000 L + $2000 D = $1000 (3500) + $2000 (0) = $3,500,000 per year.

Choice #2 (3500,0) for the Adjacent Corner Point

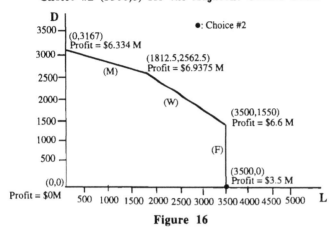

Figure 16

The Simplex Method decides on Choice #1 and moves from (L,D) = (0, 0) to (L,D) = (0,3167) since this path provides the most improvement in the objective function (yearly profit). After manipulation of the tableau associated with the corner point (0,0), the Simplex Method pivots to a new tableau (see Figure 17) associated with the corner point (0, 3167).

Simplex Method Tableau For Corner Point (0,3167)

L	D	W	M	F	RHS	
-333	0	0	667	0	**$6334000**	Profit
8/3	0	1	-10/3	0	4830	(W)
1/3	1	0	1/3	0	3167	(M)
1	0	0	0	1	3500	(F)

Figure 17

Notes: From the Simplex Method tableau (see Figure 17) after the pivot from the corner point (0,0) to the corner point (0,3167), one can read off the following:

• The basic variables correspond to the identity columns which have all zeros except for a single 1. The value of these basic (positive) variables are given in the right hand side (RHS) column:

- D = 3167 desktop computers
- W = 4830 unused workdays
- F = 3500 unused flat screens

• The identity of the non-basic (zero) variables (those variables whose columns are not identity matrix columns):

- L = 0
- M = 0

• The value of the objective function (yearly profit) $6,334,000 at the current corner point (L,D) = (0, 3167) is found in the upper right hand corner of the tableau.

Final (Optimal) Corner Point Simplex Method Solution (One Choice)

<u>Notes:</u> The single choice (1812.5, 2562.5) for the adjacent corner (see Figure 18) has the following properties:

• The Simplex Method will move along the (M) constraint edge going from corner point (L,D) = (0,3167) to corner point (L,D) = (1812.5,2562.5). The former non-basic (zero) L becomes positive (1812.5 laptops) and converts into a basic variable. The slack variable W becomes 0 (since the corner point (1812.5, 2562.5) is on the (W) constraint denoting that all of the workday resource is used up) and converts into a non-basic (zero) variable.

- The basic variables (positive) = L, D, F = 1812.5, 2562.5, 1687.5.

- The non-basic variables (zero) = W, M = 0, 0.

- The value of the slack variables W, M, and F are determined by substituting the current corner point (1812.5, 2562.5) into the (W), (M), and (F) constraints.

-- (W): $6 \cdot (L=1812.5) + 10 \cdot (D=2562.5) + W = 35000 \Rightarrow W = 35000 - 35000 = 0$ unused workdays.

-- (M): $1 \cdot (L1812.5) + 3 \cdot (D=2562.5) + M = 9500 \Rightarrow L = 9500 - 9500 = 0$ unused computer components.

-- (F): $1 \cdot (L=1812.5) + 0 \cdot (D=2562.5) + F = 3500 \Rightarrow F = 3500 - 1812.5 = 1687.5$ unused flat screens.

• The yearly profit generated = $1000 \cdot L + 2000 \cdot D = 1000 \cdot (1812.5) + 2000 \cdot (2562.5) = 6,937,500$.

Optimal GeeWhiz LP Corner Point (1812.5,2562.5)

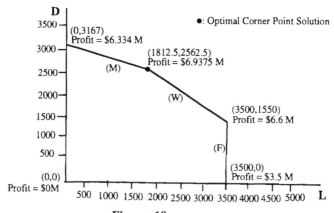

Figure 18

Note: This corner point is optimal since it is not possible to move to an adjacent corner point which *improves* the value of the yearly profit for the GeeWhiz Computer Company.

After manipulation of the tableau associated with the corner point (0,3167), the Simplex Method arrives at the final tableau (see Figure 19) associated with the corner point (1812.5,2562.5).

Final Simplex Method Tableau Corner Point (1812.5,2562.5)

L	D	W	M	F	RHS	
0	0	125	250	0	$6937500	Profit
1	0	3/8	-10/8	0	1812.5	(W)
0	1	-1/8	-2/24	0	2562.5	(M)
0	0	-3/8	10/8	1	1687.5	(F)

Figure 19

Notes: From the final Simplex Method tableau (see Figure 19) after a pivot from the corner point (0,3167) to the optimal corner point (1812.5, 2562.5), one can read off the following:

• The basic variables are those corresponding to the columns which have all zeros except for a single 1. The value of these basic (positive) variables for the optimal solution (L,D) = (1812.5,2562.5) are given in the right hand side (RHS) column:

    - L = 1812.5 laptop computers produced
    - D = 2562.5 desktop computers produced
    - F = 1687.5 unused flat screens

• The non-basic (zero) variables are those variables whose columns are not identity matrix columns.

    - W = 0 (no unused workdays at the optimal solution)
    - M = 0 (no unused computer components at the optimal solution)

• The optimal value $6,937,500 for the objective function evaluated at the optimal solution (L,D) = (1812.5,2562.5) is found in the upper right hand corner of the final tableau.

• The shadow prices for the resources W, M, and F appear in the *objective function row* of the final Simplex Method tableau (see Figure 19) in the W, M, and F columns, respectively:

    - Shadow price of resource W = $125 per workday added
    - Shadow price of resource M= $250 per computer component added
    - Shadow price of resource F = $0 per flat screen added (since there are unused flat screens at the optimal solution)

• A shadow price for a resource defines the *marginal* effect on the value of the objective function. The shadow price is valid when there a unit increase in that *one resource only*. The amount of all other resources must remain constant.

9.6 Assignment Problem

Example 7: (Forever Temps LP Assignment Problem). The manager of the employment agency Forever Temps has four jobs (Job 1, Job 2, Job 3, and Job 4) to fill and four employees (Employee 1, Employee 2, Employee 3, Employee 4) available to fill the jobs. After interviewing the four employees, the manager determined the following *subjective* scores for the compatibility of each employee's skills with each job's requirements:

Employee Job Compatibility Scores

	Job 1	Job 2	Job 3	Job 4
Employee 1	50	80	90	50
Employee 2	70	90	60	40
Employee 3	40	20	60	20
Employee 4	0	50	50	0

A score of 0 indicates the employee has no compatibility with the job's requirements while a score of 100 indicates the employee is completely compatible with the job's requirements. How should the manager assign the four employees to the four jobs to *maximize the sum* of the four compatibility scores?

Answer: There are a couple of obvious assignment methods which can be used to assign the four employees to the four jobs. Employee 1 can be assigned to his or her best job (Job 3), then Employee 2 to his or her best job (Job 2) among the jobs still available, etc. This would result in the assignment of employees to jobs provided in Figure 20.

Employees Assigned To Best Available Jobs

	Job 1	Job 2	Job 3	Job 4
Employee 1	50	80	**90**	50
Employee 2	70	**90**	60	40
Employee 3	**40**	20	60	20
Employee 4	0	50	50	**0**

Figure 20

The total compatibility score of this assignment = 90 + 90 + 40 + 0 = 220.

Another assignment method is to assign Job 1 to the best employee (Employee 2), Job 2 to the best available employee (Employee 1), etc. This would result in the assignment of employees to jobs provided in Figure 21.

Jobs Assigned To Best Available Employees

	Job 1	Job 2	Job 3	Job 4
Employee 1	50	**80**	90	50
Employee 2	**70**	90	60	40
Employee 3	40	20	**60**	20
Employee 4	0	50	50	**0**

Figure 21

The total compatibility score of this assignment = 70 + 80 + 60 + 0 = 210.

The *optimal* solution to the Forever Temps Assignment Problem (which can be found using several algorithms including the Simplex Method) is 230. There are actually two optimal solutions which provide the maximum sum of assignment compatibility scores (see Figures 22 and 23).

Forever Temps Assignment Problem Optimal Solution #1

	Job 1	Job 2	Job 3	Job 4
Employee 1	50	80	**90**	50
Employee 2	**70**	90	60	40
Employee 3	40	20	60	**20**
Employee 4	0	**50**	50	0

Figure 22

The total compatibility score of this assignment = 70 + 50 + 90 +20 = 230.

Forever Temps Assignment Problem Optimal Solution #2

	Job 1	Job 2	Job 3	Job 4
Employee 1	50	80	90	**50**
Employee 2	**70**	90	60	40
Employee 3	40	20	**60**	20
Employee 4	0	**50**	50	0

Figure 23

The total compatibility score of the this assignment = 70 + 50 + 60 + 50 = 230.

Note: There need not be a *unique* solution to an Assignment Problem. Remember that linear programs can have more than one optimal corner point with the same value of the objective function.

LP Formulation for the Forever Temps Assignment Problem

Let the index $i = 1, 2, 3, 4$ be assigned to Employees 1, 2, 3, and 4, respectively.

Let the index $j = 1, 2, 3, 4$ be assigned to Jobs 1, 2, 3, 4, respectively.

Let the decision variables $x_{ij} = 1$ (if Employee i is assigned to Job j)
0 (if Employee i is not assigned to Job j)

Note: The decision variables x_{ij} take on only the *integer* values 0 or 1 (i. e. no or yes for each possible assignment) rather than continuous values (e.g. $x_{ij} = 1/2$ or 0.66).

The linear programming formulation (which is actually an integer programming formulation) for the Forever Temps Assignment Problem is then given by:

335

Max $50 \cdot x11 + 80 \cdot x12 + 90 \cdot x13 + 50 \cdot x14 +$
$70 \cdot x21 + 90 \cdot x22 + 60 \cdot x23 + 40 \cdot x24 +$
$40 \cdot x31 + 20 \cdot x32 + 60 \cdot x33 + 20 \cdot x34 +$
$0 \cdot x41 + 50 \cdot x42 + 50 \cdot x43 + 0 \cdot x44$

(Maximize the sum of the compatibility scores)

$x11 + x21 + x31 + x41 = 1$ (Job 1)

$x12 + x22 + x32 + x42 = 1$ (Job 2)

$x13 + x23 + x33 + x43 = 1$ (Job 3)

$x14 + x24 + x34 + x44 = 1$ (Job 4)

(Each Job is assigned to exactly 1 Employee)

$x11 + x12 + x13 + x14 = 1$ (Employee 1)

$x21 + x22 + x23 + x24 = 1$ (Employee 2)

$x31 + x32 + x33 + x34 = 1$ (Employee 3)

$x41 + x42 + x43 + x44 = 1$ (Employee 4)

(Each Employee is assigned to exactly 1 Job)

$x11, x12, x13, x14, x21, x22, x23, x24, x31, x32, x33, x34, x41, x42, x43, x44$
$= 0$ or 1

(Either no or yes to each possible assignment of employees to jobs)

Linear programming using the Simplex Method (which assumes that the $x_{ij} \geq 0$) can be applied to solve the Assignment Problem. An important attribute of the peculiar structure of the constraints of the Assignment Problem given above is that when linear programming ($x_{ij} \geq 0$) is applied, the solution will *always be integer* (i.e. $x_{ij} = 0$ or 1) as required by the constraints of the Assignment Problem.

Hence the decision variables for the Assignment Problem can be changed from $x_{ij} = 0$ or 1 to $x_{ij} \geq 0$ and solved using linear programming without affecting the optimal solution (all decision variables x_{ij} will be integer 0 or 1) or the value of the optimal objective function.

Notes:

• There are several matrix manipulation algorithms available which solve the standard Assignment Problem given above even faster than the Simplex Method. Computers often use the "Hungarian Method" (see below) rather than the Simplex Method to solve the Assignment Problem.

• The Simplex Method, unlike the matrix manipulation algorithms, can be applied to variations of the above Assignment Problem as long as the constraints remain linear. Such complex linear constraints might mandate assignments or preclude assignments (as designated by the manager) of particular employees to particular jobs.

• There are two optimal solutions to the Assignment Problem given in Example 7. Given the *subjective* nature of the compatibility scores (determined by the manager after interviews), it makes sense for the manager to examine both optimal solutions (and other pertinent information) before deciding the best assignment of employees to jobs.

Hungarian Method

Example 8: (Forever Temps Hungarian Method Assignment Problem). Recall the employee job compatibility Forever Temps Assignment Problem given in Example 7. Now assume that for four different employees, the job compatibility scores are as follows:

Employee Job Compatibility Scores

	Job 1	Job 2	Job 3	Job 4
Employee 1	20	70	40	30
Employee 2	70	90	60	40
Employee 3	60	20	50	30
Employee 4	70	50	30	10

A score of 0 indicates the employee has no compatibility with the job's requirements while a score of 100 indicates the employee is completely compatible with the job's requirements. Use the Hungarian Method assign the four employees to the four jobs to *maximize the sum* of the four compatibility scores.

Answer: The Hungarian Method proceeds as follows:

Step 1: Determine the largest numerical value in each *row* . Subtract all entries in each of the rows from the maximum value in the row. After this step, there should be at least one zero in each row.

The largest numerical entries for the Employee 1, Employee 2, Employee 3, and Employee 4 rows are 70, 90, 60, and 70, respectively. After subtraction of the entries in the respective rows from these values, the job compatibility matrix becomes:

Hungarian Method (Step 1)

	Job 1	Job 2	Job 3	Job 4
Employee 1	50	0	30	40
Employee 2	20	0	30	50
Employee 3	0	40	10	30
Employee 4	0	20	40	60

Note: If the objective of this assignment is to *minimize* the total of the scores, the *smallest* numerical value in each row is selected in Step 1. For each row, the smallest numerical value is subtracted from each entry. Whether the objective is to maximize or minimize the sum of the assignment scores, there should never be negative values in the matrix after Step 1 (or any step of the Hungarian Method).

Step 2: For the matrix resulting from Step 1, subtract the smallest numerical value in each *column* from each entry in that column. After this step, there should be at least one zero in each column.

The smallest numerical values for the Job 1, Job 2, Job 3, and Job 4 columns are 0, 0, 10, and 30, respectively. After subtraction of these values from each column, the job compatibility matrix becomes:

Hungarian Method (Step 2)

	Job 1	Job 2	Job 3	Job 4
Employee 1	50	0	20	10
Employee 2	20	0	20	20
Employee 3	0	40	0	0
Employee 4	0	20	30	30

Step 3: Determine if the *minimum* number of lines (vertical and/or horizontal) needed to cover all of the zeros in the matrix following Step 2 is equal to the size of the matrix n (equal to 4, the number of Employees and the number of Jobs in this example). If so, an optimal solution has been determined. If the minimum number of lines required to cover each of the zeros is less than n, go to Step 4.

In the matrix resulting from Step 2, all the zeros in the matrix can be covered with a combination of only three lines (two vertical and one horizontal):

Hungarian Method (Step 3)

	Job 1	Job 2	Job 3	Job 4
Employee 1	50	0	20	10
Employee 2	20	0	20	20
Employee 3	0	40 ·	0	0
Employee 4	0	20	30	30

Since the minimum number of lines (three) required to cover all the zeros in the matrix is less than the size of the matrix (four), a optimal solution has not been determined.

Step 4: Determine the smallest numerical value not covered by a horizontal or vertical line for the matrix determined in Step 3. Subtract this value from itself and all values not covered by a horizontal or a vertical line. Add this value to the value at the *intersection* of each of the lines covering the zeros in the matrix. Repeat Step 3.

The smallest uncovered value in the matrix determined in Step 3 is 10. This value is subtracted from all uncovered (unshaded) values in the matrix. This value is added to the value 0 at the intersection of the lines for Employee 3 and Job 1 and the value 40 at the intersection of the lines for Employee 3 and Job 2:

Hungarian Method (Step 4)

	Job 1	Job 2	Job 3	Job 4
Employee 1	50	0	10	0
Employee 2	20	0	10	10
Employee 3	10	50	0	0
Employee 4	0	20	20	20

Step 3: Since the zeros in the matrix after Step 4 can not be covered with less than a combination of four horizontal and vertical lines, an optimal solution has been determined.

Assignment: There is only one zero in the Employee 2 row (Job 2). Employee 2 is assigned to Job 2. Also, there is only one zero in the Employee 4 row (Job 1). Employee 4 is assigned to Job 1. After Job 1 and Job 2 are eliminated from possible assignment, there is only 1 zero left in the Employee 1 row (Job 4). Employee 1 is assigned to Job 4. Employee 3 then must be assigned to Job 3.

The total compatibility score of the this assignment = $30 + 90 + 50 + 70 = 240$.

Note: The Hungarian Method need not find a unique solution. In choosing the best assignment, the decision maker is given the flexibility to use other relevant (perhaps non-quantitative) criteria in addition to the matrix (quantitative) entries.

9.7   Transportation Problem

Example 9: (Wacko Soft Drink LP Transportation Problem). The Wacko Soft Drink Company has 4 bottling plants (denoted 1, 2, 3, 4) which service 10 distributors (denoted 1, 2, $\cdots$, 10). The bottling plants and distributors are spread out in an non-uniform manner across the United States of America.

Each Wacko plant produces a weekly supply of units of soft drinks. Each Wacko distributor must satisfy a weekly demand for units of soft drinks. There is a cost to ship a unit of soft drinks from bottling plant i = 1, 2, 3, 4 to a distributor j = 1, 2, $\cdots$, 10. All this data is provided in the following table:

Unit Cost of Shipping Soft Drinks From Plants to Distributors

		1	2	3	4	5	6	7	8	9	10	Supply
P	1	40	36	34	20	20	13	10	14	10	18	400
l a	2	54	46	30	8	8	12	11	6	19	29	600
n	3	18	9	4	8	7	6	10	21	21	13	500
t	4	3	4	4	13	13	7	6	26	21	3	800
Demand		86	140	190	230	600	100	250	300	200	160	

The Wacko Soft Drink Transportation Problem seeks to determine how many units of soft drinks should be shipped each week from each bottling plant i to each distributor j in order to:

- Satisfy weekly demand at each distributor (one set of constraints)

339

- Ship no more than the weekly supply at each bottling plant (another set of constraints)
- *Minimize the total weekly shipping costs* (the objective function)

Note: The Transportation Problem minimizes the cost of *shipping*. The decision of how many units to produce each week (or each inventory cycle) in order to minimize production set-up and holding costs (see Chapter 4) are addressed outside of the Transportation Problem. The total weekly shipping cost is determined only by how many units per week are transported on each of the shipping routes in order to meet demand.

Answer: A transportation software package provides the following optimal solution:

		1	2	3	4	5	6	7	8	9	10	Supply
P						Distributor						
l	1	0	0	0	0	0	0	126	30	200	0	356 (<400)
a	2	0	0	0	230	100	0	0	270	0	0	600
n	3	0	0	0	0	500	0	0	0	0	0	500
t	4	86	140	190	0	0	100	124	0	0	160	800
Demand		86	140	190	230	600	100	250	300	200	160	

Minimum Total Shipping Cost = $126 \cdot \$10 + 30 \cdot \$14 + 200 \cdot \$10 + \cdots + 160 \cdot \$3$

$$= \$14,942.$$

LP Formulation of the Wacko Soft Drink Transportation Problem

The Wacko Soft Drink Transportation Problem will have no feasible linear programming solution unless the total amount of supply of soft drink units produced at the bottling plants each week is *greater than or equal* to the total amount of demand at the distributors each week. In this example, the total supply = 2300 units (400 + 600 + 500 + 800) each week exceeds the total demand = 2256 units (86 + 140 + ⋯ + 160) each week.

The decision variables for the Wacko Soft Drink Transportation Problem (the number of units to ship from bottling plants to distributors each week) can take on all non-negative *integer* points. This is in contrast to the Assignment Problem where the decision variables (whether or not to make an assignment) can take on only the integer values 0 or 1.

Let the decision variables X_{ij} denote the integer number of soft drink units to ship from plant i to distributor j each week.

The linear programming formulation (which is actually an integer programming formulation) for the Wacko Soft Drink Transportation Problem is then given by:

Min $40 \cdot X_{11} + 36 \cdot X_{12} + 34 \cdot X_{13} + \cdots + 18 \cdot X_{1,10} +$
      $54 \cdot X_{21} + 46 \cdot X_{22} + 30 \cdot X_{23} + \cdots + 29 \cdot X_{2,10} +$
      $18 \cdot X_{31} + \ 9 \cdot X_{32} + \ 4 \cdot X_{33} + \cdots + 13 \cdot X_{3,10} +$
      $\ 3 \cdot X_{41} + \ 4 \cdot X_{42} + \ 4 \cdot X_{43} + \cdots + \ 3 \cdot X_{4,10}$

(Minimize the total weekly soft drink shipping cost)

$$X11 + X12 + X13 + \cdots + X1,10 \leq 400$$
$$X21 + X22 + X23 + \cdots + X2,10 \leq 600$$
$$X31 + X12 + X33 + \cdots + X3,10 \leq 500$$
$$X41 + X42 + X43 + \cdots + X4,10 \leq 800$$

(Each bottling plant must ship no more soft drink units than its weekly supply)

$$X11 + X21 + X31 + X41 \geq 86$$
$$X12 + X22 + X33 + X42 \geq 140$$
$$X13 + X23 + X33 + X43 \geq 190$$

.
.
.

$$X1,10 + X2,10 + X3,10 + X4,10 \geq 160$$

(Each distributorship must satisfy its weekly demand for soft drink units)

$$X11, X12, X13, \cdots, X1,10 \text{ integer}$$
$$X21, X22, X23, \cdots, X2,10 \text{ integer}$$
$$X31, X32, X33, \cdots, X3,10 \text{ integer}$$
$$X41, X42, X43, \cdots, X4,10 \text{ integer}$$

(All number of units shipped integer)

Linear programming using the Simplex Method (which assumes that the $Xij \geq 0$) can be applied to solve the Transportation Problem. An important attribute of the peculiar structure of the constraints for the Transportation Problem is that when linear programming ($Xij \geq 0$) is applied, the solution will *always be integer* as required by the constraints of the Transportation Problem.

Hence the decision variables for the above formulation of the Transportation Problem can be changed from Xij integer to $Xij \geq 0$ without affecting the optimal solution (all decision variables will be integer) or the optimal objective function value (e.g. the minimum total weekly shipping cost).

Notes:

• A typical Transportation Problem has many supply and demand points. Computers often use matrix manipulation algorithms (see Examples 10 through 13) rather than the Simplex Method to solve the Transportation Problem. These algorithms take advantage of the matrix form of the Transportation Problem to find the optimal solution even more quickly than the Simplex Method.

• Matrix manipulation algorithms (unlike the Simplex Method) do not typically require total supply to be at least as large as total demand. These algorithms will ship whatever supply is on hand most economically (at lowest cost) and allow some demand unsatisfied.

• The Simplex Method, unlike the matrix manipulation algorithms, can be applied to complex variations of the above Transportation Problem as long as the constraints remain linear. Such complex linear constraints might mandate amounts shipped or routes used, limit amounts shipped on particular routes, preclude routes used. etc.

Northwest Corner Algorithm

Example 10: (Togs R Us Northwest Corner Transportation Problem). Togs R Us is a specialty clothing store with four supply warehouses and four retail outlets. The cost in dollars of shipping clothing units from supply warehouses to retail outlets as well as the monthly supply and demand requirements are given as follows:

Outlet Warehouse	1	2	3	4	Supply
1	5	15	16	40	15
2	35	10	25	18	6
3	20	30	6	45	14
4	40	20	46	7	11
Demand	10	12	15	9	46

Use the Northwest Corner Algorithm to determine a transportation schedule for shipping clothing units from warehouses to outlets which:

- Satisfies monthly demand at each outlet
- Does not exceed monthly supply at each warehouse

Answer: The Northwest Corner Algorithm starts at the *upper left hand corner* of the matrix (the northwest corner) and uses the supply of Warehouse 1 (row 1) to satisfy as much of the outlet demand as possible. The 15 units of supply at Warehouse 1 is used to completely satisfy the demand of Outlet 1 (10 units) and partially satisfied the demand of Outlet 2 (12 units). The demands of Outlet 3 and Outlet 4 have not been addressed yet:

Outlet Warehouse	1	2	3	4	Supply
1	5 10	15 5	16	40	15
2	35	10	25	18	6
3	20	30	6	45	14
4	40	20	46	7	11
Demand	10	12	15	9	46

The 6 units of supply of Warehouse 2 (row 2) is then used to satisfy as much *remaining* outlet demand as possible. The demand of Outlet 1 has already been satisfied so that the entire 6 units of supply is allocated to Outlet 2 (7 units of demand remaining):

Warehouse ＼ Outlet	1	2	3	4	Supply
1	5 10	15 5	16	40	15
2	35	10 6	25	18	6
3	20	30	6	45	14
4	40	20	46	7	11
Demand	10	12	15	9	46

Since Outlet 2 has been allocated only 11 units (short of its demand for 12 units), the first of the 14 units of supply from Warehouse 3 (row 3) will be allocated to Outlet 2. The remaining 13 units of supply from Warehouse 3 will be used to partially satisfy the demand requirement of Outlet 3 (15 units):

Warehouse ＼ Outlet	1	2	3	4	Supply
1	5 10	15 5	16	40	15
2	35	10 6	25	18	6
3	20	30 1	6 13	45	14
4	40	20	46	7	11
Demand	10	12	15	9	46

Since Outlet 3 is 2 units short of its demand for 15 units, the first 2 units of supply from Warehouse 4 (row 4) will be allocated to Outlet 3. Finally, the remaining 9 units of supply from Warehouse 4 is used to satisfy the demand requirement of Outlet 4 (9 units):

Warehouse ＼ Outlet	1	2	3	4	Supply
1	5 10	15 5	16	40	15
2	35	10 6	25	18	6
3	20	30 1	6 13	45	14
4	40	20	46 2	7 9	11
Demand	10	12	15	9	46

The above allocation of supply to demand represents the Northwest Corner Algorithm transportation solution.

The shipping cost for the Northwest Corner Algorithm solution = 10($5) + 5($15) + 6($10) + 1($30) + 13($6) + 2($46) + 9($7) = $448.

Notes:

• The Northwest Corner Algorithm rarely provides an optimal transportation algorithm. It allocates supply to demand points moving from the upper right left hand (northwest) corner of the matrix to the lower right hand corner without regard to the cost of shipping. Since cost is not considered, higher cost shipping routes are often assigned while lower cost shipping routes are ignored.

• However, the Northwest Corner Algorithm is easily programmable and often serves as the *starting solution* for transportation algorithms which account for cost and produce optimal or near-optimal allocations of supply to demand.

Least Cost Transportation Algorithm

Example 11: (Togs R Us Least Cost Transportation Problem). Consider again the cost in dollars of shipping clothing units from supply warehouses to retail outlets as well as the monthly supply and demand requirements given in Example 10:

Outlet Warehouse	1	2	3	4	Supply
1	5	15	16	40	15
2	35	10	25	18	6
3	20	30	6	45	14
4	40	20	46	7	11
Demand	10	12	15	9	46

Use the Least Cost Transportation Algorithm to determine a transportation schedule for shipping clothing units from warehouses to outlets which:

    - Satisfies monthly demand at each outlet
    - Does not exceed monthly supply at each warehouse

Answer: Each iteration of the Least Cost Transportation Algorithm allocates as much supply as possible (demand requirements may not be violated) to the cell with the smallest transportation cost. Ties may be broken arbitrarily.

Iteration 1: The cost of shipping from Warehouse 1 to Outlet 1 (cell 1-1) is the smallest transportation cost ($5 per unit) from any supply warehouse to any factory outlet. As much supply as possible from Warehouse 1 (10 units) is allocated to Outlet 1. Allocation of more than 10 units would violate the demand requirement (10 units) of Outlet 1:

Outlet Warehouse	1	2	3	4	Supply
1	5 10	15	16	40	15
2	35	10	25	18	6
3	20	30	6	45	14
4	40	20	46	7	11
Demand	10	12	15	9	46

Iteration 2: The cost of shipping from Warehouse 3 to Outlet 3 (cell 3-3) is the next smallest transportation cost ($6 per unit). As much supply as possible from Warehouse 3 (14 units) is allocated to the demand requirement (15 units) of Outlet 3:

Outlet Warehouse	1	2	3	4	Supply
1	5 10	15	16	40	15
2	35	10	25	18	6
3	20	30	6 14	45	14
4	40	20	46	7	11
Demand	10	12	15	9	46

Iterations of this algorithm continue using the next smallest transportation cost cell until all supply has been allocated and all demand has been met. The Least Cost Transportation Algorithm solution is as follows:

Outlet Warehouse	1	2	3	4	Supply
1	5 10	15 5	16	40	15
2	35	10 6	25	18	6
3	20	30	6 14	45	14
4	40	20 1	46 1	7 9	11
Demand	10	12	15	9	46

The shipping cost for the Least Cost Transportation Algorithm solution = 10($5) + 5($15) + 6($10) + 14($6) + 1($20) + 1($46) + 9($7) = $398 (compared to $448 using the Northwest Corner Algorithm).

345

Notes:

• The Least Cost Transportation Algorithm typically provides a lower cost solution than the Northwest Corner Algorithm (which does not account for cost in its allocation).

• The Least Cost Transportation Algorithm is not typically optimal since it examines only *one cell at a time* rather than the shipping costs of the *entire* system of supply and demand nodes. The result is that higher cost shipping routes may be assigned (due to demand requirements at lower cost cells already being satisfied) while lower cost routes are ignored

Vogel's Approximation Method (VAM)

Example 12: (Togs R Us VAM Transportation Problem). Consider again the cost in dollars of shipping clothing units from supply warehouses to retail outlets as well as the monthly supply and demand requirements given in Example 10:

Outlet / Warehouse	1	2	3	4	Supply
1	5	15	16	40	15
2	35	10	25	18	6
3	20	30	6	45	14
4	40	20	46	7	11
Demand	10	12	15	9	46

Use Vogel's Approximation Method (VAM) to determine a transportation schedule for shipping clothing units from warehouses to outlets which:

    - Satisfies monthly demand at each outlet
    - Does not exceed monthly supply at each warehouse

Answer: There are five steps involved in applying the VAM algorithm:

Step 1: Determine the difference between the two *lowest* cost cells in each row and column.

    Row 1: $15 -$ 5 = $10
    Row 2: $18 - $10 = $8
    Row 3: $20 - $6 = $14
    Row 4: $20 - $7 = $13

    Column 1: $20 - $5 = $15
    Column 2: $15 - $10 = $5
    Column 3: $16 - $6 = $10
    Column 4: $18 - $7 = $11

Step 2: Identify the row or column with the largest difference. Ties may be broken arbitrarily.

Column 1 has the largest difference ($15) between its two lowest cost cells.

Step 3: Allocate as much supply as possible (demand requirements may not be violated) to the lowest cost cell in the row or column determined to have the largest difference.

Cell 1-1 (Warehouse 1- Outlet 1) is the lowest cost cell ($5) in Column 1. As much of the 15 units of Warehouse 1 supply as possible (10 units) is allocated to Outlet 1:

Outlet Warehouse	1	2	3	4	Supply
1	5 10	15	16	40	15
2	35	10	25	18	6
3	20	30	6	45	14
4	40	20	46	7	11
Demand	10	12	15	9	46

Step 4: If all supply and demand requirements are satisfied, the VAM solution has been determined. Otherwise, go to Step 5.

Only the demand of Outlet 1 (Column 1) is met so far in this example.

Step 5: For all rows whose supply requirements have not been completely met and all columns whose demand requirements have not been completely met, calculate the difference between the two lowest cost cells. Then go back to Step 2.

The differences for the remaining rows and columns are as follows:

Row 1: 16 - 15 = 1
Row 2: 18 - 10 = 8
Row 3: 30 - 6 = 24
Row 4: 20 - 7 = 13

Column 2: 15 - 10 = 5
Column 3: 16 - 6 = 10
Column 4: 18 - 7 = 11

Step 2: The row or column with the largest difference is Row 3.

Step 3: Cell 3-3 (Warehouse 3 - Outlet 3) is the lowest cost cell ($6) in Row 3. As much of the 14 units of Warehouse 3 supply as possible (all 14 units) is allocated to Outlet 3:

Outlet Warehouse	1	2	3	4	Supply
1	5 10	15	16	40	15
2	35	10	25	18	6
3	20	30	6 14	45	14
4	40	20	46	7	11
Demand	10	12	15	9	46

Step 4: Only the demand at Outlet 1 has been completely met and the supply of Warehouse 3 has been completely allocated so far in this example. Go to Step 5.

Step 5: The differences between the two lowest cost cells for all rows and columns whose supply or demand requirements, respectively, have not been completely satisfied are as follows:

Row 1: 16 - 15 = 1
Row 2: 18 - 10 = 8
Row 4: 20 - 7 = 13

Column 2: 15 - 10 = 5
Column 3: 16 - 6 = 10
Column 4: 18 - 7 = 11

Step 2: The row or column with the largest difference is Row 4.

Step 3: Cell 4-4 (Warehouse 4 - Outlet 4) is the lowest cost ($7) in Row 4. As much of the 11 units of Warehouse 4 supply as possible (9 units) is allocated to Outlet 4:

Outlet Warehouse	1	2	3	4	Supply
1	5 10	15	16	40	15
2	35	10	25	18	6
3	20	30	6 14	45	14
4	40	20	46	7 9	11
Demand	10	12	15	9	46

Vogel's Approximation Method (VAM) algorithm will continue for 4 more iterations (7 in total) before all supply and demand requirements are met. The final transportation solution using the VAM algorithm is:

Outlet Warehouse	1	2	3	4	Supply
1	5 10	15 4	16 1	40	15
2	35	10 6	25	18	6
3	20	30	6 14	45	14
4	40	20 2	46	7 9	11
Demand	10	12	15	9	46

The shipping cost for the Vogel's Approximation Method solution = 10($5) + 4($15) + 1($16) + 6($10) + 14($6) + 2($20) + 9($7) = $373 (compared with $448 for the Northwest Corner Algorithm and $398 for the Least Cost Transportation Algorithm).

Notes:

• Vogel's Approximation Method is a heuristic method (based on common sense rules and practical experience) which determines a near-optimal transportation solution.

• In a surprisingly large percentage of transportation problems, Vogel's Approximation Method determines an optimal solution.

• To determine whether a solution is optimal, each unused cell must be examined to determine whether a shift of supply into that cell will provide a lower total shipping cost. When all cells have been examined and no shifts lowering the total shipping cost are possible, the solution is optimal. A method used to examine all cells and make the cost-saving shifts required to determine the *optimal solution* is the Stepping Stone Method.

Stepping Stone Method

Example 13: (Togs R Us Stepping Stone Method Transportation Problem). Consider again the cost in dollars of shipping clothing units from supply warehouses to retail outlets as well as the monthly supply and demand requirements given in Example 10:

Outlet Warehouse	1	2	3	4	Supply
1	5	15	16	40	15
2	35	10	25	18	6
3	20	30	6	45	14
4	40	20	46	7	11
Demand	10	12	15	9	46

Use the Stepping Stone Method to determine the optimal transportation schedule for shipping clothing units from warehouses to outlets which:

- Satisfes monthly demand at each outlet
- Does not exceed monthly supply at each warehouse
- Minimizes total shipping costs

Answer: The Stepping Stone Method is applied to an initial transportation problem solution (typically determined using the Northwest Corner Algorithm). From Example 10, the Northwest Corner Algorithm provides the following initial (non-optimal) solution for the Stepping Stone Method algorithm:

Outlet Warehouse	1	2	3	4	Supply
1	5 10	15 5	16	40	15
2	35	10 6	25	18	6
3	20	30 1	6 13	45	14
4	40	20	46 2	7 9	11
Demand	10	12	15	9	46

The total shipping cost of this initial (non-optimal) solution = 10($5) + 5($15) + 6($10) + 1($30) + 13($6) + 2($46) + 9($7) = $448.

Step 1: Choose any empty cell (where no supply has been allocated). Identify a closed path leading to that cell.

Examine the empty cell 2-1 denoting the units shipped from Warehouse 2 to Outlet 1. A closed path (see matrix below) from this cell back to itself would proceed to cell 1-1 (Warehouse 1 - Outlet 1), cell 1-2 (Warehouse 1 - Outlet 2), cell 2-2 (Warehouse 2 - Cell 2) and back to cell 2-1 (Warehouse 2 - Outlet 1). Note that there are no empty cells on this path other than the original empty cell 2-1.

Outlet Warehouse	1	2	3	4	Supply
1	5 10	15 5	16	40	15
2	35	10 6	25	18	6
3	20	30 1	6 13	45	14
4	40	20	46 2	7 9	11
Demand	10	12	15	9	46

Note: A closed path is a path made up of a combination of vertical and horizontal lines at 90 degree angles leading from an empty cell back to itself. On a closed path, there may be no other empty cell other than the starting empty cell. Thus, the path must proceed at right angles from an empty cell through a series of filled cells and back to the original empty cell

Step 2: Shift one unit into the empty cell and modify the remaining cells on the corners of the path so that the supply and demand requirements are not violated. This requires that when one unit is added to or subtracted from a cell in a given row or column, one unit is subtracted from or added to, respectively, another cell in that row or column.

The following addition and subtractions along the closed path from cell 2-1 must take place:

One unit is added to cell 2-1 (Warehouse 2 - Outlet 1)
One unit is subtracted from cell 1-1 (Warehouse 1 - Outlet 1)
One unit is added to cell 1-2 (Warehouse 1 - Outlet 2)
One unit is subtracted from cell 2-2 (Warehouse 2 - Outlet 2)

Step 3: Determine whether the shift lowers the total shipping cost. If so, the shift is desirable. Move as many units as possible into the empty cell.

The additional shipping costs of this shift of one unit into cell 2-1 are as follows:

Cell 2-1: +$35
Cell 1-2: +$15
+$50

The shipping cost savings of this shift of one unit into cell 2-1 are as follows:

Cell 1-1: -$5
Cell 2-1: -$10
-$15

Hence, a shift of one unit of supply into cell 2-1 (from Warehouse 2 to Outlet 1) would increase total shipping costs by $50 - $15 = $35. This is not a desirable shift. Typically, a plus sign is placed in such an empty cell (see matrix below). This denotes that the cell has been examined and the shift results in an increase in the total shipping cost.

Outlet Warehouse	1	2	3	4	Supply
1	5 10	15 5	16	40	15
2	35 +	10 6	25	18	6
3	20	30 1	6 13	45	14
4	40	20	46 2	7 9	11
Demand	10	12	15	9	46

On the other hand, examine the empty cell 4-2 (Warehouse 4 - Outlet 2). A closed path (see matrix above) from cell 4-2 back to itself would pass through cell 3-2, cell 3-3, and cell 4-3.

One unit is added to cell 4-2. In order not to violate supply and demand requirements, one unit is subtracted from cell 3-2, one unit is added to cell 3-3, and one unit is subtracted from cell 4-3.

The additional shipping costs of this shift of one unit into cell 4-2 are as follows:

    Cell 4-2: +$20
    Cell 3-3: +$6
              +$26

The shipping cost savings of this shift of one unit into cell 4-2 are as follows:

    Cell 3-2: -$30
    Cell 2-1: -$46
              -$76

Hence, a shift of one unit into the empty cell 4-2 results in a decrease in total shipping costs of $76 - $26 = $50. This is a desirable shift. One would like to shift as many units as possible into cell 4-2 due to the cost advantage of $50 for every unit shifted.

However, the most that can be shifted without violating the demand and supply requirements is 1 unit. In general, the most that can be shifted in an empty cell where there is found to be a cost advantage is the amount of the smallest allocation to a cell from which a subtraction is made (1 unit for cell 4-2 on the closed path in the above matrix). The transportation matrix updated by an iteration of the Stepping Stone Algorithm is:

Updated Transportation Matrix (Stepping Stone Algorithm)

Outlet Warehouse	1	2	3	4	Supply
1	5 10	15 5	16	40	15
2	35 +	10 6	25	18	6
3	20	30 +	6 14	45	14
4	40	20 1	46 1	7 9	11
Demand	10	12	15	9	46

The total shipping cost for the updated solution = 10($5) + 5($15) + 6($10) + 14($6) + 1($20) + 1($46) + 9($7) = $398 (a decrease of $50 from the previous cost of $448).

Step 4: Repeat Steps 1 through 3 until each empty cell has been examined to determine if a shift of supply into the cell lowers the total shipping cost.

Examine empty cell 1-3 (Warehouse 1 - Outlet 3). A closed path (see matrix above) from cell 1-3 back to itself would pass through cell 4-3, cell 4-2, and cell 1-2.

One unit is added to cell 1-3. In order not to violate supply and demand requirements, one unit is subtracted from cell 4-3, one unit is added to cell 4-2 and one unit is subtracted from cell 1-2.

The additional shipping costs of this shift of one unit in cell 1-3 are as follows:

    Cell 1-3: +$16
    Cell 4-2: +$20
              +$36

The shipping cost savings of this shift of one unit into cell 1-3 are as follows:

    Cell 4-3: -$46
    Cell 1-2: -$15
              -$61

There is a total shipping cost savings of $61 - $36 = $25 for each unit shifted into empty cell 1-3. This is a desirable shift . One would like to shift as many units as possible into cell 1-3 due to the savings of $25 for every unit shifted.

However, the most that can be shifted without violating the demand and supply requirements is 1 unit since this is the amount of the smallest allocation to a cell from which a subtraction is made (cell 4-3 on the above closed path). The updated transportation matrix (which will be determined to be optimal) using the Stepping Stone Algorithm is:

Optimal Transportation Matrix (Stepping Stone Algorithm)

Outlet Warehouse	1	2	3	4	Supply
1	5 10	15 4	16 1	40 +	15
2	35 +	10 6	25 +	18 +	6
3	20 +	30 +	6 14	45 +	14
4	40 +	20 2	46 +	7 9	11
Demand	10	12	15	9	46

The total shipping cost of this updated (optimal) allocation = 10($5) + 4($15) + 1($16) + 6($10) + 14($6) + 2($20) + 9($7) = $373 (a decrease of $25 from the previous total shipping cost of $398).

One can examine all other empty cells and determine that none provide a decrease in the total shipping costs. The above matrix (where each empty cell has a plus sign) provides the optimal solution to the transportation problem. Note that this is the same solution determined by Vogel's Approximation Method (see Example 12).

9.8 Integer Programming

Most real world problems involve decision variables which must take on integer values. That is, it is impossible to build 0.47 hospitals or to ship 3517.65 units to an outlet. We have seen that there are special integer programming problems (i.e. the Assignment Problem and the Transportation Problem) whose structure allows linear programming (requiring only continuous decision variables) to determine the optimal integer programming solution (requiring integer decision variables). In general, however, it is not possible to solve integer programming problems directly using linear programming.

It is tempting to approximate the solution to integer programs by rounding off its linear programming solution to the nearest integer values (with the expectation that the resulting integer solution will be close to optimal). For example, if the linear programming solution to an integer program determines that it is optimal to ship 3517.65 units to an outlet, it is expected that if 3517 or 3518 units are shipped, the value of the integer program objective function will be close to its optimal value.

However, in many cases such an approach may not provide a near optimal or even a feasible solution to an integer programming problem. For example, if a linear programming solution to an integer hospital building program determines that it is optimal to build 0.47 hospitals, it may not be feasible to build no hospitals (due to demand constraints). Or it may be that the decision whether to build 0 or 1 hospitals is the critical decision to be made and is not amenable to an approximation.

The most widely used technique used to solve integer programming problems is the Branch and Bound Method. This technique involves the systematic application of linear programming (easy to solve) to subsproblems of the integer program until an optimal integer solution is determined.

Branch and Bound Method

Example 14: (Integer Program Branch and Bound Method Solution). Consider the following integer program (linear objective function and constraints, integer decision variables):

$$Max\ 170 \cdot x_1 + 100 \cdot x_2$$
$$x_1 + 0.75 \cdot x_2 \leq 10$$
$$x_1 + 0.27 \cdot x_2 \leq 8$$
$$0.5 \cdot x_1 + 0.75 \cdot x_2 \leq 8$$
$$x_1, x_2\ integer$$

Determine the optimal solution using the Branch and Bound Method.

Answer: Consider the corresponding linear programming solution to the above integer program. This linear program would have the same objective function and constraints but would require only that the decision variables be non-negative. This linear program would consider not only the feasible integer values considered by the above integer program but also other feasible non-negative values.

Since this linear program would consider all the feasible integer values considered by the integer program *plus* other continuous, non-negative values, the value of the objective function for this corresponding linear program must be at least as large as the value of the integer program. Thus the value of this corresponding linear program represents an *upper bound* on the value of the above integer program.

The notion that a linear program can provide an upper bound on the maximum value of the corresponding integer program is at the core of the Branch and Bound Method.

Step 1: Determine the *upper bound* on the value of the integer program by solving the corresponding linear program (all decision variables non-negative):

Node 1

$$Max\ 170 \cdot x_1\ +\ 100 \cdot x_2$$
$$x_1\ +\ 0.75 \cdot x_2 \le 10$$
$$x_1\ +\ 0.27 \cdot x_2 \le 8$$
$$0.5 \cdot x_1\ +\ 0.75 \cdot x_2 \le 8$$
$$x_1, x_2 \ge 0$$

The solution using the Simplex Method for this LP (designated Node 1) is:

$$Max\ = 1585.42$$
$$x_1\ = 6.875$$
$$x_2\ = 4.167$$

The optimal value of the integer program objective function can not be greater than 1585.42.

Step 2: Examine the upper bound solution. If all the decision variables are integer, the optimal integer program solution has been determined. Otherwise, go to Step 3.

Since neither 6.875 nor 4.167 is integer, the upper bound solution determined by linear program is not a feasible solution to the integer program.

Step 3: Determine a feasible *lower bound* on the value of the integer program by branching a linear programming node into two linear programming subproblem nodes using two mutually exclusive constraints on a non-integer (basic) decision variable.

Each non-integer decision variable can be described by an integer component k plus a fractional component Δ. The non-integer decision variable chosen for branching is the one with the largest fractional component ($x_1\ = 6.875 = k + \Delta = 6 + 0.875$ for Node 1). The mutually exclusive constraints added to the linear programming subproblems are:

$$x_1 \le k = 6$$
$$x_1 \ge k + 1 = 7$$

since k = 6 for the initial linear program (Node 1).

The two linear programming subproblem nodes (Nodes 2 and 3) become (see Figure 24):

	Node 2		Node 3

$$\text{Node 2}$$

$$Max\ 170 \cdot x_1 \ +\ 100 \cdot x_2$$
$$x_1 \ +\ 0.75 \cdot x_2 \leq 10$$
$$x_1 \ +\ 0.27 \cdot x_2 \leq 8$$
$$0.5 \cdot x_1 \ +\ 0.75 \cdot x_2 \leq 8$$
$$x_1 \leq 6$$

$$\text{Node 3}$$

$$Max\ 170 \cdot x_1 \ +\ 100 \cdot x_2$$
$$x_1 \ +\ 0.75 \cdot x_2 \leq 10$$
$$x_1 \ +\ 0.27 \cdot x_2 \leq 8$$
$$0.5 \cdot x_1 \ +\ 0.75 \cdot x_2 \leq 8$$
$$x_1 \geq 7$$

Solving the above linear programs yields the following solutions:

$$\text{Node 2}$$

$$Max\ = 1553.34$$
$$x_1 = 6$$
$$x_2 = 5.333$$

$$\text{Node 3}$$

$$Max\ = 1560.38$$
$$x_1 = 7$$
$$x_2 = 3.704$$

Note that the linear programs solved at Node 2 and Node 3 consist of the linear program solved at Node 1 plus an extra constraint for each node. Since an extra constraint has been added (and the feasible region is smaller), the optimal value of the objective functions for Node 2 (1553.34) and Node 3 (1560.38) is less than or equal to the optimal value for Node 1 (1585.42).

The solutions to the linear programs associated with Node 2 and Node 3 have non-integer values for a decision variable and thus are not feasible for the integer program. But these solutions provide new upper bounds. That is, no integer solution found branching off of Node 2 can be larger than 1553.34 while no integer solution found branching off of Node 3 can be larger than 1560.38 (see Figure 24).

Notes:

• When a node provides a feasible integer solution (a lower bound on the maximum value of the integer program), no further branching occurs from that node. Branching at this node is said to be *terminated*.

• When a node provides a non-integer solution (an integer program upper bound for that branch) with an objective function value less than or equal to a lower bound value already provided by another node with a integer solution, branching at this node is terminated.

• The optimal solution to the integer program occurs when a feasible integer solution is found and it is no longer possible to find an integer solution with a larger objective function value (all branches have been terminated).

Branching is continued for both Node 2 and Node 3 since neither node provides a feasible integer solution (which would have provided a lower bound on the objective function of the integer program), and neither node has an objective function value less than or equal to that of a node with an integer solution.

Node 2 (which has only the one non-integer decision variable x_2) is split into Node 4 and Node 5 by placing an additional constraint on x_2 (see Figure 24):

Node 4	**Node 5**
$Max\ 170 \cdot x_1\ +\ 100 \cdot x_2$	$Max\ 170 \cdot x_1\ +\ 100 \cdot x_2$
$x_1\ +\ 0.75 \cdot x_2\ \leq 10$	$x_1\ +\ 0.75 \cdot x_2\ \leq 10$
$x_1\ +\ 0.27 \cdot x_2\ \leq\ 8$	$x_1\ +\ 0.27 \cdot x_2\ \leq\ 8$
$0.5 \cdot x_1\ +\ 0.75 \cdot x_2\ \leq\ 8$	$0.5 \cdot x_1\ +\ 0.75 \cdot x_2\ \leq\ 8$
$x_1 \leq 6$	$x_1 \leq 6$
$x_2 \leq 5$	$x_2 \geq 6$

Solving the linear programs for Node 4 and Node 5 yields the following solutions:

Node 4	**Node 5**
$Max\ = 1520$	$Max\ = 1535$
$x_1\ = 6$	$x_1\ = 5.5$
$x_2\ = 5$	$x_2\ = 6$

Since Node 4 and Node 5 each has an extra constraint compared to Node 2, the value of the objective functions for Node 4 (1520) and Node 5 (1535) is less than that of Node 2 (1553.34).

Node 4 provides a feasible integer solution. Since this is only *one* point in the feasible region for the integer program, the *optimal* value for the integer program must be at least as large as 1520. Thus Node 4 provides a *lower bound* on the optimal value of the integer program and further branching from Node 4 is terminated (see Figure 24).

Node 5 provides a non-integer solution with a value of 1535. No solution to the integer program with a value larger than 1535 will be found branching from Node 5. However, it is possible to find an integer solution with a objection function value greater than 1520 (the current best integer objective function value at Node 4) branching from Node 5. Hence, branching from Node 5 will be done to search for a larger lower bound on the integer program than that provided by Node 4.

Node 5 (which has only the one non-integer decision variable x_1) is split into Node 6 and Node 7 by placing an additional constraint on x_1 (see Figure 24):

357

Node 6	Node 7
$Max\ 170 \cdot x_1 + 100 \cdot x_2$	$Max\ 170 \cdot x_1 + 100 \cdot x_2$
$x_1 + 0.75 \cdot x_2 \le 10$	$x_1 + 0.75 \cdot x_2 \le 10$
$x_1 + 0.27 \cdot x_2 \le 8$	$x_1 + 0.27 \cdot x_2 \le 8$
$0.5 \cdot x_1 + 0.75 \cdot x_2 \le 8$	$0.5 \cdot x_1 + 0.75 \cdot x_2 \le 8$
$x_1 \le 6$	$x_1 \le 6$
$x_2 \ge 6$	$x_2 \ge 6$
$x_1 \le 5$	$x_1 \ge 6$

Note that the constraints on x_1 at Node 6 can be replace by $x_1 \le 5$ while the constraints on x_1 at Node 7 could be replaced by $x_1 = 6$. Solving the linear programs for Node 6 and Node 7 yields the following solutions:

Node 6	Node 7
$Max = 1450$	Infeasible
$x_1 = 5$	
$x_2 = 6$	

Node 6 provides a feasible integer solution. The objective function value for this feasible integer solution provides another lower bound for the maximum value of the integer program. However, since the value 1450 is smaller than the value 1520 found at Node 4, this branch is terminated (since adding additional constraints would only lower the value of the objective function).

Node 7 results in an infeasible solution. It is not possible to satisfy constraint number 1 when x_1 is 6 and x_2 is at least 6. This branch is terminated.

To this point in the Branch and Bound Method (see Figure 24), the largest lower bound value for the objective function of the integer program is found at Node 4 (1520). Only Node 3 (with a value of 1560.38) has the possibility of providing an integer solution with a larger objective function value than 1520.

Node 3 is split into Node 8 and Node 9 by placing additional constraints on the only non-integer decision variable x_2 (see Figure 24):

Node 8	Node 9
$Max\ 170 \cdot x_1 + 100 \cdot x_2$	$Max\ 170 \cdot x_1 + 100 \cdot x_2$
$x_1 + 0.75 \cdot x_2 \le 10$	$x_1 + 0.75 \cdot x_2 \le 10$
$x_1 + 0.27 \cdot x_2 \le 8$	$x_1 + 0.27 \cdot x_2 \le 8$
$0.5 \cdot x_1 + 0.75 \cdot x_2 \le 8$	$0.5 \cdot x_1 + 0.75 \cdot x_2 \le 8$
$x_1 \ge 7$	$x_1 \ge 7$
$x_2 \le 3$	$x_2 \ge 4$

Solving the linear programs for Node 8 and Node 9 yields the following solutions:

Node 8	*Node 9*
$Max = 1490$	*Infeasible*
$x_1 = 7$	
$x_2 = 3$	

Node 8 provides a feasible integer solution and branching from Node 8 is terminated (see Figure 24). Note that the objective function value for Node 8 (1490) is smaller than the current largest lower bound provided by Node 4 (1520).

Node 9 results in an infeasible solution. It is not possible to satisfy constraint number 1 or 2 when x_1 is at least 7 and x_2 is at least 4. This branch is terminated.

The Branch and Bound Method process is finished since no branch offers the possibility of an integer solution with a larger objective function value than the 1520 provided by Node 4 (see Figure 24).

Integer Program Branch and Bound Method Solution

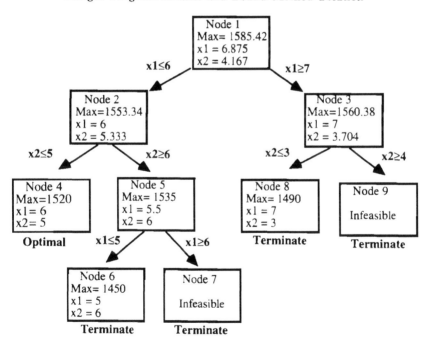

Figure 24

359

Notes:

• The real world application of integer programming is very limited due to the complexity of solving integer programs. Even the simple two variable integer program in Example 14 required that nine separate linear programs be solved. For large integer programs, the number of iterations quickly becomes unwieldy.

• When the magnitude of the decision variables allows, integer programs are often solved by rounding off the solution to the corresponding linear program. The resulting inaccuracy is absorbed to avoid the difficulty of solving the integer program.

9.9 Duality

Section 9.2 discussed the two types of LP formulations: Type 1 (where the objective is to maximize production output given a fixed level of input resources) and Type 2 (where the objective is to minimize input resources used to achieve a fixed level of production output).

Type 1 and Type 2 LPs can be cast into two distinct forms: the *primal* form (which has been used exclusively to this point in the treatment of LP) and the *dual* form. It is essential to understand that the primal and dual forms of an LP provided the *exact same optimal solution*. Why then even examine the dual form of an LP? The answer is two-fold:

1) **Insight.** The dual form provides an opposite but equivalent perspective on the optimization problem. This opposite point of view can provide added insight to the decision maker as to resource levels that would be advantageous to change.

2) **Computational Efficiency.** The dual form is sometimes a significantly easier problem to solve (in terms of iterations required by the Simplex Method or in terms of computation time for a computer software package) than the primal form.

Example 15: (Type 1 Dual Form). Recall the Type 1 GeeWhiz Computer Company LP from Example 2:

Decision Variables	L (# *laptops*) , D (# *desktops*)
Objective Function	*Max* $\$1000 \cdot L + \$2000 \cdot D$
Resource Constraints	$6 \cdot L + 10 \cdot D \leq 36500$ (W)
	$1 \cdot L + 3 \cdot D \leq 9500$ (M)
	$1.1 \cdot L + 2.1 \cdot D \leq 9000$ (I)
	$1 \cdot L + 0 \cdot D \leq 3500$ (F)
Non-Negative Decision Variables	$L \geq 0$, $D \geq 0$

Provide the dual form of the Type 1 GeeWhiz Computer Company LP.

Answer: The general steps to convert a primal form LP into a dual form LP are as follows:

Step 1: Convert the primal maximization objective function to a minimization objective function (or covert the primal minimization objective function to a maximization).

The dual form of the GeeWhiz LP will have a minimization objective function.

Step 2: If the primal LP is a maximization problem, all the constraints should have a ≤ sense. Likewise, if the primal LP is a minimization problem, all the constraints should have a ≥ sense. If the sense of the inequality is backwards (e.g. a ≥ constraint for a maximization problem), multiply the constraint through by -1 to obtain the correct sense of the inequality. If a constraint is an equality, split the constraint into two inequalities (one ≤ and one ≥). Multiply the inequality with the wrong sense through by -1 so that both inequalities have the correct sense.

All the constraints for the primal GeeWhiz LP (a maximization problem) all have the correct ≤ sense. There is no need to split an equality constraint into two inequalities or to multiply an inequality constraint with the wrong sense by -1.

Step 3: Each constraint of the primal LP becomes a decision variable for the dual LP. The right hand side value for each constraint in the primal LP become the coefficient of the corresponding decision variable of the objective function for the dual LP.

The resources workdays (W), computer components (M), inspection hours (I), and flat screens (F) become the decision variables. The objective function for the dual LP becomes:

$$Min\ 36500 \cdot W + 9500 \cdot M + 9000 \cdot I + 3500 \cdot F$$

Step 4: Each decision variable of the primal LP becomes a constraint for the dual LP. The coefficient of each decision variable in the objective function of the primal LP becomes the right hand side value of the corresponding constraint for the dual LP.

There are two decision variables, laptops (L) and desktops (D), in the primal LP. These become the two constraints of the dual LP. The right hand side values for the laptop (L) constraint and the desktop (D) constraint become:

 RHS

 (L): 1000

 (D): 2000

Step 5: The resource constraint coefficients in the nth *column* of the primal LP become the resource constraint coefficients of the nth *row* of the dual LP.

The dual form of the Type 1 GeeWhiz Computer Company LP is then:

Decision Variables: (W) , (M) , (I) , (F)

Objective Function: $Min\ 36500 \cdot W + 9500 \cdot M + 9000 \cdot I + 3500 \cdot F$

Resource Constraints: $6 \cdot W + 1 \cdot M + 1.1 \cdot I + 1 \cdot F \geq 1000$ (L)

                                 $10 \cdot W + 3 \cdot M + 2.1 \cdot I + 0 \cdot F \geq 2000$ (D)

Non-Negative Decision Variables: $W \geq 0$, $M \geq 0$, $I \geq 0$, $F \geq 0$

Notes:

• These same five steps can be used to convert a dual LP into a primal LP.

• Whereas the primal GeeWhiz LP has two decision variables and four constraints, its dual LP has four decision variables and two constraints.

Let's examine the insight provided by the dual form of the GeeWhiz LP:

1) Examine the *decision variables* of the dual LP. Whereas the primal LP is concerned finding the optimal value for the number of laptops (L) and desktops (D) to manufacture, the dual LP is concerned with the optimal value for the resources workdays (W), computer components (M), inspection hours (I), and flat screens (F). The optimal value for the dual LP decision variables provides the decision maker with the value of more of each type of resource. Note that the optimal value of each decision variable of the dual LP is the same as the *shadow price* for that resource determined by the primal LP.

2) Examine the *objective function* of the dual LP. Whereas the primal LP is concerned with maximization of the profit from the manufacture of laptops (L) and desktops (D), the dual LP is concerned with the efficient (minimum) usage of the available workdays (W), computer components (M), inspection hours (I), and flat screens (F) resources.

3) Examine the *resource* constraints of the dual LP. Whereas the primal LP is concerned that the laptops (L) and desktops (D) manufactured do not use more than the available resources, the dual LP is concerned that the sum of the marginal values of the resources required to manufacture each product be at least as large as the profit for that product.

For example, constraint (L) requires that the 6 workdays, 1 computer component, 1.1 inspection hours, and 1 flat screen required to produce a laptop times their corresponding shadow (marginal) prices be at least equal to the profit for a laptop. Otherwise, it is not the most efficient (minimum) use of resources to make another laptop.

The dual LP minimizes usage of resources by ensuring that resources are used to produce laptops (L) or a desktop (D) only as long as the marginal value of producing a laptop or a desktop is as least as large as the profit for that product.

Let's examine the computational efficiency of the dual form of the GeeWhiz LP:

1) Remember that the number of decision variables and constraints for the primal LP equals the number of constraints and decision variables, respectively, for the dual LP. It is typically significantly faster for a computer to solve an LP with a large number of decision

variables compared with the number of constraints than to solve an LP with a large number of constraints compared with the number of decision variables.

2) The dual Gee Whiz LP has only two constraints (and four decision variables) while the primal LP has four constraints (and two decision variables). One should expect that a computer will be able to solve the dual LP somewhat faster. However, when the primal LP has a large number of constraints versus the number of decision variables, there can be significant computational advantages in solving the dual LP rather than the primal LP.

Example 16: (Type 2 Dual Form). Recall from Example 3 the Roots N' Berries LP:

Decision Variables	$X_1, X_2, X_3, \cdots, X_8$ *(lbs. of each trail mix item in a package)*
Objective Function	*Minimize* $\$0.49 \cdot X_1 + \$1.25 \cdot X_2 + \$2.50 \cdot X_3 + \cdots + \$1.60 \cdot X_8$

Output Constraints

$$350 \cdot X_1 + 690 \cdot X_2 + 860 \cdot X_3 + \cdots + 420 \cdot X_8 \leq 350 \qquad (C)$$

$$20 \cdot X_1 + 85 \cdot X_2 + 72 \cdot X_3 + \cdots + 55 \cdot X_8 \geq 125 \qquad (P)$$

$$25 \cdot X_1 + 100 \cdot X_2 + 90 \cdot X_3 + \cdots + 30 \cdot X_8 \leq 50 \qquad (F)$$

$$6 \cdot X_1 + 10 \cdot X_2 + 12 \cdot X_3 + \cdots + 16 \cdot X_8 \geq 20 \qquad (I)$$

Non-Negative Decision Variables	$X_1 \geq 0$, $X_2 \geq 0$, $X_3 \geq 0$, $\cdots$, $X_8 \geq 0$

Provide the dual form of this Type 2 Roots N' Berries LP.

Answer: The dual form of the Type 2 Roots N' Berries LP is given by:

Decision Variables	C, P, F, I *(shadow prices for cal, protein, fat, and iron)*
Objective Function	*Maximize* $-350 \cdot C + 125 \cdot P - 50 \cdot F + 20 \cdot I$

Output Constraints

$$-350 \cdot C + 20 \cdot P - 25 \cdot F + 6 \cdot I \leq \$0.49 \qquad (Raisins)$$

$$-690 \cdot C + 85 \cdot P - 100 \cdot F + 10 \cdot I \leq \$1.25 \qquad (Peanuts)$$

$$-860 \cdot C + 72 \cdot P - 90 \cdot F + 12 \cdot I \leq \$2.50 \qquad (Almonds)$$

$$\bullet$$
$$\bullet$$
$$\bullet$$

$$-420 \cdot C + 55 \cdot P - 30 \cdot F + 16 \cdot I \leq \$1.60 \qquad (Dried\ Figs)$$

Non-Negative Decision Variables	$C \geq 0$, $P \geq 0$, $F \geq 0$, $I \geq 0$

Exercise 9

Linear Programming (LP)

Problem #1 (Linear Programming)

The Glitz Photography Studio produces portfolios for adults (A) and children (C) aspiring to acting or modeling careers. The services offered by Glitz consist of the photo session and the follow-up development procedures to produce a portfolio of pictures. On the average, the photo session for an adult client takes 1.5 hours and development procedures take 30 hours of staff time. On the average, the photo session for children takes 3 hours and development procedures take 10 hours of staff time. The Glitz Photography Studio has resources to handle up to 60 hours per week of photo sessions and up to 360 hours per week of development procedures. The studio charges a fixed fee of $1000 per adult and $200 per child to produce a portfolio.

a) Formulate a linear program (LP) to derive the number of adults and child to schedule each week in order to maximize the weekly revenue for the Glitz Photography Studio.

b) Solve the LP using the Graphical Method. The graphical solution may be checked using a LP software package, if desired.

c) Trace the path of the Simplex Method (from corner point to adjacent corner point until the optimal corner point is determined) for the LP to maximize the weekly revenue.

d) Suppose Glitz charged the same fixed fee for adults and children. The objective function then would be to maximize the number of clients served each week. Formulate a new LP to maximize the number of clients (number of adults plus number of children) per week.

e) Solve this new LP using the Graphical Method. The graphical solution may be checked using a LP software package, if desired.

f) Trace the path of the Simplex Method for this new LP to maximize the number of clients per week starting at the origin corner point.

g) Since only the objective function changed for this new LP to maximize the number of clients (the constraints and thus the feasible region remain the same as the previously solved LP to maximize weekly revenue), which corner point of the unchanged feasible region should be used as the initial corner point for this new LP.

Problem #2 (Linear Programming)

The Green Pastures nursing home can admit only three types of patients to its facility: post-surgical patients (P), hip fracture patients (H) and stroke patients (S). The administrator of the Green Pastures wishes to determine what mix of patients would maximize the total yearly revenue given the constraints on the amount of resources available at the nursing home. The most important resources consumed by these patients are: bed days (BD), nursing hours (NH), and physical therapy (PT) hours. Based on past patient admissions, the forecast for the amount of these resources used by each type of patient is as follows:

	Post-Surgical (P)	Hip Fracture (H)	Stroke (S)
Bed Days/Admission	20	80	140
Nursing Hours/Admission	35	170	360
PT Hours/Admission	4	100	100

In addition, the revenue provided to Green Pastures for each patient type admission is forecasted to be:

Post-Surgical	$3,000
Hip	$15,300
Stroke	$26,600

Finally, the limits on the amount of these three resources available at the Green Pastures nursing home each year are:

Bed days/year	36,500
Nursing hours/year	75,100
PT hours/year	39,000

2a) Formulate an LP to find the number of admissions per year of each patient type which maximizes total yearly revenue for the Green Pastures nursing home.

2b) Solve the LP formulated using an LP software package. Determine the optimal number of each patient types and the maximum total yearly revenue. What are the basic variables and non-basic variables for this LP? What are the shadow prices for the resources bed days/year, nursing hours/year, and PT hour/year?

2c) Would it be worthwhile to hire an additional nurse (at a cost of $32,000/year) in order to increase the amount of nursing hours available to 77,100/year? (Resolve the linear program with the new amount of nursing hours available).

2d) Assume that the additional nurse is NOT hired (nursing hours are again 75,100/year). Suppose that criterion for optimality is maximum *net* yearly revenue instead of maximum total yearly revenue and that the contributions to net revenue for each patient type are:

Patient Type	Net Revenue
Post-Surgical	$600
Hip	$1700
Stroke	$5500

Does the optimal number of each patient type change when maximum net revenue is used instead of maximum total revenue is used as the objective function? Do the binding constraints change?

Problem #3 (Simplex Method)

Consider an LP to maximize revenue under resource constraints. The final (optimal) simplex tableau from a software LP package is given as follows:

Decision Variables		Slack Variables				
$X1$	$X2$	$S1$	$S2$	$S3$		
0	0	25	0	40	650	
1	0	2	0	3	12	Resource 1
0	0	4	1	4	20	Resource 2
0	1	0	0	6	15	Resource 3

3a) What are the basic variables and their values in the optimal solution? What are the non-basic variables and their values?

3b) Which resources have binding and non-binding constraints at the optimal solution?

3c) What would be the effect on the value of maximum revenue if a small number of units of each resource could be added?

Problem #4 (Assignment Problem)

Ten students (names given below) in the Slippery Rock University intern program must be assigned as administrative residents to one of 10 hospitals (A through J). Each student has interviewed at four of the ten facilities and has given their ranking for their first through fourth choice to the university. Each hospital interviewed exactly four students and has also provided rankings of the students who interviewed at their facility to the university. The two tables below provide the students' rankings of the hospitals and the hospitals' rankings of the students. An "X" indicates that a student did not interview at that hospital.

Student Ranking of Hospitals

	A	B	C	D	E	F	G	H	I	J
Al	1	X	2	X	X	3	X	X	X	4
Betty	X	3	X	1	X	4	2	X	X	X
Carol	4	X	X	2	X	X	X	3	1	X
Dave	X	X	4	X	2	X	X	X	1	3
Eph	X	2	X	4	X	X	3	1	X	X
Fran	4	X	X	X	2	1	X	3	X	X
Gert	1	X	X	X	2	X	3	X	4	X
Hal	X	1	X	4	2	X	X	X	3	X
Irma	X	3	2	X	X	X	1	X	X	4
Jack	X	X	3	X	X	2	X	4	X	1

Hospital Ranking of Students

	A	B	C	D	E	F	G	H	I	J
Al	1	X	1	X	X	1	X	X	X	2
Betty	X	2	X	2	X	4	2	X	X	X
Carol	2	X	X	1	X	X	X	1	1	X
Dave	X	X	2	X	1	X	X	X	2	3
Eph	X	3	X	3	X	X	1	3	X	X
Fran	3	X	X	X	4	2	X	2	X	X
Gert	4	X	X	X	3	X	4	X	4	X
Hal	X	1	X	4	2	X	X	X	3	X
Irma	X	4	4	X	X	X	3	X	X	4
Jack	X	X	3	X	X	3	X	4	X	1

Use an Assignment Problem software package to answer the following questions:

4a) What is the best matching of students to hospitals using the student rankings?

4b) What is the best matching of students to hospitals using the hospital rankings?

Note: It must be decided how to handle the ranking for matches between a hospital and a student where the student did not interview. These cases are indicated by an "X" in the tables above. Since one is attempting to minimize the sum of the above rankings, one could place a high number in place of each X to insure that a student is not matched to that hospital. For example, one can replace all X's with a "5" or other number greater than 4.

4c) Compare the sum of the rankings using the student preferences in 4a) and for the university preferences in 4b). Is Slippery Rock University able to satisfy the students or the hospitals better if one of these matching criteria are used to implement the matching process?

Problem #5 (Transportation Problem)

The MegaMedical Corporation of America owns 6 hospitals on Wilshire Boulevard in West Los Angeles, each of which contracts out its blood tests. The blood tests for the 6 hospitals are contracted to one of three nearby labs which are also owned by MegaMedical. The costs to MegaMedical for the blood tests are identical at each of the three labs. The quality of the blood testing at each of the labs is also comparable.

The weekly volume of blood tests performed at each of the 6 hospitals is as follows:

Agony Hospital	2500
Bland Hospital	5400
Costly Hospital	1900
Dow Hospital	3800
Eager Hospital	4400
Farr Hospital	1300
Total	19,300

The maximum weekly volume of blood tests that can be analyzed by each of the three labs is as follows:

Lab #1	10000
Lab #2	5000
Lab #3	8000
Total	23,000

The table below provides the average shipping costs (in dollars) per blood sample routed from a hospital to a lab:

	Lab #1	Lab #2	Lab #3
Agony	1.20	0.60	1.00
Bland	1.30	0.40	0.20
Costly	0.40	1.00	1.20
Dow	0.60	1.10	0.40
Eager	0.80	0.50	0.50
Farr	0.20	1.50	0.70

Use a software Transportation Problem module to answer the following questions:

5a) What is the optimal routing of blood tests to labs which minimizes the total weekly shipping costs?

5b) By relocating resources, MegaMedical could increase the maximum volume for Lab #2 by 2500 tests/week while decreasing the volume at Lab #1 by the same amount. This move of equipment and manpower will increase the overall cost to MegaMedical for the lab tests by $1500/week. Is this relocation of resources economical for MegaMedical to consider?

Chapter 9

Linear Programming (LP) Additional Solved Problems

Problem #1 (Linear Programming)

The Women's Guild at Sea of Troubles Hospital sells packages of trail mix during the hospital's annual Spring fundraising event. These trail mix packages are quite popular and sell out each year. The Guild sells two types of packaged trail mixes at the fundraising event:

Blend A: Contains 1 ounce of almonds & 3 ounces of raisins
Blend B: Contains 2 ounces of sunflower seeds & 2 ounces of raisins

Through donations from local vendors, the guild has collected 40 ounces of almonds, 180 ounces of raisins, and 120 ounces of sunflower seeds from which to assemble Blend A and Blend B trail mix packages for sale at the fundraising event. The Guild wishes to maximize its sales revenue from packages of Blend A trail mix (sold for $1 per package) and Blend B trail mix (sold for $2 per package).

a) Formulate the Linear Program (LP) to determine the optimal number of Blend A and Blend B trail mix packages to be made from the available almonds, raisins, and sunflower seeds. Identify the decision variables, the constraints, and the objective function.

Answer: The decision variables are A = # packages of Blend A made and B = # packages of Blend B made. Since the # packages made must be non-negative, the following holds:

$A \geq 0, B \geq 0$

There are three constraints corresponding to the fixed amount of almonds, raisins, and sunflower seeds donated by local vendors.

The amount of packages of Blend A and Blend B made must use less than the 40 ounces of almonds available. Since Blend A uses 1 ounce of almonds and Blend B uses 0 ounces of almonds, the almond constraint (A) holds:

$1 \cdot A + 0 \cdot B \leq 40$ (A)

The amount of packages of Blend A and Blend B made must use less than the 180 ounces of raisins available. Since Blend A uses 3 ounces of raisins and Blend B uses 2 ounces of raisins, the raisin constraint (R) holds:

$3 \cdot A + 2 \cdot B \leq 180$ (R)

The amount of packages of Blend A and Blend B made must use less than the 120 ounces of sunflower seeds available. Since Blend A uses 0 ounces of sunflower seeds and Blend B uses 2 ounces of sunflower seeds, the sunflower seed constraint (S) holds:

$0 \cdot A + 2 \cdot B \leq 120 \ (S)$

The objective is to maximize the sales revenue. Since there is a sales revenue of $1 per package of Blend A sold and a sales revenue of $2 per package of Blend B sold, the objective function is given by:

$Max \ \$1 \cdot A + \$2 \cdot B$

The Linear Program (LP) to maximize the sales revenue from selling packages of Blend A and Blend B trail mix given the available amount of almonds, raisins, and sunflower seeds is given by:

Sea of Troubles LP Formulation

$$Max \ \$1 \cdot A + \$2 \cdot B$$
$$1 \cdot A + 0 \cdot B \leq \ \ 40 \ (A)$$
$$3 \cdot A + 2 \cdot B \leq 180 \ (R)$$
$$0 \cdot A + 2 \cdot B \leq 120 \ (S)$$
$$A \geq 0, \quad B \geq 0$$

b) Solve the LP using the Graphical Method.

Answer: The feasible region for the Sea of Troubles Hospital LP with corner points labeled is graphed as follows:

Sea of Troubles Hospital LP Feasible Region

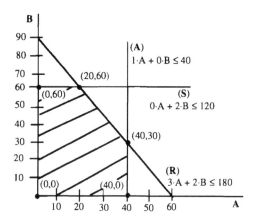

The optimal solution may be determined by evaluating the corner points of the feasible region:

Linear Programming (LP) Additional Solved Problems

Corner Point Evaluation

Corner Point (A,B)	Sales Revenue $1·A + $2·B
(0,0)	$0
(0,60)	$120
(20,60)	**$140**
(40,30)	$100
(40,0)	$40

The optimal corner point is (A,B) = (20,60) with an objective function (sales revenue) value of $140.

c) Trace the path of the Simplex Method would take in solving the LP starting at the origin (0,0) corner point and moving to the optimal corner point solution.

Answer: The Simplex Method starting at the origin corner point (0,0) with sales revenue $0 will move to the adjacent corner point (0,60) with a sales revenue $120 (since moving to the adjacent corner point (40,0) will increase the sales revenue only to $40). From the corner point (0,60), the only adjacent corner point that will increase the sales revenue is the corner point (20,60) with sales revenue $140. Since there is no adjacent corner point which will increase the sales revenue, the corner point (20,60) is optimal.

Simplex Method Corner Point Movement

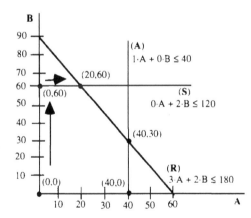

d) What are the binding and non-binding constraints at the optimal solution?

Answer: The optimal corner point solution (A,B) = (20,60) is at the intersection of the raisin (R) constraint and the sunflower seed (S) constraint. These two constraints are binding constraints. All available raisins and sunflower seeds are used up at the optimal solution.

The almond (A) constraint is not binding. At the optimal solution (20, 60), only 20 ounces out of the available 40 ounces of almonds is used up. Adding more almonds will not increase the value of the maximum sales revenue for trail mix packages.

Problem #2 Linear Programming

a) The John Gopher Company in Illinois, USA produces four different types of golf course lawn mowers: light-duty (L) mowers for the greens, medium-duty (M) mowers for the fairways, heavy-duty (H) mowers for the roughs, and all-purpose (A) mowers capable of cutting all types of grass.

The production of each light duty mower requires 5 hours of welding, 12 hours of assembly, and 2 hours of inspection; each medium-duty mower requires 3 hours of welding, 15 hours of assembly and 3 hours of inspection; each heavy-duty mower requires 2 hours of welding, 18 hours of assembly, and 5 hours of inspection; and each all-purpose mower requires 5 hours of welding, 20 hours of assembly, and 6 hours of inspection.

Total resources available each month to John Gopher are 300 hours of welding, 400 hours of assembly, and 100 hours of inspection. Each light-duty mower produced generates a profit of $200; each medium-duty mower generates a profit of $300; each heavy-duty mower generates a profit of $500; and each all-purpose mower generates a profit of $750. It is assumed that John Gopher can sell all equipment it can produce each month.

Formulate as a linear program (LP) the determination of the optimal number of each type mower for John Gopher to produce each month in order to maximize total monthly profit.

Answer: The decision variables are L = # light-duty mowers produced, M = # medium-duty mowers produced, H = # heavy-duty mowers produced, and A = # all-purpose mowers produced. Since the # mowers of all types produced must be non-negative, the following holds:

$$L \geq 0, \; M \geq 0, \; H \geq 0, \; A \geq 0$$

There are three constraints corresponding to the fixed amount of hours of welding (W), assembly (S), and Inspection (I) available each month.

The amount of hours of welding used by the four types of mowers must not exceed the 300 hours available each month. Since mower type L uses 5 hours, type M uses 3 hours, type H uses 2 hours, and type A uses 5 hours of welding, respectively, the welding constraint (W) holds:

$$5 \cdot L + 3 \cdot M + 2 \cdot H + 5 \cdot A \leq 300 \qquad (W)$$

The amount of hours of assembly used by the four types of mowers must not exceed the 400 hours available each month. Since mower type L uses 12 hours, type M uses 15 hours, type H uses 18 hours, and type A uses 20 hours of assembly, respectively, the assembly constraint (S) holds:

$$12 \cdot L + 15 \cdot M + 18 \cdot H + 20 \cdot A \leq 400 \quad (S)$$

The amount of hours of inspection used by the four types of mowers must not exceed the 100 hours available each month. Since mower type L uses 2 hours, type M uses 3 hours, type H uses 5 hours, and type A uses 6 hours of inspection, respectively, the inspection constraint (I) holds:

$$2 \cdot L + 3 \cdot M + 5 \cdot H + 6 \cdot A \leq 100 \quad (I)$$

The objective is to maximize the monthly profit. Since there is a profit of $200 for each light-duty mower produced, $300 for each medium-duty mower produced, $500 for each heavy-duty mower produced, and $750 for each all-purpose mower produced, the objective function is given by:

$$Max \ \$200 \cdot L + \$300 \cdot M + \$500 \cdot H + \$750 \cdot A$$

The linear program (LP) to maximize the monthly profit from producing the four types of mowers given the available amount of welding, assembly, and inspection hours is given by:

John Gopher LP Formulation

$$
\begin{aligned}
Max \ \$200 \cdot L &+ \$300 \cdot M + \$500 \cdot H + \$750 \cdot A \\
5 \cdot L &+ 3 \cdot M + 2 \cdot H + 5 \cdot A \leq 300 \quad (W) \\
12 \cdot L &+ 15 \cdot M + 18 \cdot H + 20 \cdot A \leq 400 \quad (S) \\
2 \cdot L &+ 3 \cdot M + 5 \cdot H + 6 \cdot A \leq 100 \quad (I) \\
L \geq 0, \quad &M \geq 0, \quad H \geq 0, \quad A \geq 0
\end{aligned}
$$

b) Solve the following linear program (LP) using the Graphical Method. For the feasible region, identify the corner points, the optimal solution, and the path the Simplex Method will take to arrive at the optimal solution if the initial feasible solution is the origin (0,0). Are the constraints on Resource 1 and Resource 2 binding on the optimal solution?

$$
\begin{aligned}
Max \ \$3 \cdot x_1 &+ \$2 \cdot x_2 \\
x_1 &+ x_2 \leq 5 \quad (Resource \ 1) \\
2 \cdot x_1 &+ x_2 \leq 6 \quad (Resource \ 2) \\
x_1 \geq 0, \quad &x_2 \geq 0
\end{aligned}
$$

Answer: The feasible region for the linear program (with the corner points labeled as A, B, C, and D) is as follows:

Linear Program Feasible Region

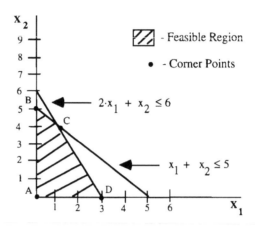

The optimal solution may be determined by evaluating the corner points of the LP feasible region:

LP Corner Point Evaluation

	Corner Point (x_1, x_2)	Objective Function $\$3 \cdot x_1 + \$2 \cdot x_2$
A:	(0,0)	$ 0
B:	(0,5)	$10
C:	(1,4)	$11
D:	(3,0)	$9

The optimal corner point is $(x_1, x_2) = (1,4)$ with an objective function value of $11.

The Simplex Method starting at the origin corner point (0,0) with objective function value $0 will move to the adjacent corner point (0,5) with a objective function $10 (since moving to the adjacent corner point (3,0) will increase the objective function only to $9). From the corner point (0,5), the only adjacent corner point that will increase the objective function is the corner point (1,4) with a value of $11. Since there is no adjacent corner point which will increase the objective function, the corner point (1,4) is optimal.

Simplex Method Corner Point Movement

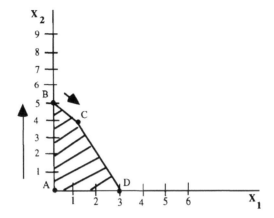

The optimal corner point C located at $(x_1, x_2) = (1,4)$ is at the intersection of the Resource 1 constraint and the Resource 2 constraint. All available Resource 1 and Resource 2 are used up at the optimal solution. These two resource constraints are binding constraints.

> *"The most important questions of life are, for the most part, really only problems of probability."*
>
> Laplace, *Theorie Analytique des Probabilites*

10 | Queueing

10.1 Introduction

Complex stochastic processes provide the closest approximation to the real world but are typically quite difficult to solve. Two powerful techniques are useful in analyzing a complex stochastic decision process:

> 1) Queueing which can solve for the *average performance* of a complex stochastic process analytically (i.e. provide an expression for the solution as a function of the problem parameters). However, application of queueing requires several simplifying assumptions on the parameters of the stochastic decision problem. To the extent that these assumptions are not a reasonable approximation to the stochastic process, the usefulness of the queueing solution will diminish.

> 2) Simulation (see Appendix A) which can examine the *intricate details* of a complex stochastic process not possible using queueing. Simulation allows the simplifying assumptions required by queueing to be relaxed. However, simulation of a complex stochastic process is *expensive*. Manpower is required to build a model which captures the underlying stochastic behavior adequately. Time is required to write a computer program, perform the simulation runs, and analyze the output data. Manpower and time translate directly into cost and schedule.

The tradeoff between queueing (loss of accuracy due to the required simplifying assumptions on the parameters) and simulation (cost and schedule impacts due to manpower and time demands) is typically resolved as follows:

> - Initially, queueing is used to provide the decision maker with a quick approximate ("ball park") analytical solution. If a more detailed examination of the complex stochastic process is required, the resources required to initiate a simulation may then be allocated.

Example 1: (Moribund Airlines Queueing Formulation). Moribund Airlines needs to decide how many service counters will be required for customers at the new hub terminal it plans to build in Catchcan, Alaska. Customers wait in a single line (queue) for service at one of the multiple service counters. Costs associated with the decision of how many service counters to provide are as follows:

- There is an investment cost to build and staff each service counter.

- There is an opportunity cost (loss of future revenue) to the airline associated with the expected time customers must wait in the queue. This loss of revenue is due to a loss of customer goodwill.

- There is an opportunity cost (loss of future revenue) to the airline associated with the expected queue length when a customer arrives at Moribund Airlines. This loss of revenue is due to customers switching airlines rather than joining the queue.

Provide a queueing formulation to determine the number of service counters which minimizes the total annual cost to Moribund Airlines.

Answer: The decision sought by Moribund Airlines is the number of service counters c which minimizes the total annual cost objective function $TC(c)$ defined by:

$$TC(c) = C(c) + u \cdot W(c) + v \cdot L(c)$$

where:

$C(c)$ = Annual investment cost (amortized) to build and staff c service counters.

$W(c)$ = *Expected* waiting line time before being served if c service counters are available. Queueing is used to derive this average waiting line time.

u = Annual opportunity cost per unit of expected customer waiting line time at Moribund Airlines. This loss of future revenue is due to loss of customer goodwill.

$L(c)$ = *Expected* waiting line length when a new customer arrives at Moribund Airlines if c service counters are available. Queueing is used to derive this average waiting line length.

v = Annual opportunity cost for each customer in the waiting line when a new customer arrives at Moribund Airlines. This loss of future revenue is due to customers who switch to other airlines rather than joining the queue.

Notes:

• The total annual cost objective function TC(c) has a classic *cost-benefit* form (see Figure 1). Increasing the number of service counters incurs the cost of more investment (due to the increased building and staffing required) but the benefit of less loss of revenue (due to the maintenance of customer goodwill and the avoidance of customer switches to other airlines).

Total Annual Cost For Moribund Airlines Service Counter Decision

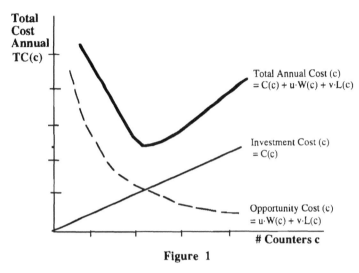

Figure 1

• The expected waiting line length L(c) and the expected waiting line time W(c) are a function of the number of service counters c *only*. L(c) and W(c) are not functions of *time* of day/week/etc. It is assumed that the queue has reached a steady-state or average performance mode which is *constant* over some time period of interest.

• The steady-state for a queueing system is typically achieved some time after the anomalies associated with startup (e.g. there may be no customers or an unusually large number of customers in line when a business first opens) have passed and the queueing system has settled into an *average performance* mode.

• Most stochastic processes (e.g. this airline customer service process) do not behave uniformly over time. There are usually peak and low volume times of the day, week. etc. Queueing solves for the average characteristics of the queue (e.g. expected waiting line length, expected waiting line time, etc.) analytically during the steady-state period. More detailed or non-steady-state characteristics of the queue (e.g. queue length soon after startup, wait times at peak times of the day, etc.) must be examined by simulation.

Quantification of the Queueing Parameters

The annual investment cost C(c) of building c service counters and maintaining the necessary staff can be estimated by engineers, architects, managers, etc. based on historical data from past projects.

The annual opportunity costs (loss of revenue) u and v to Moribund Airlines for loss of customer goodwill (due to long waits in line) and for customer airline switches (due to unacceptably long lines at the time of arrival), respectively, are based on *subjective* estimates. Annual opportunity costs due to loss of customer goodwill and due to customer airline switches are determined in one of two ways:

(1) Performing a *survey* of potential Moribund Airlines customers. This survey would determine the effect of a wait in line prior to service on a customer's future actions in choosing an airline and the effect of the length of the line upon arrival at Moribund Airlines on a customer's likelihood of switching to another airline.

(2) Relying on *expert opinion* (people with considerable experience in the airline service industry) to determine the effect the wait in line and the length of the line upon arrival at Moribund Airlines will have on future customer actions.

The determination using queueing of the expected waiting line time W(c) and the expected waiting line length L(c) requires an examination of the flow of a single line queueing system (see Figure 2).

Single Line Queueing System Flow

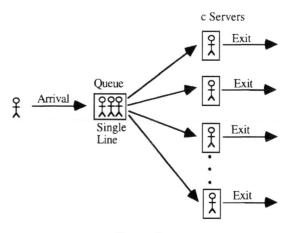

Figure 2

The flow in a single line queueing system (assuming c servers) is as follows:

- A customer seeking service arrives at the queueing system (the arrival rate for customers is specified by an arrival distribution) and proceeds to a server if one is free.

- If all c servers are already busy, the customer joins the end of the queue (single line) and waits his or her turn for service.

- When a customer at one of the servers completes service (the service completion time for customers is specified by a service distribution), the server becomes free and the person at the head of the queue will begin service with the free server.

The three principal elements of a queueing system are as follows:

• **Arrival Process**: Number of customers arriving over time.

• **Service Process**: Service completion times for customers.

• **Queue Discipline**: Manner in which next customer is selected from the queue.

  - First in, first out (FIFO) (e.g. banks)
  - Shortest expected service time (e.g. computer operating system)
  - Most critical customer next (e.g. triage at a hospital)

An important assumption for a queueing process to reach a steady-state mode and to have an analytical solution is the *independence* of the arrival and service processes. That is, a large or small number of arrivals in a given unit of time has no effect on the average service completion time. For example, servers do not hurry when there is a long line or slow down when there is a short line.

Note: If this independence assumption is not a reasonable assumption, simulation can be used to model the *interaction* between the number of arrivals and the service completion time (or any other complexity which affects the capability of the queueing system to reach a steady-state mode). However, remember that the quick, inexpensive, "ball park" solution provided by queueing can sometimes provide greater utility to a decision maker than the detailed but expensive and time-consuming analysis provided by simulation.

10.2 Arrival Process

The arrival process for a queueing system is stochastic and must be modeled by an appropriate probability distribution. The arrival distribution for queueing is typically defined by an expected arrival rate or, *equivalently*, an expected interarrival time.

Example 2: (Arrival Rate vs. Interarrival Time). The customer service facility at Moribund Airlines has an expected arrival rate of 10 customers per hour. What is expected customer interarrival time?

Answer: If customers arrive at an expected rate of 10 per hour, the expected time between arrivals will be 1/10 hours or 6 minutes. Any single time between customer arrivals may vary from zero to a large positive value. However, the mean (average) of the interarrival times will approach six minutes as the total number of customer arrivals increases.

Suppose Moribund Airlines has surveyed customer arrivals at an already operating airline customer service facility that is *considered similar* in size and clientele to the facility that Moribund Airlines is contemplating building. This survey indicates that on the average 8 customers per minute will arrive at the similar facility. Then the average customer interarrival time is 1/8 of a minute or 0.125 minutes.

The arrival process of 8 customers per minute on the average may be explained by one of a large number of probability distributions or may not conform to any theoretical probability distribution. If the arrival process does not conform satisfactorily to a theoretical probability distribution, queueing will not lead to an analytical solution.

Note: The queueing system can still be analyzed using simulation. Simulation can handle *any* characterization of the arrival process. Often *empirical* data (observed arrival data rather than data generated from a theoretical arrival distribution) is used in a simulation.

Certain arrival distributions which can be easily manipulated mathematically are required in order to solve a queueing system analytically. The <u>Poisson distribution</u> is such a distribution. The Poisson distribution can be used when the following are reasonable assumptions:

- The arrivals are *random* (referred to as <u>Markovian</u>). That is, a customer's arrival is independent of all other customers and is based only on his or her own needs for service.

- The expected number of arrivals per unit of time λ is constant (i. e. *independent of time*). That is, customers arrive randomly but at a constant expected rate λ over the steady-state time period (e.g. hours, day, week, etc.) being examined.

<u>Note</u>: Of the two assumptions, arrivals are more likely to violate the assumption of being independent of time over the steady-state time period. To the extent that either of these two assumptions is not reasonable, the accuracy of the analytical queueing solution will suffer.

Poisson Distribution

The Poisson distribution takes on *discrete* non-negative integer values $k = 0, 1, 2, 3, \cdots,$ ∞. The Poisson distribution is completely specified by a single parameter λ, the constant expected number of arrivals per unit of time.

The probability of k arrivals during a unit of time for an arrival process specified by a Poisson distribution with expected value λ is given by:

$$P \ (k \ arrivals) = \lambda^k e^{-\lambda} / k! \qquad k = 0, 1, 2, 3, \cdots, \infty$$

<u>Notes</u>:

• The expected arrival rate λ is not limited to non-negative *integer* values but may be any non-negative value ($\lambda \geq 0$). That is, the expected value of the Poisson distribution need not be a possible value of the Poisson distribution.

• The variance of the Poisson distribution is also λ. The standard deviation for the Poisson distribution is $\lambda^{1/2}$.

• k! is called "k factorial" and is computed as $k! = k \cdot (k-1) \cdot (k-2) \cdots (2) \cdot (1)$. For example, $3! = 3 \cdot 2 \cdot 1$. The term 0! is defined to be equal to 1.

• The Poisson distribution probabilities for k arrivals within a given time assuming an expected arrival rate λ are available in Table 4 (λ = equal 0.1 to 6). Poisson distribution probabilities for larger values of λ are easily computed using a calculator.

• If λ is defined for one unit of time (e.g. 2 arrivals per minute) but the time period of interest is different (e.g. hours), λ can be scaled linearly to this time period (e.g. 120 arrivals per hour).

Before proceeding with the analysis of the queueing system, one should test whether the observed arrival frequency data collected by Moribund Airlines fits a discrete Poisson distribution with satisfactory confidence.

Example 3: (Arrival Distribution Chi-Square Test). Moribund Airlines surveyed arrivals at the similar airline facility for 100 minutes and found that the number of arrivals during a minute ranged from 0 to 18 customers. The following data was collected on the frequency of the number of arrivals per minute over the 100 minutes examined.

Observed Arrival Data (100 Minutes)

# Arrivals Per Minute	Frequency
0	1
1	1
2	1
3	3
4	6
5	8
6	13
7	13
8	15
9	9
10	10
11	8
12	3
13	4
14	2
15	1
16	0
17	1
18+	1
	100

Use the Chi -Square Test to test the hypothesis that the above observed arrival frequency data is generated by a Poisson distribution.

Answer: The sample mean $\bar{X}$ for the above arrival frequency data is computed as follows:

$$\bar{X} = \frac{0 \cdot 1 + 1 \cdot 1 + 2 \cdot 1 + 3 \cdot 3 + \cdots + 15 \cdot 1 + 16 \cdot 0 + 17 \cdot 1 + 18 \cdot 1}{100}$$

$$= 8 \; arrivals \; per \; minute$$

The hypothesis that the arrival frequency data collected at the similar airline facility is generated by a Poisson distribution is tested using the Chi-Square (x^2) Test (see Section 1.11) for goodness of fit. For the Chi-Square (x^2) Test, the actual (observed) and expected arrival frequency data are compared:

- The *actual* data is the above arrival frequency data collected by Moribund Airlines at the similar facility.

- The *expected* data is computed using the Poisson distribution ($\lambda^k e^{-\lambda} / k!$) with the sample mean $\bar{X}$ = 8 arrivals per minute for the actual (observed) arrival frequency data used to estimate the expected arrival rate λ.

Due to the randomness in any stochastic process and the limited sampling performed (only 100 minutes of data collected) at the similar facility, one would expect the actual data to have deviation from the expected data even if the similar facility did have an arrival process generated by a Poisson distribution with $\lambda = 8$ arrivals per minute. The Chi-Square Test computes the sum of the squared deviation between the actual (observed) arrival frequency data and the expected arrival frequency data to allow inferences to be made on the following hypotheses:

- Null hypothesis: The arrival data for the similar facility is generated by a Poisson distribution. The deviation between the actual arrival frequency data and the expected arrival frequency data is consistent with the randomness of the Poisson arrival process and the limited sampling.

- Alternative hypothesis: The arrival data for the similar facility is not generated by a Poisson distribution.

Recall that the computation for the Chi-Square statistic, x^2, is given by:

$$x^2 = \Sigma \, (A_i - E_i)^2 / E_i \qquad\qquad i = 1, 2, 3, \cdots, N$$

where A_i =   *Actual* number of outcomes (from the collected data) in cell i
        E_i =   *Expected* number of outcomes (estimated from the theoretical distribution) in cell i
        i =   Index of the cells (from cell 1 to cell N)

For the theoretical Poisson distribution ($\lambda^k e^{-\lambda} / k!$) with the sample mean $\bar{X} = 8$ arrivals per minute used as an estimate for λ, the probability of 0, 1, 2, 3, $\cdots$ *discrete* arrivals in a minute is given by:

$$P(0 \; arrivals) \quad = \frac{8^0 \cdot e^{-8}}{0!} = 0.0003$$

$$P(1 \; arrival) \quad = \frac{8^1 \cdot e^{-8}}{1!} = 0.0027$$

$$P(2 \; arrivals) \quad = \frac{8^2 \cdot e^{-8}}{2!} = 0.0107$$

$$\bullet$$
$$\bullet$$
$$\bullet$$

$$P(17 \; arrivals) \quad = \frac{8^{17} \cdot e^{-8}}{17!} = 0.0021$$

Note: There is some probability of 18, 19, 20, 21, etc. arrivals in a minute but the probabilities for these numbers of arrivals in a minute are quite small. All these probabilities are included in the P(18+ arrivals). The P(18+ arrivals) is computed using the property of complement sets (see Section 1.4):

$$P(E) = 1 - P(\hat{E})$$
$$E = \left\{18+\right\}, \; \hat{E} = \left\{0, 1, 2, 3, \cdots, 17\right\}$$

$P(18+ \ arrivals) = 1 - P(0 \ arrrivals) - P(1 \ arrivals) - \cdots - P(17 \ arrivals)$

$= 0.0009$

Then the expected number of outcomes 0, 1, 2, $\cdots$, 18+ arrivals per minute for the N=100 observations (minutes of data) is simply 100 times the individual outcome probabilities:

$E(Number \ of \ 0 \ arrivals \ in \ 100 \ minutes \ of \ data) \ = 100 \cdot 0.0003 \ = \ 0.03$

$E(Number \ of \ 1 \ arrival \ in \ 100 \ minutes \ of \ data) \ = 100 \cdot 0.0027 \ = \ 0.27$

$E(Number \ of \ 2 \ arrivals \ in \ 100 \ minutes \ of \ data) \ = 100 \cdot 0.0107 \ = \ 1.07$

·
·
·

$E(Number \ of \ 18+ \ arrivals \ in \ 100 \ minutes \ of \ data) = 100 \cdot 0.0009 = \underline{0.09}$

100

The probabilities of k arrivals per minute using the Poisson distribution ($\lambda = 8$) and the expected frequency of k arrivals per minute during 100 minutes of data are added to the table in Example 3 to obtain the data for the computation of the Chi-Square statistic (see Figure 4).

Data For Computation for Chi-Square Statistic

# Arrivals per Minute	Actual Frequency (Observed)	Cell Probability ($\lambda = 8$)	Expected Frequency (Prob x 100)
0	1	0.0003	0.03
1	1	0.0027	0.27
2	1	0.0107	1.07
3	3	0.0286	2.86
4	6	0.0573	5.73
5	8	0.0916	9.16
6	13	0.1221	12.21
7	13	0.1396	13.96
8	15	0.1396	13.96
9	9	0.1241	12.41
10	10	0.0993	9.93
11	8	0.0772	7.72
12	3	0.0481	4.81
13	4	0.0296	2.96
14	2	0.0169	1.69
15	1	0.0090	0.90
16	0	0.0045	0.45
17	1	0.0021	0.21
18+	1	0.0009	0.09
	100	1.0000	100.00

Figure 4

Note: The Chi-Square Test requires that the values of the cell $E_i's$ (the *expected* number of outcomes in each cell) must be at least 5 in order for the cell to be statistically significant. It is sometimes necessary to *pool* data (combine contiguous cells) and thereby reduce the total number of cells in order to get each of the cell $E_i's$ to be at least 5.

For this example, the outcomes associated with between 0 and 4 arrivals per minute are pooled into one cell to obtain 9.96 expected arrivals in that cell. The outcomes associated with between 13 and 18+ arrivals per minute are pooled into one cell to obtain 6.30 expected arrivals in that cell. Finally, the outcomes associated with 11 and 12 arrivals per minute are pooled into one cell to obtain 12.03 expected arrivals in that cell. The data for computation of the Chi-Square statistic after pooling is given in Figure 5.

Pooled Data for Computation of the Chi-Square Statistic

# Arrivals Per Minute	Actual Frequency (Observed)	Cell Probability ($\lambda=8$)	Expected Frequency (Prob x 100)
0 - 4	12	0.0996	9.96
5	8	0.0916	9.16
6	13	0.1221	12.21
7	13	0.1396	13.96
8	15	0.1396	13.96
9	9	0.1241	12.41
10	10	0.0993	9.93
11-12	11	0.1203	12.03
13-18+	9	0.0630	6.30
	100	**1.0000**	**100.00**

Figure 5

For the 9 cells remaining after pooling, the Chi-Square statistic may now be computed (see Figure 6).

Chi-Square Statistic Calculation Table

Cell #	# Arr Per Min	Actual Frequency (Observed)	Expected Frequency (Probx100)	Actual Minus Expected	$(Act-Exp)^2$	$\dfrac{(Act-Exp)^2}{Exp}$
1	0 - 4	12	9.96	2.04	4.16	0.42
2	5	8	9.16	-1.16	1.35	0.15
3	6	13	12.21	0.79	0.62	0.05
4	7	13	13.96	-0.96	0.92	0.07
5	8	15	13.96	1.04	1.08	0.08
6	9	9	12.41	-3.41	11.63	0.94
7	10	10	9.93	0.07	0.005	0.00
8	11-12	11	12.03	-1.03	1.06	0.09
9	13-18+	9	6.30	2.70	7.29	1.16
		100	**100.00**			$\chi^2=2.94$

Figure 6

Note: The Chi-Square statistic follows the Chi-Square distribution with N-k-1 degrees of freedom where:

N = Number of cells (after pooling to obtain at least 5 *expected* outcomes in each cell).

k = Number of parameters that needed to be *estimated* to find the expected outcomes in each cell. For the Poisson distribution, k=1 (λ); for the Normal distribution, k=2 (μ, σ); etc.

1 = A degree of freedom is lost since the total number of outcomes (100) is known. If the total number of outcomes is known and the number of outcomes in *all but one cell* is known, then the number of outcomes in the last remaining cell is no longer random but may be computed

For Example 3, the number of degrees of freedom is 9 - 1 - 1 = 7 since the number of cells after pooling is 9 and the expected value λ of the Poisson distribution had to be estimated using the sample mean arrival rate $\bar{X}$. Recall that the smaller the value of the Chi-Square statistic, the better the fit of the actual data collected is to the theoretical Poisson distribution. One must be *smaller* than a percentile value of the Chi-Square distribution to claim confidence to that percentile.

The value of the Chi-Square statistic computed in Figure 6 is 2.94. From the percentiles of the Chi-Square distribution given in Table 3 for 7 degrees of freedom, it is noted that:

- If the Chi-Square statistic is 2.83 or less, one is 90% confident that the actual (observed) data is generated by a Poisson distribution (λ=8). That is, even when the data is governed by a Poisson distribution (λ=8), the probability of a Chi-square statistic 2.83 or greater due to randomness is 0.90. This amount of deviation between the actual data and the expected data is not troublesome and will probably lead one not to reject the null hypothesis.

- If the Chi-Square statistic is 4.25 or less, one is 75% confident that the actual (observed) data is generated a Poisson distribution (λ=8). That is, even when the data is governed by a Poisson distribution (λ=8), the probability of a Chi-square statistic 4.25 or greater is only 0.75. This amount of deviation between the actual and the expected data is troublesome and may lead one to reject the null hypothesis.

Since the value of the Chi-Square statistic is 2.94, one is almost 90% confident that the actual arrival data is generated by a Poisson distribution with λ estimated by the sample mean $\bar{X}$ = 8 arrivals per minute. The null hypothesis that the arrival frequency data in Example 3 is generated by a Poisson distribution is not rejected.

Note: A *subjective* argument has been made that the arrival frequency data at the surveyed airline facility will be similar to that of the new Moribund Airline facility to be built. A *statistical* argument has been made that the observed arrival frequency data at the surveyed facility is generated by a Poisson distribution. However, it has not been *proved* that the new Moribund Airline facility will have arrival frequency data that is generated by a Poisson distribution (and can't be proved until actual data is collected at the new Moribund Airlines facility).

10.3 Service Process

Suppose Moribund Airlines has also surveyed customer service completion times at the similar customer service facility and has estimated that a customer requires 20 minutes on the average to complete service (once the customer leaves the single line and begins being served at one of the service counters).

The service process of 20 minutes on the average may be explained by one of a large number of probability distributions or may not conform to any theoretical probability distribution. If the service process does not conform satisfactorily to a theoretical distribution, queuing will not lead to an analytical solution.

Note: The queueing system can still be solved using simulation. Simulation can handle *any* characterization of the service process. Often *empirical* data (observed service completion time data rather than data generated from a theoretical service completion time distribution) is used in a simulation.

Certain service completion time distributions which can be easily manipulated mathematically are required in order to solve a queueing system analytically. The Exponential distribution is such a distribution. It can be used when the following are reasonable assumptions:

- Service completion times are random (referred to as Markovian). That is, a customer's service completion time is independent of all other customers and is based only on his or her own needs.

- The expected service completion time $1/\mu$ per customer is constant (independent of the starting time of the service). That is, customers have random service completion times but with a constant expected service completion time $1/\mu$ over the steady-state time period (e.g. hours, day, week, etc.) being examined.

Note: Service completion times for a queueing system are more likely to violate the assumption of being independent of starting time over the steady state time period. To the extent that either of these two assumptions is not reasonable, the accuracy of the analytically derived queueing solution will suffer.

The Poisson and Exponential distributions are related. Both are Markovian distributions (i.e. Poisson arrivals and Exponential service completion times are completely independent (random) of each other and independent of time). In fact, the Poisson and Exponential distributions often provide a discrete and continuous characterization, respectively, of the *same* stochastic process. For example, if the arrival rate (discrete) has a Poisson distribution, the interarrival times (continuous) have an Exponential distribution. Likewise, if service completion times (continuous) have an Exponential distribution, the service completion rate (discrete) has a Poisson distribution.

Exponential Distribution

The Exponential distribution takes on *continuous*, non-negative values $t \geq 0$. The Exponential distribution is completely specified by a single parameter $1/\mu$, the constant expected service completion time. The *probability distribution* f(t) for the service completion time is given by:

$$f(t) = P(Service\ completion\ time = t)$$
$$= \mu e^{-\mu t} \qquad\qquad t \geq 0$$

The *cumulative distribution function* F(t) for the service completion time is given by:

$$F(t) = P(Service\ completion\ time \leq t)$$
$$= 1 - e^{-\mu t} \qquad\qquad t \geq 0$$

Notes:

• The inverse of the constant expected service completion *time* $1/\mu$ is the expected service completion *rate* μ. That is, if the expected service completion time $1/\mu$ for the Exponential distribution is 20 minutes (1/3 hours) per customer, then the expected service completion rate μ is 3 customers per hour.

• The variance of the Exponential distribution is $1/\mu^2$. The standard deviation for the Exponential distribution is then $1/\mu$.

• Values of e^{-x} are given in Table 5 (x = equal 0.0 to 9.95). Exponential distribution probabilities for larger values of x are easily computed using a calculator.

Before proceeding with the analysis of the queueing system, one should test whether the observed service completion time data (collected by Moribund Airlines) fits a continuous Exponential distribution with satisfactory confidence.

Example 4: (Service Distribution Chi-Square Test). Moribund Airlines has also surveyed customer service completion times at the similar facility for 100 customers and observed the following data on the frequency of service completion times (in 5 minute intervals).

Observed Service Completion Times (100 Customers)

Service Completion Time Interval	Frequency
0 - 5	26
5 - 10	18
10 - 15	13
15 - 20	11
20 - 25	6
25 - 30	5
30 - 35	4
35 - 40	4
40 - 45	2
45 - 50	2
50 - 55	2
55 - 60	2
60 - 65	1
65 - 70	1
70 - 75	0
75 - 80+	3
	100

Use the Chi-Square Test to test the hypothesis that the above observed customer service completion time data is generated by an Exponential distribution.

Answer: The sample mean $\bar{X}$ for the above data (conservatively using the upper bound of each service completion time interval) is computed as follows:

$$\bar{X} = \frac{5 \cdot 26 + 10 \cdot 18 + 15 \cdot 13 + \cdots + 70 \cdot 1 + 75 \cdot 0 + 80 \cdot 3}{100}$$

$= 20$ *minutes per customer*

The hypothesis that the service completion time data collected at the similar airline facility is generated by an Exponential distribution is tested using the Chi-Square (x^2) Test for goodness of fit. For the Chi-Square (x^2) Test, the actual (observed) and expected service completion time data are compared:

- The *actual* data is the above service completion time data collected by Moribund Airlines at the similar facility.

- The *expected* data is computed using the Exponential cumulative distribution $(1 - e^{-\mu t})$ using the sample mean $\bar{X} = 20$ minutes for the actual (observed) service completion time data as an estimate for the expected service completion time $1/\mu$.

Due to the randomness in any stochastic process and the limited sampling performed (service completion time data for only 100 customers) at the similar facility, one would expect the actual data to have some deviation from the expected data even if the similar facility did have a service process generated by an Exponential distribution with $1/\mu = 20$ minutes per customer. The Chi-Square Test computes the sum of the squared deviation between the actual (observed) service completion times and the expected service completion times to allow inferences to be made on the following hypotheses:

- Null hypothesis: The service completion time data for the similar facility is generated by an Exponential distribution. The deviation between the actual arrival frequency data and the expected arrival frequency data is consistent with the randomness of the Exponential service process and the limited sampling.

- Alternative hypothesis: The service completion time data for the similar facility is not generated by an Exponential distribution.

Recall that the computation for the Chi-Square statistic, x^2, is given by:

$$x^2 = \Sigma \, (A_i - E_i)^2 \, / \, E_i \qquad\qquad i = 1, 2, 3, \cdots, N$$

where A_i = *Actual* number of outcomes (from the collected data) in cell i
E_i = *Expected* number of outcomes (estimated from the theoretical distribution) in cell i
i = Index of the cells (from cell 1 to cell N)

To apply the Chi-Square test to the *continuous* Exponential distribution, one must compute the expected number of service completion time outcomes in each 5 minute time interval cell using the cumulative distribution.

The Exponential cumulative distribution function for the similar airline facility where $1/\mu$ = 20 minutes = 1/3 hour per customer (and hence μ = 1/20 minutes = 3 customers per hour) is given by:

$$F(t) = 1 - e^{-\mu t} \qquad\qquad t \geq 0$$

$$= 1 - e^{-t/20} \qquad\qquad t \geq 0 \text{ (t measured in minutes)}$$

$$= 1 - e^{-3t} \qquad\qquad t \geq 0 \text{ (t measured in hours)}$$

The expected number of service time completion outcomes in each 5 minute cell is equal to the expected probability of residing in that cell times 100. The expected probabilities of residing in each cell are computed using the Exponential cumulative distribution function:

$$
\begin{aligned}
P(0 \leq t \leq 5 \text{ minutes}) &= P(t \leq 5) - P(t \leq 0) \\
&= F(5) - F(0) \\
&= F(5) - 0 \\
&= 1 - e^{-5/20} \\
&= 0.221
\end{aligned}
$$

Expected # outcomes = 22.1 service completions in cell 0 to 5 minutes.

$$
\begin{aligned}
P(5 \leq t \leq 10 \text{ minutes}) &= P(t \leq 10) - P(t \leq 5) \\
&= F(10) - F(5) \\
&= (1 - e^{-10/20}) - (1 - e^{-5/20}) \\
&= 0.396 - 0.221 \\
&= 0.175
\end{aligned}
$$

Expected # outcomes = 17.5 service completions in cell 5 to 10 minutes.

The expected probabilities and number of service completions for each cell are added to the table in Example 4 (see Figure 7).

Data For Computation of Chi-Square Statistic

Completion Time Interval (Minutes)	Actual Frequency (Observed)	Cell Probability ($1/\mu$=20 min)	Expected Frequency (Prob x 100)
0 - 5	26	0.221	22.1
5 - 10	18	0.175	17.5
10 - 15	13	0.132	13.2
15 - 20	11	0.104	10.4
20 - 25	6	0.081	8.1
25 - 30	5	0.064	6.4
30 - 35	4	0.050	5.0
35 - 40	4	0.038	3.8
40 - 45	2	0.030	3.0
45 - 50	2	0.023	2.3
50 - 55	2	0.018	1.8
55 - 60	2	0.014	1.4
60 - 65	1	0.011	1.1
65 - 70	1	0.009	0.9
70 - 75	0	0.007	0.7
75 - 80+	3	0.023	2.3
	100		100.00

Figure 7

Note: The Chi-Square Test requires that the expected number of outcomes in each cell must be at least 5 in order for the cell to be statistically significant. Pooling of data in contiguous cells is required.

For the expected outcomes given in Figure 7, the service completions between 60 and 80+ minutes are pooled to obtain 5.0 expected outcomes in that cell; between 45 and 60 minutes are pooled to obtain 5.5 expected outcomes in that cell; and between 35 and 45 minutes are pooled to obtain 6.8 expected outcomes in that cell. Figure 7 is updated to provide the pooled data for computation of the Chi-Square statistic (see Figure 8).

Pooled Data for Computation of the Chi-Square Statistic

Completion Time Interval (Minutes)	Actual Frequency (Observed)	Expected Frequency (Prob x 100)
0 - 5	26	22.1
5 - 10	18	17.5
10 - 15	13	13.2
15 - 20	11	10.4
20 - 25	6	8.1
25 - 30	5	6.4
30 - 35	4	5.0
35 - 45	6	6.8
45 - 60	6	5.5
60 - 80+	5	5.0
	100	100.00

Figure 8

For the 10 cells remaining after pooling, the Chi-Square statistic may now be computed (see Figure 9).

Chi-Square Calculation Table

Cell #	Completion Time Interval (Minutes)	Actual Frequency (Observed)	Expected Frequency (Prob x 100)	$\frac{(Act-Exp)^2}{Exp}$
1	0 - 5	26	22.1	0.688
2	5 - 10	18	17.5	0.004
3	10 - 15	13	13.2	0.003
4	15 - 20	11	10.4	0.035
5	20 - 25	6	8.1	0.544
6	25 - 30	5	6.4	0.306
7	30 - 35	4	5.0	0.200
8	35 - 45	6	6.8	0.094
9	45 - 60	6	5.5	0.045
10	60 - 80+	5	5.0	0.000
		100	**100.00**	$\chi^2 = 1.92$

Figure 9

Notes: The Chi-Square statistic follows the Chi-Square distribution with N-k-1 degrees of freedom where:

- N = Number of cells (after pooling to obtain at least 5 *expected* outcomes in each cell).

- k = Number of parameters that needed to be *estimated* to find the expected outcomes in each cell. For the Exponential distribution, k=1 (1μ); for the Normal distribution, k=2 (μ, σ); etc.

- 1 = A degree of freedom is lost since the total number of outcomes (100) is known. If the total number of outcomes is known and the number of outcomes in *all but one cell* is known, then the number of outcomes in the last remaining cell is no longer random but may be computed.

For Example 4, the number of degrees of freedom is 10 - 1 - 1 = 8 since the number of cells after pooling is 10 and the expected value $1/\mu$ of the Exponential distribution had to be estimated using the sample mean $\bar{X}$. Recall that the smaller the value of the Chi-Square statistic, the better the fit of the actual data collected is to the theoretical Exponential distribution. One must be *smaller* than a percentile value of the Chi-Square distribution to claim confidence to that percentile.

The value of the Chi-Square statistic computed in Figure 9 is 1.92. From the percentiles of the Chi-Square distribution given in Table 3 for 8 degrees of freedom, it is noted that:

- If the Chi-Square statistic is 2.18 or less, one is 97.5% confident that the actual (observed) data is generated by an Exponential distribution ($1/\mu=20$ minutes). That is, even if the data is governed by an Exponential distribution ($1/\mu$=20 minutes), the probability of getting a Chi-square statistic 2.18 or greater due to randomness is 0.975. This amount of deviation between the actual and the expected data is not troublesome and will probably lead one not to reject the null hypothesis.

Since the value of the Chi-Square statistic is 1.92, one is over 97.5% confident that the collected service completion time data is generated by an Exponential distribution with $1/\mu$ estimated by the sample mean service completion time $\bar{X}$ = 20 minutes per customer. The null hypothesis that the service completion time data is generated by an Exponential distribution is not rejected.

Note: The same *subjective* argument has been made that the service completion time data at the surveyed airline facility will be similar to new facility to be built by Moribund Airlines. The same *statistical* argument has been made that the service completion time data at the surveyed facility is generated by an Exponential distribution. However, it has not been *proved* that the new Moribund Airlines facility will have service completion time data that is generated by an Exponential distribution (and can't be proved until actual data is collected at the new Moribund Airlines facility).

10.4 Solving the Queueing Problem

Queue Discipline

Queue discipline is the method used to select the next person (from all those waiting in the queue) to be served. For example, at a health clinic, people in the queue may include those who have appointments to see an available staff doctor; those who have appointments to see a particular doctor; those who are emergency cases requiring quick service; those who are walk-ins; etc. The method used to select the person from the queue to serve next may be quite complex.

Simulation can deal with all such queue discipline complexities. However, in order for queueing to solve the queueing system *analytically*, the simplifying assumption of a first in - first out (FIFO) queue discipline is required. For a first in - first out queue discipline, the person in line the longest will be the next served when a server becomes free.

Queueing Notation

The notation "M/M/c" is used to denote a single line queueing system with the following three elements:

- A Markovian (random) arrival process.
- A Markovian (random) service process.
- A constant number c servers.

Recall that the Poisson and Exponential distributions are Markovian distributions. That is, the number of arrivals per unit of time (described by a Poisson distribution) and the service completion times (described by an Exponential distribution) are random and have an expected value which remains constant over time.

The M/M/1 Single Server Queue

Initially, the analytical solution to the M/M/1 single server queueing system will be presented. Later, results will be presented for the general M/M/c multiple server queueing system.

Assumptions which allow an analytical solution of the single line M/M/1 single server queueing system are as follows:

- Poisson arrivals with constant expected arrival rate λ (Markovian).
- Exponential service with constant expected completion time $1/\mu$ (Markovian).
- Single server (constant c = 1).

Examples of a single line single server queueing system include a doctor's office with a single doctor, a convenience store with a single clerk, a single automated teller machine, etc.

The steady state expected waiting line length L (sometimes called L(1) for the single server queueing system) is given by:

$$L = \lambda^2 / \mu(\mu - \lambda)$$

Notes:

• The expected waiting line length L for the *steady state* single server queue is a function of the expected arrival rate λ and the expected service completion rate μ (which is the inverse of the expected service completion time $1/\mu$). L is *not* a function of time during the steady-state period.

• The equation for the steady state expected waiting line length L has physical meaning only when $\lambda < \mu$ (the expected arrival rate must be less than the expected service completion rate). If the arrival rate λ is greater than or equal to the service completion rate μ, the line length will (theoretically) grow *without bound* to infinity and a steady state for the queue will never be attained.

The steady state expected waiting line time W (sometimes called W(1) for the single server queueing system) is given by:

$$W = L / \lambda$$

Notes:

• If the steady state expected waiting line length L = 10 customers and the expected arrival rate λ = 5 customers per minute, then the expected waiting line time is 10/5 or 2 minutes. The expected waiting line time W does *not* include the expected service completion time $1/\mu$.

• A simplified expected value characterization of the steady state queue helps to understand the equation for expected waiting line time W = L/λ :

- At any given time, the expected waiting line length L is 10 customers. Within L/λ (= 10/5 = 2) minutes, L (= 10) new customers are expected to arrive. And in L/λ (= 2) more minutes, 10 additional customers are expected to arrive and replace the 10 original new customers. Thus, within L/λ plus L/λ time (= 4 minutes), the L (= 10) new customers entered the line and exited the line. The average or expected waiting line time W is then L/λ (= 2 minutes).

The steady-state probability of *exactly* N customers in the single server queueing *system* (including customers in the waiting line and being served) is given by:

P(exactly N customers in the queueing system) = $(1 - \lambda/\mu) \cdot (\lambda/\mu)^N$

where N = 0, 1, 2, 3, $\cdots$

<u>Notes</u>:

• The probability that the single server is busy = λ/μ. That is, if $\lambda = 2$ and $\mu = 3$, then the steady-state probability that the server is busy is 2/3. This probability can be derived using complement sets:

$$E = \left\{ Server\ busy \right\} = \left\{ At\ least\ 1\ customer\ in\ the\ queueing\ system \right\}$$

$$\hat{E} = \left\{ Server\ not\ busy \right\} = \left\{ Nobody\ in\ the\ queuing\ system \right\}$$

Then the steady-state probability that the server is busy is computed as follows:

$$P\left\{ Server\ busy \right\} = 1 - P\left\{ Server\ not\ busy \right\}$$

$$= 1 - P\left\{ Nobody\ in\ the\ queuing\ system \right\}$$

$$= 1 - (1 - \lambda/\mu) \cdot (\lambda/\mu)^0$$

$$= 1 - (1 - \lambda/\mu)$$

$$= \lambda/\mu$$

• The quantity ρ (rho) = λ/μ for the single server queue is called the <u>server utilization rate</u>. The server utilization rate ρ which is the *proportion* of the time the server is busy during the steady-state period. This proportion must be less than 1 (i.e. $\lambda < \mu$); a server utilization rate ρ greater than or equal to 1 ($\lambda \geq \mu$) would result in the queue growing without bound and never attaining a steady state.

• For the single server queueing system, the expected number of customers being served by the single server at any given time is equal to the server utilization rate λ/μ. This is demonstrated as follows:

$E(Number\ being\ served\ by\ the\ single\ server) = 1 \cdot (\lambda/\mu) + 0 \cdot (1 - \lambda/\mu) = \lambda/\mu$

The expected number of customers in the queueing system is then the expected waiting line length L plus the expected number of customers being served:

$$E(Number\ in\ the\ queueing\ system) = E(Number\ in\ line) + Server\ utilization\ rate$$
$$= L + \lambda/\mu$$

The waiting line time distribution is described by an Exponential distribution with parameter W:

- The Exponential distribution for the waiting line time in the single server queue has expected value $1/\mu$ equal to W. Hence $\mu = 1/W$.

- The probability that a customer will have a waiting line time t or less is computed using the cumulative distribution function for the Exponential distribution (denoted by F(t)):

$$F(t) = P(Waiting\ line\ time \leq t)$$

$$F(t) = 1 - e^{-\mu t} \qquad\qquad t \geq 0$$

$$F(t) = 1 - e^{-t/W} \qquad\qquad t \geq 0$$

Notes:

• The Exponential waiting line time distribution (describing the time spent in the waiting line) is *different* from the Exponential service completion time distribution (describing the time spent being served after leaving the waiting line).

• The total expected time in the queuing system equals the expected waiting line time W plus the expected service completion time $1/\mu$.

$$E(Time\ in\ queueing\ system) = E(Waiting\ line\ time) + E(Service\ time)$$
$$= W + 1/\mu$$

Example 5: (Dr. Quackenbush M/M/1 Single Server Queue). The physician Dr. Quackenbush has an office which includes a waiting room and a single examination room. There is an expected patient arrival rate of 2 patients per hour and an expected service completion time of 20 minutes per patient. Assume a Poisson arrival process and an Exponential service process.

Note: The expected arrival rate $\lambda = 2$ patients per hour. The expected service completion time $1/\mu = 20$ minutes = 1/3 hour so that μ = the expected service completion rate = 20 minutes per customer = 3 patients per hour. Since $\lambda < \mu$, it is expected that the queue will attain a steady-state.

1) What is the expected number of patients waiting to see Dr. Quackenbush (the expected waiting room line length)?

Expected waiting line length L $= \lambda^2 / \mu(\mu - \lambda)$

$$= 2^2 / 3(3 - 2)$$

$$= 4/3 \ patients$$

2) What is the doctor utilization rate (proportion of the time Dr. Quackenbush is busy)?

P(Doctor is busy) $= \rho$

$$= \lambda/\mu$$

$$= 2/3$$

3) What is the expected number of patients in Dr. Quackenbush's office (in the waiting room and being served in the examination room)?

E(Number in queueing system) $=$ *E(Number in line)* $+$ *Doctor utilization rate*

$$= L + \lambda/\mu$$

$$= 4/3 + 2/3$$

$$= 2 \ patients$$

4) What is the expected waiting line time (in the waiting room) before seeing Dr. Quackenbush?

Expected waiting line time W $= L / \lambda$

$$= (4/3 \ patients) / (2 \ patients \ per \ hour)$$

$$= 2/3 \ hours$$

5) What is the expected time a patient spends in Dr. Quackenbush's office (the queueing system) either in the waiting room or being served in the examination?

E(Time in queueing system) $=$ *E(Waiting line time)* $+$ *E(Service time)*

$$= W + 1/\mu$$

$$= 2/3 \ hour + 1/3 \ hour$$

$$= 1 \ hour$$

6) What is the probability that a patient waits less than 1 hour in the waiting room prior to being served?

- The waiting line time has an Exponential distribution with expected value $1/\mu = W = 2/3$ hours. Hence, $\mu = 1/W = 3/2$ per hour. The cumulative Exponential distribution function is used:

$P(Waiting\ line\ time < t) = 1 - e^{-\mu t}$

$\qquad\qquad = 1 - e^{-t/W}$

$\qquad\qquad = 1 - e^{-3t/2}$ $\qquad\qquad$ (*t measured in hours*)

$P(Waiting\ line\ time < 1\ hour) = 1 - e^{-3/2}$

$\qquad\qquad = 0.7769$

7) What is the probability of exactly 2 patients in Dr. Quackenbush's office (in the queueing system)?

$P(exactly\ N\ customers\ in\ the\ queueing\ system) = (1 - \lambda/\mu)\cdot(\lambda/\mu)^N$

$P(exactly\ 2\ patients\ in\ the\ doctor's\ office) = (1 - \lambda/\mu)\cdot(\lambda/\mu)^2$

$\qquad\qquad = (1 - 2/3)\cdot(2/3)^2$

$\qquad\qquad = 4/27$

8) What is the probability of at least 1 patient in Dr. Quackenbush's office (either in the waiting room or being served in the examination room)?

<u>Note</u>: The probability of at least 1 patient in the office is also the probability that Dr. Quackenbush is busy (λ/μ). For the single server queue, the server is busy as long as there is at least one customer in the system.

$P(exactly\ N\ patients\ in\ the\ queueing\ system) = (1 - \lambda/\mu)\cdot(\lambda/\mu)^N$

$P(at\ least\ 1\ patient\ in\ the\ office) = 1 - P(0\ patients\ in\ the\ office)$

$\qquad\qquad = 1 - (1 - \lambda/\mu)\cdot(\lambda/\mu)^0$

$\qquad\qquad = 1 - (1 - \lambda/\mu)$

$\qquad\qquad = \lambda/\mu$

$$= 2/3$$

The M/M/c Multiple Server Queue

Assumptions which allow an analytical solution of the single line M/M/c multiple server queueing problem are as follows:

- Poisson arrivals with constant expected arrival rate λ (Markovian).
- Exponential service with constant expected completion time $1/\mu$ (Markovian).
- Multiple servers (constant number c)

Examples of a single line multiple server queueing system include an airline service line with multiple service agents, a bank with multiple tellers, etc.

For the M/M/c queueing system, define $\rho = \lambda/c\mu$ (where c is the number of servers) as the server utilization rate (probability *each* of the c servers is busy):

- For the M/M/1 single server queue (c = 1), $\rho = \lambda/\mu$.
- For the M/M/c multiple server queue, $\rho = \lambda/c\mu$.

For the multiple server queue to have an analytical solution, the server utilization rate ρ must be strictly less than 1 (i.e. $\lambda < c\mu$). That is, the *total* service rate ($c\mu$) of the c servers must be strictly greater than the arrival rate (λ). Otherwise, the queue will grow without bound and never attain a steady state.

Example 6: (Minimum Number of Servers for a M/M/c Queueing System). Consider a bank with an expected arrival rate of 15 customers per hour. Each teller can be expected to complete all transactions for 3 customers per hour. What is the minimum number of tellers required to ensure that the line will not grow without bound and will attain a steady state?

Answer: The server utilization rate $\rho = \lambda/c\mu$ must be strictly less than 1 for the line not to grow without bound. Since $\lambda = 15$ and $\mu = 3$, the minimum number of tellers is determined as follows:

$$\lambda / c\mu < 1$$

$$\Rightarrow c > \lambda/\mu$$

$$\Rightarrow c > 15/3$$

$$\Rightarrow c > 5$$

At least 6 tellers are required in order for the queue not to grow without bound and to achieve a steady state.

The steady state expected waiting line time $W(c)$ for the multiple server queue with c servers is given by the following complicated expression:

$$W(c) = \frac{\dfrac{\rho^c}{(c!)(1 - \rho/c)} \cdot \dfrac{1}{\displaystyle\sum_{j=0}^{c-1} \rho^j/j! + \dfrac{\rho^c}{(c!)\cdot(1 - \rho/c)}}}{\mu c - \lambda}$$

The expression for the expected waiting line time $W(c)$ for the multiple server queue is quite cumbersome to compute by hand. Queueing software packages will compute and report the steady state expected waiting line time $W(c)$ for the M/M/c multiple server queue.

The steady state <u>expected waiting line length</u> $L(c)$ for the multiple server queue with c servers is given by:

$$L(c) = \lambda \cdot W(c)$$

Notes:

• The same relationship between expected waiting line time W and expected line length L hold for the multiple server queue (i.e. $L(c) = \lambda \cdot W(c)$) as for the single server queue (i.e. $L = \lambda \cdot W$).

• A simplified expected value characterization of the steady state queue helps to understand the equation for the expected waiting line length:

- Assume the arrival rate λ is 2 customers per minute and the expected waiting line time W is 5 minutes. Then from the time a given customer joins the line until the time the customer leaves the line and begins service (an expected time of 5 minutes), it is expected that 10 more customers have joined the line (i.e. 2 customers per minute times 5 minutes). Since the line is in steady-state (i.e. the same expected number of customers in line when a customer arrives and leaves), it is expected that there must be 10 customers in line on the average when the given customer (or any customer) joins the line.

• Similar to the M/M/1 single server queue, the waiting line time distribution for the M/M/c multiple server queue is described by an Exponential distribution with expected value $W(c)$:

- The Exponential distribution has expected value $1/\mu$ equal to $W(c)$. Hence, $\mu = 1/W(c)$.

- The probability that a customer will a waiting line time t or less is computed the cumulative distribution function for the Exponential distribution (denoted by $F(t)$):

$$F(t) = P(Waiting\ line\ time \leq t)$$
$$= 1 - e^{-\mu t} \qquad\qquad t \geq 0$$
$$= 1 - e^{-t/W(c)} \qquad\qquad t \geq 0$$

10.5 Final Notes

Recall that the objective of the Moribund Airlines multiple server queueing decision was to find the number of service counters c which minimized the total annual cost (including the building cost as well as the opportunity cost due to loss of customer goodwill and customer switches to other airlines). The total annual cost objective function TC(c) for c service counters was given by:

$$TC(c) = C(c) + u \cdot W(c) + v \cdot L(c)$$

Notes:

• The annual investment cost C(c) of building c service counters and maintaining the staff is typically estimated by engineers, architects, managers, etc. based on historical data from past projects.

• Analytical formulas for the expected waiting line time W(c) and the expected waiting line length L(c) at the time of customer arrival have been provided for the M/M/c multiple server queue (see Section 10.4).

• The annual opportunity costs (loss of revenue) to Moribund Airlines μ (due to loss of goodwill for each unit of customer waiting line time) and υ (due to customer airline switches for each person in the waiting line at the time of customer arrival), respectively, are *subjective* estimates. These costs may be estimated through expert opinion or through surveys of future customer actions (see Section 10.1).

One can now plot TC(c) vs. the number of service counters c (see Figure 10) to determine the *integer* number of service counters which provides the minimum total annual cost to Moribund Airlines. Naturally, the decision maker will also take into account other considerations (e.g. long range strategic plans for Moribund Airlines, space availability, local demographic projections, etc.) in the final choice of the number of service counters c to build.

Selection of the Number of Service Counters for Moribund Airlines

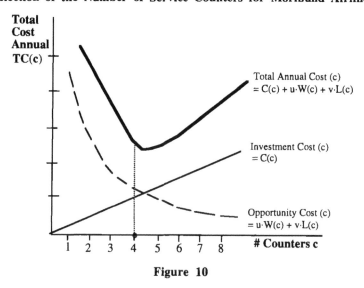

Figure 10

402

Limitations Inherent in the Analytical Queueing Solution Process

A number of restrictive assumptions have been made in order to allow an analytical solution for the Moribund Airlines queueing system during its steady-state period. To the extent that these assumptions are not reasonable, the analytical queueing solution derived will suffer:

> - Poisson (constant rate) arrival process: What if large groups of customers arrive at known times of the day (arrivals are not at a constant expected rate)?

> - Exponential (constant rate) service process: What if service agents hurry or provide less service when a line forms (there is an interaction between the arrival process and the service process)?

> - Constant number of servers c at all times: What if the number of service agents during the steady-state period fluctuates due to staffing difficulties?

> - First in - first out queue discipline: What if some customers are handled out of arrival sequence based on how soon their plane leaves or their frequent flyer status?

> - Steady-state solution: What if a decision maker is interested in the queue characteristics at Moribund Airlines during an extremely busy period generated by temporary discount fares (a time which is not steady-state)?

Simulation can be used to characterize a queueing system despite any or all of the queueing analytical solution assumptions being violated. However, simulation is expensive in terms of the manpower and time required to build the simulation and analyze the output.

Queueing can be effectively used to provide a "ball park" solution. If it becomes necessary to refine this solution, resources can then be allocated to develop a more realistic and detailed simulation.

Exercise 10

Queueing

Problem #1 (Queueing Assumptions)

The Yet Another Frozen Yogurt Shoppe planned for a shopping mall must decide how many employees are required to serve yogurt to its walk-in customers.

a) Define all input parameters and assumptions needed to formulate this business as a queueing process that can be solved analytically for the optimum number of employees.

b) Where appropriate, indicate what data could be collected or estimated to quantify this queueing problem.

Note: No calculations are required. Only a list of applicable parameters and assumptions is required.

Problem #2 (M/M/1 Hand Calculation)

Copy Spot has set up a copying kiosk in a business district. Assume that the copying kiosk has a Poisson arrival process (mean $\lambda = 10$ customers per hour), an Exponential service time (mean $1/\mu = 4$ minutes per customer = $1/15$ hour per customer), and a single server ($c=1$). A first in-first out (FIFO) queue discipline is used. Calculate the following:

a) What is the expected ("steady state") waiting line length L ?

b) What is the expected ("steady state") waiting line time W in the queue before service?

c) What is the probability that the ("steady state") waiting line time W (prior to receiving service) is less that 6 minutes?

d) What is the ("steady state") probability of exactly 1 customer in the Copy Spot queueing system (in the waiting line or being served)? 2 customers?

e) What is the expected time in the Copy Spot queueing system (including time in the waiting line and time being served)?

f) What is the ("steady state") probability of at least 1 customer in the Copy Spot queueing system (at the copying kiosk). That is, what is the probability that the server is busy? Compute this two ways.

Problem #3 (M/M/1 Queue)

Use a queueing software package to verify the answers calculated by hand in Problem #2.

Problem #4 (M/M/c Queue)

The switchboard at Oliver Twist Hospital has 7 lines. Calls arrive at the switchboard according to a Poisson distribution with a constant rate $\lambda = 1$ call/minute. The service completion time per call is Exponentially distributed with a constant time $1/\mu = 5$ minutes. Calls may be placed on hold so that all customers have the opportunity to wait for their calls to be answered.

Use a queueing software package to answer the following questions:

4a) What is the probability that a given line is being utilized (i.e. what is the server utilization rate)?

4b) What is the average number of calls on hold (i.e. what is the average waiting line length)?

4c) What is the average number of calls in the switchboard system either on hold or being answered (i. e. what is the expected number of customers in the queueing system)?

4d) What is the probability that a call into the hospital switchboard system will be put on hold (i.e. what is the probability that an arriving customer calling in must wait in the queue)?

4e) What is the expected time a call will spend on hold (i.e. what is the expected waiting line time)?

4f) What is the expected time a call into the Oliver Twist Hospital switchboard will spend waiting on hold or being answered (i.e. what is the expected time a customer will spend in the queueing system)?

Chapter 10

Queueing Additional Solved Problems

Problem #1 (Queueing)

The small outpatient clinic at A. Ling Community Hospital is set up as an M/M/1 (single doctor) queue. Patients arrive at the clinic according to a Poisson process with rate $\lambda = 10$ patients per hour. Patient service completion time is described by an Exponential process with $1/\mu = 0.067$ hours per patient. Assume that the queueing system has settled into a steady-state mode.

a) What is the doctor (server) utilization rate?

Answer: The doctor (server) utilization rate is given as follows:

Server utilization rate $= \rho = \lambda/\mu = 10/15 = 2/3$

b) What is the expected number of patients waiting for service (in the queue)? What is the expected number of patients at the clinic (in the queueing system)?

Answer: The expected number of patients waiting in line for service is given as follows:

$$L = \lambda^2 / u \cdot (u - \lambda) = (10)^2 / (15) \cdot (15 - 10) = 100/75 = 4/3 \; patients$$

The expected number of patients in the clinic (the queueing system) is the expected number of patients in line L plus the expected number of patients being served (given by the doctor utilization rate ρ):

$$\begin{aligned} Expected\; number\; of\; patients\; in\; the\; clinic &= L + \rho \\ &= 4/3 + 2/3 \\ &= 2\; patients \end{aligned}$$

c) What is the expected waiting time for patients prior to service (in the queue)? What is expected time patients spend at the clinic (in the queueing system)?

Answer: The expected length of time a patient must wait prior to service (in line) is given by:

$$\begin{aligned} W &= L/\lambda \\ &= 4/3\; patients\, /\, 10\; patients\; per\; hour \\ &= 1.333\, /\, 10 \\ &= 0.133\; hours \end{aligned}$$

The expected length of time in the clinic (in line or in service) is given by W plus $1/\mu$:

Expected length of time at the clinic $= W + 1/\mu$

$$= 0.133 + 0.067$$
$$= 0.20 \ hours$$

d) What is the probability that the wait for a patient in the queue prior to service is less than 0.20 hours?

Answer: The waiting line time is Exponential with expected value $1/\mu = W = 0.1333$ hours. The probability that the waiting time is less than 0.20 hours is given by:

$$F(t) \quad = 1 - e^{-\mu t}$$
$$F(0.20) = 1 - e^{-0.20/0.1333}$$
$$= 1 - e^{-1.5}$$
$$= 0.78$$

Problem #2 Queueing

An automated teller machine (ATM) is set up as a single line M/M/1 queueing system. Customers arrive at the ATM according to a Poisson process with expected arrival rate $\lambda = 15$ customers per hour. Customer service is described by an Exponential process with expected service completion time $1/\mu = 3$ minutes per customer. Assume that the queueing system has settled into a steady-state mode.

a) What is the ATM (server) utilization rate?

Answer: The ATM (server) utilization rate is given as follows:

$$ATM \ utilization \ rate = \rho = \lambda/\mu = 15/20 = 3/4$$

b) What is the expected number of customers waiting for the ATM (in the waiting line)? What is the expected number of customers at the ATM site (in the queueing system)?

Answer: The expected number of patients waiting in line at the ATM for service is given as follows:

$$L = \lambda^2 / u \cdot (u - \lambda) = (15)^2 / (20) \cdot (20 - 15) = 225/100 = 2.25 \ customers$$

The expected number of customers at the ATM site (the queueing system) is the expected number of customers in line L plus the expected number of customers being served (given by the ATM utilization rate ρ):

Expected number of customers at the ATM $= L + \rho$
$$= 2.25 + 3/4$$
$$= 3 \ customers$$

c) What is the expected waiting time for customers prior to service (in the waiting line)? What is expected time customers spend at the ATM site (in the queueing system)?

Answer: The expected length of time a customer must wait prior to service (in the waiting line) is given by:

$$W = L/\lambda$$
$$= 2.25 \ customers \ / \ 15 \ customers \ per \ hour$$
$$= 0.15 \ hours$$

The expected length of time at the ATM site (in line or in service) is given by W plus $1/\mu$:

$$Expected \ length \ of \ time \ at \ the \ ATM = W + 1/\mu$$
$$= 0.15 + 0.05$$
$$= 0.20 \ hours$$

d) What is the probability that the wait for a customer in the waiting line prior to service is less than 0.15 hours?

Answer: The waiting line time is Exponential with expected value $1/\mu = W = 0.15$ hours (i.e. $\mu = 1/W$) The probability that the waiting line time is less than 0.15 hours is given by:

$$P(W < t) = F(t) = 1 - e^{-\mu t}$$
$$= 1 - e^{-t/W}$$

$$F(0.15) = 1 - e^{-0.15/0.15}$$
$$= 1 - e^{-1}$$
$$= 0.6321$$

e) Consider a group of c ATM's in a shopping mall set up as a single line M/M/c queueing system where the Poisson arrival rate is 30 per hour. If the Exponential service completion rate of each ATM is 5 customers per hour, what is the minimum number of ATM's required so that this M/M/c queueing system will settle into a steady-state?

Answer: For a M/M/c queue, $\lambda/c\mu$ must be < 1 to achieve a steady-state mode. Since $\lambda = 30$ and $\mu = 5$:

30/5c < 1
$\Rightarrow c > 6$
$\Rightarrow$ at least 7 ATM's (servers)

"It is a truth very certain that when it is not in our power to determine what is true we ought to follow what is most probable."

Descartes, *Discourse on Method*

11 | Simulation

11.1 Definitions

System Definitions

A <u>system</u> consists a set of related *entities*. A system has *states* whose values are described by a set of *relationships* using the *parameters* (fixed values) of the system. An example of all these elements for a system defined by a single doctor's office is given in Figure 1.

Example System: Single Doctor's Office

System	A single doctor's office (M/M/1 queuing system)
Entities	Waiting room Examination room
States	Waiting line length (L) Waiting line time (W)
Relationships	$P(N \text{ patients in the office}) = (1 - \lambda/\mu)\cdot(\lambda/\mu)^s$ $P(\text{Waiting line time} < t) = 1 - e^{-t/w}$ (These relationships defined by the fixed parameters λ, μ of this queueing system are valid for the steady-state period)

Figure 1

The states of a system (e.g. waiting line length or waiting line time for a customer entering the doctor's office) will change over time. During a given interval of time, a state can be in a steady-state mode or in a transient state mode defined as follows:

- A state of a system is in a <u>steady-state</u> mode if its probability distribution converges to a fixed probability distribution (e.g. the probability a patient spends

409

less than t time in the waiting line converges to $1 - e^{-t/W}$ while the probability of exactly N patients in the queueing system converges to $(1 - \lambda/\mu) \cdot (\lambda/\mu)^N$.

- A state of a system is in a <u>transient state</u> mode if its probability distribution does not converge to a fixed probability distribution.

The behavior of the state waiting line length is displayed in Figure 2 for a M/M/1 queueing system defined by a convenience store with a single cashier. During the first hour, the waiting line length is in a transient state mode. Later, the waiting line length settles into a steady state mode (where the number of customers in line and being served converges to the probability distribution $(1 - \lambda/\mu) \cdot (\lambda/\mu)^N$).

Transient vs. Steady-State Behavior at a Convenience Store

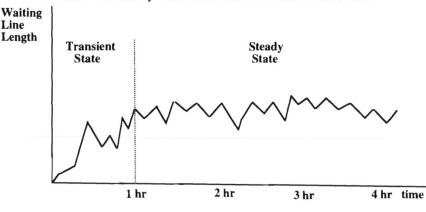

Figure 2

Notes:

• For a M/M/1 queueing system with $\lambda/\mu < 1$, the state waiting line length will settle into a steady state mode after a startup period. When $\lambda/\mu \geq 1$, the state waiting line length will not settle into a steady state mode and will remain in a transient state mode.

• For a general queueing system, a state in the transient mode may have explosive or non-explosive behavior:

- A queueing system with $\lambda/\mu \geq 1$ will have the state waiting line length increase without bound theoretically. Realistically, the size of the waiting room and patient turnaways will mitigate the unbounded increase in the waiting line length. Such a state is said to have *explosive* transient behavior (see Figure 3).

- A queueing system with arrival and service rates which *vary* over time but which maintains $\lambda/\mu < 1$ will have the state waiting line length remain finite over time. However, the expected waiting line length will vary over time and the queueing system will never attain a steady state. Such a state is said to have *non-explosive* transient behavior (see the oscillating system in Figure 4).

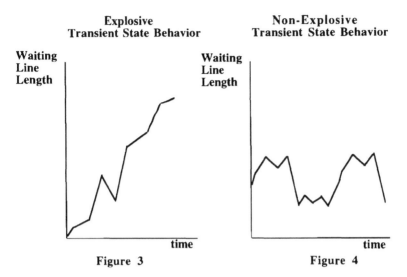

Figure 3   Figure 4

• A state in the steady-state mode can also be classified as stable and non-stable. This state is said to be non-stable (see Figure 5) if following an anomalous event (e.g. the doctor is called away for a couple of hours to attend to an emergency at a nearby hospital and the waiting line grows unusually long), the state does not returns to its steady-state (e.g. the waiting line length begins to grow without bound or else settles into a *different* steady-state behavior). A state is said to be stable (see Figure 6) if following an anomalous event, the state returns to its previous steady-state behavior (e.g. the waiting line length returns to the steady-state mode which existed prior to the anomalous event).

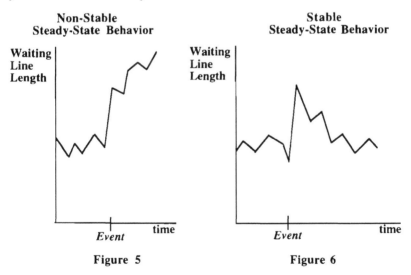

Figure 5   Figure 6

411

Model Definition

A model is a representation used to characterize a real world system. Two common classifications of models are deterministic and stochastic:

- Deterministic models are often solvable by quantitative methods and provide analytical *solutions* (e.g. the economic order quantity for a deterministic inventory model, the task start times for a deterministic network, etc.) for the system.

- Stochastic models are typically not solvable in closed form by quantitative methods and are only able to provide system performance *data* (e.g. waiting line times at the startup of a queuing system, project completion times for a stochastic network) for which *statistics* (e.g. sample mean of the waiting line times, sample standard deviation of the project completion times) can be computed and *hypotheses* (e.g. the waiting line time data is generated by an Exponential distribution) tested.

The *advantages* of building a model to represent a real world system are evident:

- Models are simpler and more convenient to manipulate than the system itself. For example, estimation of the effect of changing coefficients of the constraints for an LP model of a production process is much simpler than changing the actual production process.

- Models often lead to an improved understanding of the system. For example, formulation of an LP model for a production process (specifying the input resources, the constraints on these resources, the objective function to be maximized, etc.) helps one to understand the current production process and the sensitivities of the process to its constraints. This itself can lead to an improved production process.

- Models often point in the direction of a solution technique. For example, determination of a model for a production process requires examination of whether the constraints and objective function are non-linear ($\Rightarrow$ non-linear programming), whether the coefficients of the decision variables and slack variables are Random Variables ($\Rightarrow$ stochastic programming), whether the decision variables are integer ($\Rightarrow$ integer programming), etc.

The *disadvantages* of building a model to represent a real world system must be weighed before committing the required resources (e.g. personnel, time, money, etc.):

- The model may never provide the benefits to justify the cost of building the model. For example, a regression model (e.g. for earthquake prediction) may never satisfactorily forecast the future behavior of the system.

- The model may be taken to mirror the real world rather than to provide only a *representation* of the real world. For example, when real world data does not conform to a regression model forecast, it is often the data which is massaged rather than the regression model.

- The model has the potential to provide spurious results. For example, when regression models are used to forecast demand far outside of the range of the data used to build the regression.

Note: When a model of a system is built based on a set of input and output data, the model often predicts this data quite well. However, the general utility of a model is not *validated* until the model demonstrates success in predicting system performance using *independent* data (i. e. actual input and output data not related to the data used to build the model).

11.2 Simulation Overview

Simulation is the process of conducting "experiments" on a deterministic or stochastic *model* of a complex *system* and of making statistical inferences using the performance data for the system. Simulation brings together three powerful quantitative techniques:

- Model building
- Computer science
- Statistics

The use of simulation to examine simultaneous interactions of the many parts of a complex system became popular following advances in large electronic computers in the early 1950's. Today, once the model is developed, the simulation of a complex deterministic or stochastic process often can be performed using a desktop or handheld computer.

Alternatives to simulation for characterizing the behavior of a complex system are:

1) Experimenting with the actual system (e.g. changing the types of input resources used or the total input resources available for a production process).

2) Representing the system by a simplified model which will lead to a loss of accuracy but which will be amenable to a quantitative technique (e.g. queueing where the average performance of the system can be derived analytically).

Simulation does not require experimentation with the actual inputs and outputs of the system. This *non-interruption* of the on-going activities of a system is an attractive feature of simulation. For example, imagine if the only way to predict the accuracy of the United States nuclear forces was to exercise the actual system (i. e. launch an attack on another country).

The *advantages* of a simulation are many and lead to its popularity as a "solution" technique:

- A system can be modeled to the desired level of detail rather than simplified in order to allow its solution by an analytical quantitative technique. For example, a complex queue discipline representative of a large hospital's emergency room may be modeled to the desired fidelity rather than the simplistic assumption of a first in-first out (FIFO) queue discipline.

- Time can be *mathematically* expanded or compressed to help analyze the behavior of a system with greater precision. For example, thousands of years of detailed planetary data can be compressed into minutes for quick examination of cosmic interactions; microseconds of nuclear fission performance data can be expanded to hours in order to capture all of the detailed particle interactions; etc.

- Experimentation on the simulated performance of a product (e. g. computer chip throughput speed) can be made prior to making the considerable investment required for a full production process.

- Sensitivity analysis is easily performed once the baseline simulation model has been developed and validated. Parameters of the system can be quickly changed in a simulation and new performance data generated. For example, once the simulation for a queueing system has been developed, the arrival rates, service rates, # servers, etc. can easily be changed and the effects of these changes on the performance of the queueing system can be quickly determined.

The *disadvantages* of simulation are its potential for high cost, lack of general applicability, and lack of definitive results:

- The cost of modeling and computer programming can be quite large. For example, complex simulations used for policy analysis (e.g. health, education, crime, etc.) can involve many man-years of effort to build the simulation and to validate the simulation. This can cost millions of dollars.

- The cost of collecting data to characterize the behavior of the parameters of a system can be large. For example, the cost to collect data on the behavior of the insured in simulating health care plan expenditures with different deductibles, co-payments, maximum yearly payments, etc. can amount to millions of dollars.

- A system may be quite unique (e. g. a queueing system with unusual arrival and service processes over time) so that the applicability of the simulation model to more general systems without major revision and revalidation may be limited.

- A "solution" is not provided. For example, a simulation may provide initial indication that building 2 service counters at an airline customer service facility provides the minimum total annual cost (construction costs and opportunity costs due to customer dissatisfaction) solution. Does the decision maker settle on 2 service counters, collect more data to refine the input cost parameters, perform more simulation trials to provide higher statistical confidence, etc.?

Note: While the applicable quantitative method provides a *solution*, a simulation only provides *data* for statistical inference on the solution.

11.3 Simulation of Complex Models

Simulation is applied both to complex deterministic models and complex stochastic models. Simulation is often used to examine those complex *deterministic* models that require a great deal of bookkeeping since computers are able to perform efficiently this bookkeeping that is so tedious to perform by hand. Simulation of complex deterministic models are also performed when intermediate results and not just the final analytical solution are of interest.

For example, a deterministic simulation could be used to predict the steady-state (equilibrium) population of two opposing animal species in the wild. Consider a population of foxes that is introduced into a national park where a large population of rabbits already exists:

- As the number of foxes in the national park increases, rabbits within the park are killed for food and the number of rabbits decreases.

- As the number of rabbits decreases, there is not enough food to support the fox population and the number of foxes starts to decrease (i.e. the foxes will either migrate out of the park or starve).

- With a smaller number of predator foxes, the rabbit population starts to increase supporting a larger fox population. And so on.

This rise and fall of the fox and rabbit population can be modeled using simultaneous differential equations whose solution determines the equilibrium population of foxes and rabbits in the national park. Simulation of this survival process provides the bookkeeping to examine the detail of the rise and fall over time of the population of the two species (i.e. intermediate results rather than only the final equilibrium point determined when the deterministic differential equations are solved). Simulation also allows the initial number of rabbits or the initial number of predator foxes introduced into the national park to be varied. This is important since different initial fox and rabbit populations can lead to different equilibrium points for the two species!

The remainder of this section examines the simulation of complex *stochastic* processes. Simulation of complex stochastic processes is of great interest since such processes often can not be solved analytically using quantitative techniques.

Example 1: (Bank M/M/1 Single Line Queue). Consider the following flow for a customer arriving at a bank with single waiting line (queue) and a single teller (server):

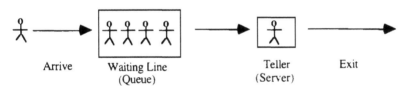

| Arrive | Waiting Line (Queue) | Teller (Server) | Exit |

Assume the following stochastic queueing model for the customer arrival and service processes and for the queue discipline:

- Arrival rate: Every 10 minutes there is a 50% chance of no customer arrivals to the bank and a 50% chance of exactly one customer arrival to the bank.

- Arrivals are independent of each other and the expected arrival rate is constant throughout the day (Markovian).

- Service completion time: 50% of the customers require 10 minutes service completion time from the teller while the other 50% of the customers require 20 minutes service completion time from the teller.

- Service completion times are independent of each other and the expected service completion time is constant throughout the day (Markovian).

- Queue discipline: First in - first out (FIFO).

Describe a possible methodology to simulate this stochastic queueing system.

Answer: A simple methodology to simulate this stochastic M/M/1 queueing system is as follows:

- Arrival Rate: During each 10 minutes of *simulated* time, a fair coin (50% Heads, 50% Tails) is flipped. If Heads, there are no customer arrivals in that 10 minute period. If Tails, there is one customer arrival at the start of that 10 minute period.

- Service Completion Time: When each new customer arrives, a fair coin is flipped. If Heads, the customer service completion time will be 10 minutes. If Tails, the customer service completion time will be 20 minutes.

- Queue Discipline: When a new customer arrives and the teller is free, the customer immediately begins service. When a customer arrives and the teller is busy, the customer joins the end of the line. When the busy teller completes service on a customer, the customer at the head of the line is served next.

By advancing time and bookkeeping the flow of customers through the bank, this stochastic queueing process can be simulated and performance data can be generated. Queue performance statistics (e.g. the mean waiting line length, the proportion of time the teller is busy, etc.) can then be computed from this simulation output performance data.

For example, bank single server queue simulation data for 80 minutes of arrivals (in 10 minute simulated time intervals) is provided in Figure 7. Each new customers arriving at the bank is numbered and tracked through the bank M/M/1 queueing system.

Bank M/M/1 Single Server Queue Simulation Data
(Single Simulation Trial)

Simulated Time Intervals (Minutes)	Customer Arrivals (Coin Flip)	Service Completion Time (Coin Flip)	Waiting Line	Teller Utilization
0-10	Tails (Customer 1)	Tails (20 min)	----	Customer 1
10-20	Tails (Customer 2)	Heads (10 min)	Customer 2	Customer 1
20-30	Tails (Customer 3)	Heads (10 min)	Customer 3	Customer 2
30-40	Heads (No Arrivals)	----	----	Customer 3
40-50	Tails (Customer 4)	Heads (10 min)	----	Customer 4
50-60	Heads (No Arrivals)	----	----	----
60-70	Heads (No Arrivals)	----	----	----
70-80	Tails (Customer 5)	Tails (20 min)	----	Customer 5
80-90				Customer 5

Figure 7

416

Statistics using the bank M/M/1 single server queueing system simulation performance data over the 80 minutes of customer arrivals (provided in Figure 7) can now be computed:

- 5 customers arrived in 80 minutes.
- The teller worked 90 minutes to service all customers arriving in the 80 minutes.
- Mean waiting line time = 20 minutes total / 5 customers = 4 minutes/customer.
- Mean waiting line length = $(1) \cdot (20/80) + (0) \cdot (60/80) = 1/4$ customers.
- Mean teller utilization rate = 70 min busy / 90 min on the job = 0.77.

The simulation performance data in Figure 7 represents only a single simulation *trial* for 80 minutes of arrivals to the bank. Typically, a simulation would be generated for a longer period of interest (e.g. an entire day) and would consist of many trials (e.g. N = 30 or more trials with independent coin flips to allow the assumption of Normal output data). Statistics such as the sample mean, the sample standard deviation, confidence intervals, goodness of fits, etc. would then computed for the bank waiting line length, waiting line time, waiting line time distribution, etc. using the output data from the N simulation trials.

Random Numbers

It is necessary to simulate arrival processes more complex than that given in Example 1 (i.e. zero or one arrivals to the bank each 10 minute interval). For example, consider a Poisson arrival process with $\lambda = 0.5$ arrivals per hour. Figure 8 provides the Poisson probability and cumulative distributions for k = 0, 1, 2, 3, and 4 or more customer arrivals during an hour when $\lambda = 0.5$ arrivals per hour.

Poisson Arrival Process ($\lambda = 0.5$)

Arrivals Per Hour k	Probability Distribution P(k) $P(X=k)$	Cumulative Distribution F(k) $P(X \leq k)$
0	0.607	0.607
1	0.303	0.910
2	0.076	0.986
3	0.012	0.998
4+	0.002	1.000

Figure 8

Note: For a Poisson Random Variable X, the discrete probability distribution $P(X = k)$ is given by:

$$P(X = k) = \lambda^k e^{-\lambda} / k! \qquad k = 0, 1, 2, 3, \cdots$$

To simulate a Poisson arrival process ($\lambda = 0.5$), a scheme which randomly generates 0, 1, 2, 3, and 4 or more arrivals per hour in the proper proportions 0.607, 0.303, 0.076, 0.012, and 0.002, respectively, is required. This will be accomplished using a Random Number between 1 and 1000 and the Poisson ($\lambda = 0.5$) *cumulative* distribution. A Random Number is a number that is *equally likely* to take on any of its possible values.

There are several methods to generate a Random Number (often abbreviated by its acronym RN) between 1 and 1000 for each hour of simulated time. For example:

- Place 1000 equal sized balls numbered with the integers 1 to 1000 into a hopper. After spinning the hopper (i.e. randomizing the process), pull out one ball and note the number; then return the ball to the hopper. Repeat this process at the start of each hour of simulated time.

- Place 1000 equal sized strips of paper numbered with the integers 1 to 1000 in a bag. After shaking the bag (i.e. randomizing the process), pull out one strip of paper and note the number; then return the strip of paper to the bag. Repeat this process at the start of each hour of simulated time.

Each ball or strip of paper chosen (which provides a Random Number between 1 to 1000) can then be used (see Example 2) to generate 0, 1, 2, 3, or 4 or more arrivals for that hour of simulated time using the cumulative Poisson distribution ($\lambda = 0.5$).

Note: The term "hour of simulated time" is not to be taken literally. It merely refers to a step in a computer simulation where a Random Number and it corresponding stochastic outcome are generated. These steps occur very fast (e.g. millions of steps per second).

Example 2: (Random Numbers Used To Generate Discrete Poisson Outcomes). A ball numbered 697 has been randomly pulled out of the hopper containing equal sized balls numbered from 1 to 1000. Use this Random Number 697 to determine the number of arrivals during an hour of simulated time for the *discrete* Poisson arrival process with $\lambda = 0.5$ arrivals per hour.

Answer: The Random Number RN = 697 selected at the start of the hour of simulated time is mapped into the cumulative Poisson distribution (see Figure 9) to obtain the discrete Poisson outcome of 1 arrival during the hour. That is, any RN between 1 and 607 (probability 0.607) would generate 0 arrivals during the hour; any RN between 608 and 910 (probability 0.303) would generate 1 arrival during the hour; etc.

Discrete Stochastic Poisson ($\lambda = 0.5$) Arrival Generation

Random Number Interval	Probability	# Arrivals
$1 \leq RN \leq 607$	0.607	0
$608 \leq RN \leq 910$	0.303	1
$911 \leq RN \leq 986$	0.076	2
$987 \leq RN \leq 998$	0.012	3
$999 \leq RN \leq 1000$	0.002	4+

Figure 9

Graphically, the Random Number 697 (considered as the decimal 0.697 since probabilities are always between 0 and 1) can be mapped into the discrete cumulative Poisson ($\lambda = 0.5$) distribution F(k) (also a decimal between 0 and 1) to determine the Poisson outcome of 1 arrival during the hour (see Figure 10).

Mapping of the Random Number 697 into the
Discrete Poisson ($\lambda = 0.5$) Cumulative Distribution

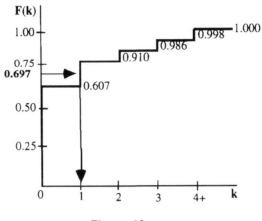

Figure 10

A computer-based simulation typically accesses Random Numbers by one of two methods:

Method #1: (Computer Stored Table of Random Numbers). A previously generated and published table of Random Numbers (see Figure 11) is stored in computer memory. Such a table of digits has the property that each single digit is a Random Number. Also, a consecutive *set* of two, three, four, etc. digits is again a Random Number.

Sample Table of Random Numbers

9822	4931	7892	8765	7846	0986	9035	9467
2013	3508	2929	5999	9633	3805	6103	2531
9813	4079	6154	0207	5858	3282	6970	2280
0405	5350	8450	2547	9232	2471	1731	8148
0703	2050	5726	5902	3876	0134	5921	4314
0061	4653	7369	2337	7699	5161	1672	3582
2566	7074	9346	2412	0355	0415	9267	2044
1824	0202	4794	6942	8501	8529	4662	1208
4391	2612	6152	3400	5031	9586	2236	7135
1348	5205	9094	7718	4459	7286	3583	5927

Figure 11

Method #2: (Computer Generated Random Numbers). Computer routines (e.g. a numerical expansion of π to a large number of decimal places) can *generate* Random Numbers *as needed* rather than having to *store* large tables of Random Numbers. The consecutive digits corresponding to the decimal places of π are considered random (i. e. no repeating pattern for the decimal value of π has been determined by mathematicians).

Notes: Properties of a Random Number include:

• On a given draw of a Random Number stream, all possible outcomes are equally likely.

• The Random Number streams (2222) and (4795) are equally likely four-digit streams even though the second stream seems "more random" than the first stream.

• The definition of probability (i.e. proportion in the long run) holds. That is, as the total number N Random Number digits (chosen from the integers 0 to 9) grows large, the proportions of 0's, 1's, 2's, etc. will approach the value 1/10:

$$\frac{\#0's}{N} \approx \frac{\#1's}{N} \approx \cdots \approx \frac{\#9's}{N} \approx \frac{1}{10}$$

The Random Numbers in a stored table in computer memory or generated as needed by computer routines are sometimes called <u>Pseudo</u> <u>Random</u> <u>Numbers</u>. To understand why the term "pseudo" *(false)* is applied, consider the following simulation scenario:

- One wishes to generate 30 independent trials for a simulation of a stochastic process. Each individual trial requires 100 Random Number streams to determine the performance of the scholastic process. One starts at the same position in a Random Number table or at the same decimal place of π for each trial. This will lead to 30 *identical* streams of 100 Random Numbers and the exact same performance of the stochastic process for each of the 30 simulation trials. The 30 simulation trials will not be independent and, in fact, will be identical!

This potential lack of independent trials using stored tables or computer generated Random Numbers is overcome by using a different seed number for each simulation trial. Each different <u>seed</u> <u>number</u> points to a different (and thus independent) location in the stored table or a different decimal place of π in which to start identifying Random Numbers. In this way, each simulation trial will be *independent* of all other trials. For example, the seed number 12345 (e.g. the 123rd row and 45th column in a table or the 12345th decimal position for π) and the seed number 13425 will provide independent simulation outputs.

Now suppose that the customer service process for the bank M/M/1 queueing system described in Example 1 is best characterized by the *continuous* Exponential distribution with expected value $1/\mu$. That is, the customer service completion time is defined by the Exponential probability distribution:

$$f(t) = \mu e^{-\mu t} \qquad\qquad t \geq 0$$

and an Exponential cumulative distribution:

$$F(t) = 1 - e^{-\mu t} \qquad\qquad t \geq 0.$$

Then the value of service completion time t can be derived as a closed form expression of the cumulative distribution function F(t) as follows:

$$1 - F(t) = e^{-\mu t}$$

$$ln(1 - F(t)) = -\mu t$$

$$t = \frac{-\ln(1 - F(t))}{\mu} \qquad \text{where } 0 \le F(t) \le 1.$$

This equation represents the mapping of a cumulative distribution function value F(t) into a continuous Exponential service completion time value t. Remember that the term F(t) is a decimal which varies between 0 and 1. Values for the service completion times t can be determined by using Random Numbers to determine equally likely decimal values between 0 and 1 for F(t) (see Example 3).

Example 3: (Random Numbers Used To Generate Continuous Exponential Outcomes). A bank customer service completion time is Exponential with mean $1/\mu = 1.5$ hours ($\mu = 0.667$ service completions per hour). Use the Random Number 328 to determine the customer service completion time for this *continuous* Exponential service process.

Answer: The Random Number 328 is considered as the decimal 0.328. This decimal value which is equally likely to be between 0 and 1 is substituted for the value F(t) in the equation mapping F(t) into a continuous Exponential service completion time t:

$$F(t) = 1 - e^{-0.667 \ t}$$

$$t = \frac{-\ln(1 - F(t))}{0.667}$$

$$= \frac{-\ln(1 - 0.328)}{0.667}$$

$$= \frac{-\ln(0.672)}{0.667}$$

$$= 0.595 \ hrs$$

Graphically, the mapping of the Random Number 328 which is considered as the decimal value 0.328 (between 0 and 1) into the continuous cumulative Exponential ($1/\mu = 1.5$ hours) distribution F(t) (also a decimal between 0 and 1) is demonstrated in Figure 12.

Mapping of the Random Number 328 into the Continuous Exponential ($1/\mu = 1.5$) Cumulative Distribution

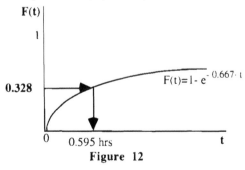

Figure 12

It is common in simulation that one is required to generate a random stochastic outcome for a continuous Normal Random Variable X with expected value μ and standard deviation σ (i.e. $X \sim N(\mu, \sigma^2)$). Values for a Normal Random Variable X can be determined using

Random Numbers to determine equally likely decimal values between 0 and 1 for the Standard Normal distribution Z ~ N(0,1) cumulative distribution (see Example 4).

Example 4: (Random Numbers Used to Generate Continuous Normal Outcomes). Consider a Normal Random Variable X with expected value 5 and standard deviation 2 (i. e. X ~ N(5,4)). Use the Random Number 195 to determine a continuous Normal outcome.

Answer: The Random Number 195 is considered as the decimal 0.195. This decimal value which is equally likely to be between 0 and 1 is substituted for the value of the cumulative distribution F(z). Table 1b is used to determine the value $z_{.195}$ for the N(0,1) Standard Normal cumulative distribution F(z) such that:

$$F(z_{.195}) \ = \ P(Z \ \le \ z_{.195}) \ = \ 0.195 \qquad\qquad Z \sim N(0,1) \,.$$

Table 1b indicates that 0.195 of the probability (area) under the curve of the Standard Normal distribution Z ~ N(0,1) is to the left of the value z = -0.86 where z can vary from $-\infty$ to $+\infty$ (but with 0.997 probability is between -3 and +3). That is,

$$F(-0.86) \ = \ P(Z \ \le \ -0.86) \ = \ 0.195 \qquad\qquad Z \sim N(0,1) \,.$$

Then the outcome from the Normal Random Variable X ~ N(5,4) associated with the Random Number 195 is given by:

$$\mu \ + \ z_{.195} \cdot \sigma \qquad\qquad P(X < \mu + z_{.195} \cdot \sigma) = 0.195$$

$$5 \ + \ (-0.86) \cdot 2 \qquad\qquad P(X < 5 + (-0.86) \cdot 2) = 0.195$$

$$3.28 \qquad\qquad P(X < 3.28) = 0.195$$

Graphically, the Random Number 195 which is considered as the decimal value 0.195 (between 0 and 1) is mapped into the continuous Standard Normal Distribution to obtain a variate z = -0.86 between $-\infty$ and $+\infty$ (see Figure 13). This Standard Normal N(0,1) variate is then used to determine a outcome from the continuous Normal N(5,4) distribution.

Mapping of the Random Number 195 into the Continuous Normal X~N(5,4) Distribution

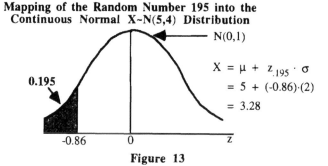

Figure 13

Random Numbers can be used to characterize queueing systems of much greater complexity than the single line M/M/1 queueing system with a Poisson arrival rate and an Exponential service completion time. Additional complexity in a queueing system (e.g.

different *types* of customers each having different arrival rates and service completion times, a changing number of servers over time, a complicated non-FIFO queue discipline, etc.) can all be handled in a straightforward manner using simulation. All that is required is more bookkeeping to keep track of elements such as the arrival rate and the service time for each customer type, the number of servers at any given time, what criteria is used to choose the next customer from the queue, etc.

The bookkeeping of simulation results and the generation of statistics for even simple simulations represent difficult tasks to be accomplished by hand. Fortunately, computers allow large numbers of independent simulation trials to be run and statistics to be computed on a quick turnaround basis even for models of highly detailed stochastic processes.

Simulation Validation

Simulation validation occurs when there is an acceptable level of *statistical* confidence that the simulation provides correct performance data for the stochastic process. Simulation validation is performed by comparing simulated output data to actual output data (known results) under the same input conditions. The Chi-Square test is often useful in determining the goodness of fit of the simulated (i.e. expected) data to the actual data.

Notes:

• The actual data used to construct the simulation should not also be used to validate the simulation model! Independent data must be used to validate the simulation.

• A *complete* validation of a simulation may not be possible. A complicated simulation will have thousands of different input conditions. Possible values for the stochastic inputs can lead to millions of paths through the computer simulation program. Obtaining both actual performance data and simulated performance data for all possible input conditions and paths through the simulation is typically not possible or cost-effective.

• It is critical to validate the simulation for at least the *most likely* input conditions and paths through the computer simulation program. The stochastic conditions under which the simulation has not been validated should be noted as a possible source of any anomalous performance data generated by the simulation.

Determination of the Required Number of Simulation Trials

The size of the simulation *run* (i.e. the number of simulation trials) which is required to achieve the desired accuracy of the simulation performance data has *no analytical solution*. Two important considerations which affect the number of simulation trials required are:

- How quickly the simulation statistic of interest (e.g. mean customer waiting line time, the maximum waiting line length at any hour, etc.) converges to its true value.

- What accuracy is desired for the simulation statistic of interest (e.g. the size of the 95th percentile confidence interval around its mean value of a parameter, the confidence level provided by the Chi-Square statistic, etc.).

While it is typically not possible to determine the required number of trials prior to the start of a simulation run, information on the rate of convergence of an outcome (e.g. how quickly an outcome sample mean settles into a small interval) is gained as the simulation progresses. Consequently, an estimate of the number of simulation trials to be run is sometimes determined in two steps (i.e. the "learn-as-you-go" technique):

- Step 1: An initial nominal run size to determine the convergence rate of the desired simulation outcome.

- Step 2: A final production simulation run with the number trials estimated using the information gathered in Step 1.

Another method to determine the number of simulation trials is to allow the simulation to terminate itself when the current confidence interval or the current Chi-Square statistic (which can be computed after each simulation trial) is reduced to a specified size.

The convergence of the mean of a simulation outcome to its true value may be *quite slow*. If the sample standard deviation of the mean of a simulation outcome is s, then the standard deviation after N trials will be reduced to $s/N^{1/2}$. That is, to reduce the standard deviation around the mean of an outcome of a simulation by a factor of 10, the number of trials must be increased by a factor of 100. Thus, achieving even a reasonably sized sample deviation for the mean of a simulation outcome may require a surprisingly large number of trials.

Variance Reduction Techniques

There is great interest in reducing the variance of a simulation outcome using techniques other than increasing the number of simulation trials. Several sampling techniques exist to reduce the variance of a simulation outcome using a *fixed number of trials*. A popular variance reduction technique is the sampling method called antithetic variates. The method of antithetic variates uses *pairs* of Random Number sets to generate simulation performance data. Two simulation trials are performed at a time using a set of Random Numbers and the *opposite* set of Random Numbers (See Example 4).

Example 5: (Antithetic Poisson Arrival Data). Consider the following set of simulated customer arrival data during an 8 hour day at the Hole in the Donut Shoppe. These arrivals have been generated using the set of eight three-digit Random Numbers from the beginning of the Table of Random Numbers (see Figure 11) applied to the cumulative Poisson distribution with $\lambda=0.5$ arrivals per hour (see Figure 9).

Hour	Random Number	# Arrivals
1	982	2
2	249	0
3	317	0
4	892	1
5	876	1
6	578	0
7	460	0
8	986	3

Provide the *antithetic* set of Random Numbers and the resulting antithetic simulated input arrival data for 8 hours of customer arrivals at the Hole in the Donut Shoppe.

Answer: The antithetic simulated input sample of arrivals is determined as follows:

1) Determine the antithetic set of 8 three-digit antithetic Random Numbers by subtracting each of the above Random Numbers from it highest possible value (i.e. 999 for a three-digit Random Number).

2) Use this antithetic set of 8 three-digit antithetic Random Number to generate antithetic simulated input data for 8 hours of customer arrivals (see Figure 9).

Each of the above 8 three-digit Random Numbers is subtracted from the value 999 to obtain the set of 8 antithetic three-digit Random Numbers 017, 750, 682, ⋯, and 013, respectively. This set of antithetic Random Numbers is mapped into the cumulative distribution F(k) for a Poisson distribution with $\lambda = 0.5$ (see Figure 9) to generate the antithetic simulated input sample of 8 hours of customer arrivals at the Hole in the Donut Shoppe.

The antithetic set of simulated input arrival data to the above set of simulated input arrival data (which are used to perform two simulation trials at a time) is derived in Figure 14.

Arrival Data vs. Antithetic Arrival Data

Hour	Random Number RN	Arrival Data	Antithetic Random Number (999 - RN)	Antithetic Arrival Data
1	982	2	017	0
2	249	0	750	1
3	317	0	682	1
4	892	1	107	0
5	876	1	123	0
6	578	0	421	0
7	460	0	539	0
8	986	3	013	0
		7 arrivals in 8 hours		2 arrivals in 8 hours

Figure 14

Notes: Generating paired sets of simulated input data and antithetic simulated input data has the effect of *reducing* the variance of the simulation outcome data given a fixed number of simulation trials. Consider two simulation outcome trials denoted by X_1 and X_2:

• The covariance (correlation) between the a simulation trial outcome X_1 (using the sample input data Z) and a simulation trial outcome X_2 (using the antithetic sample input data Z^*) is likely to be negative (i.e. $Cov(X_1, X_2) < 0$). For example, the sample input data Z (i.e. 7 arrivals in 8 hours > 4 arrivals expected in 8 hours) will likely lead to a waiting line length above the expected value in the simulation trial X_1. Conversely, the antithetic sample input data Z^* (i.e. 2 arrivals in 8 hours < 4 arrivals expected in 8 hours) will likely lead to a waiting line length below the expected value in the simulation trial X_2.

• Using the laws of probability, the variance of the sum of the simulation trial outcomes X_1 and X_2, written as $Var(X_1 + X_2)$, is given as follows:

$$Var(X_1 + X_2) = Var(X_1) + Var(X_2) + 2 \cdot Cov(X_1, X_2)$$

- Using *independent* input data Z and Z^* (e.g. with different seed numbers), the covariance of the simulation outcomes X_1 and X_2 equals zero (i.e. $Cov(X_1, X_2) = 0$). Then the variance of the sum of the simulation trial outcomes X_1 and X_2, written as $Var(X_1 + X_2)$, is given as follows:

$$Var(X_1 + X_2) = Var(X_1) + Var(X_2) + 2 \cdot Cov(X_1, X_2)$$

$$Var(X_1 + X_2) = Var(X_1) + Var(X_2)$$

- Using *antithetic* input data Z and Z^* (i.e. using opposite sets of Random Numbers), the covariance of the outcomes X_1 and X_2 is less than zero (i.e. $Cov(X_1, X_2) < 0$). Then the variance of the sum of the simulation trial outcomes, written as $Var(X_1 + X_2)$, is given as follows:

$$Var(X_1 + X_2) = Var(X_1) + Var(X_2) + 2 \cdot Cov(X_1, X_2)$$

$$Var(X_1 + X_2) < Var(X_1) + Var(X_2)$$

• The variance of the *sum* of the two simulation outcomes associated with the antithetic inputs is less than the variance of the sum of the two simulation outcomes associated with the independent inputs. Hence the variance of the *mean* (i.e. the sum divided by the constant 2) of the two antithetic simulation outcomes, written as $Var(X_1 + X_2 / 2)$, is also less than the variance of the mean of the two independent simulation outcomes.

• In general, the variance of the mean of a simulation outcome using a *fixed* number of trials N can be reduced by using input data (e.g. number of arrivals, service completion times, etc.) determined by N/2 sets of Random Numbers and N/2 sets of antithetic Random Numbers.

11. 4 Summary

For many real world problems, it is difficult or impossible to find a deterministic model of the system solvable by quantitative methods. A stochastic model relating the random and deterministic aspects of the system may be set up but such models are not typically solvable in closed form by quantitative methods. Building a physical model of the system and experimenting to determine the performance of the physical model is too time-consuming and expensive in most cases.

The most promising alternative is simulation where a mathematical model of the system is constructed and experiments are performed with the mathematical model. A simulation generates sample data that depicts the behavior of a system. The simulation is advanced through time to determine what performance can be expected from the system under a given set of conditions. After the simulation is observed under a set of conditions for a period of time, the system may be altered in an attempt to improve its performance.

Advantages of Simulation

Simulation is a powerful technique for examination of the details of complex stochastic processes. Some stochastic processes are sufficiently intricate that no simplifying assumptions (which allow the stochastic process to be solved by quantitative methods) can reasonably be made. In such instances, simulation is the only option which can determine the behavior of a system to a high level of accuracy.

Simulation can be a valuable training tool. It allows one to gain a better understanding of how a system performs than one can get merely by solving an equation or applying an algorithm.

Simulation allows one to study not just the average performance of the system (predicted by a quantitative method) but also the detailed and non-steady-state performance of the system. It may be precisely this non-steady-state performance (e.g. waiting line times at the motor vehicle registry office during the time of peak customer arrival) that is of interest.

Simulation allows conditions of the system to be tightly controlled. It has been found that the performance of people in a system is modified from their normal mode when they understand they are being studied.

Simulation is typically less expensive and less risky than experimentation on the actual system. A set of simulation conditions which lead theoretically to disastrous results for a business may simply be discarded from future consideration without actual harm to the business.

Simulation can expand time or compress time to allow a detailed examination of the processes of a system. Some processes (e.g. chemical process) may be impossible to examine in depth in real time while other processes (e.g. 100 year flood water height) may be impractical to examine in real time.

Limitations Of Simulation

Simulation can hide some of the critical assumptions of the mathematical model. Faulty assumptions (which may not even be documented) can easily lead to spurious results. Also, there may not be enough real world data to build the mathematical model and then to validate the simulation.

Simulation does not typically provide an optimal solution. It only provides system performance under a chosen set of conditions. One must search for the set of conditions under which the system would perform optimally through both intuition and trial and error.

Simulation may not lead to a definitive conclusion on the performance of the system. Statistical inferences made on the simulation data may not be conclusive. More simulation data will be required but definitive results will not be guaranteed.

Simulation is sometimes too expensive to be cost-effective. The time and cost to build a mathematical model, to validate the simulation, to collect the data under sets of conditions, and to analyze the simulation data may not be warranted by the system performance insight provided by the simulation. There may not be reasonable assurance that the benefits of the simulation will outweigh the resources required to implement the simulation.

Conclusions

There is a tradeoff whether to use a quick and inexpensive quantitative method to determine an approximate analytical solution for the average performance of a system or to use a time-consuming and expensive simulation to determine the detailed performance of a system.

The decision to allocate the resources required for a simulation of a complex stochastic or deterministic process is typically made on the basis a positive response to the following three criteria:

1) Enough real world data exists to build the simulation and then to validate the simulation.

2) The simulation results are likely to be definitive rather than to suggest the generation of further simulation data.

3) The cost of a simulation relative to the application of a quantitative technique is acceptable when the benefit of the increased accuracy is considered.

That is, simulation is the proper choice when a simulation is expected to lead to results which are accurate and definitive and whose utility compares favorably with applicable quantitative techniques.

Exercise 11

Simulation

Problem #1 (Seed Numbers)

Explain the role of seed numbers in the application of simulation.

Problem #4 (Generating Random Values)

The number of arrivals at the Payne Hospital emergency room in any hour has the following discrete distribution:

# Arrivals	Probability
0	0.2
1	0.4
2	0.3
3	0.1

Use the following set of 10 four-digit Random Numbers to generate the number of arrivals in the first 10 hours of the day:

9822	4931	7892	8765	7846
2013	3508	2929	5999	1633

Problem #3 (Antithetic Values)

Use the four-digit Random Number 2389 and its antithetic Random Number to generate the corresponding outcome and its antithetic outcome. respectively. for a Normal Random Variable X with Expected Value 10 and Standard Deviation 3 (i.e. $X \sim N(10,9)$).

Problem #4 (Simulation of Fair Coin)

a) Use the following sequence of 10 four-digit Random Numbers to determine the outcomes ("Heads" or "Tails") of 10 flips of a fair coin:

9822	4931	7892	8765	7846
2013	3508	2929	5999	9633

b) If the coin is fair. will the outcomes of 10 flips using 10 Random Numbers always produce the expected number of 5 "Heads" and 5 "Tails"?

c) Use the following sequence of 10 four-digit Random Numbers and their 10 antithetic values (9999 - RN) to determine the outcomes of 20 flips of a fair coin:

9822	4931	7892	8765	7846
2013	3508	2929	5999	9633

0177	5068	2107	1234	2153
7986	6491	7070	4000	0366

d) If the coin is fair, will the outcomes of 20 flips using 10 Random Numbers and their 10 antithetic Random Numbers always produce the expected number of 10 "Heads" and 10 "Tails"?

e) What is the value of using the sequence of antithetic Random Numbers along with a sequence of Random Numbers?

Chapter 11

Simulation Additional Solved Problems

Problem #1 (Simulation)

Consider the following probabilities for the discrete Poisson distribution with $\lambda = 2$ arrivals per minute:

# Arrivals	Probability
0	0.135
1	0.271
2	0.271
3	0.180
4	0.090
5	0.036
6	0.012
7	0.004
8	0.001

Use the following table of 10 four-digit Random Numbers to generate the number of arrivals in the next 10 minutes:

9822	4931	7892	8765	7846
2013	3508	2929	5999	9633

Answer: First, a cumulative distribution is determined for the Poisson distribution with $\lambda = 2$ arrivals per minute:

# Arrivals	Cumulative Probability
0	0.135
≤ 1	0.406
≤ 2	0.677
≤ 3	0.857
≤ 4	0.947
≤ 5	0.983
≤ 6	0.995
≤ 7	0.999
≤ 8	1.000

Note that a four-digit random variable ranges from 0000 to 9999. The above cumulative probabilities can be used when 0.0001 is subtracted from the above cumulative probability value. A four-digit Random Number RN can be then used to generate arrivals for the cumulative Poisson distribution with $\lambda = 2$ arrivals per minute as follows:

$0000 \leq RN \leq 0134 \Rightarrow \# \textit{Arrivals} = 0$

$0135 \leq RN \leq 4059 \Rightarrow \# \textit{Arrivals} = 1$

$4060 \leq RN \leq 6769 \Rightarrow \# \textit{Arrivals} = 2$

$6770 \leq RN \leq 8569 \Rightarrow \# \textit{Arrivals} = 3$

$8570 \leq RN \leq 9469 \Rightarrow \# \textit{Arrivals} = 4$

$9470 \leq RN \leq 9829 \Rightarrow \# \textit{Arrivals} = 5$

$9830 \leq RN \leq 9949 \Rightarrow \# \textit{Arrivals} = 6$

$9950 \leq RN \leq 9998 \Rightarrow \# \textit{Arrivals} = 7$

$9999 \leq RN \leq 9999 \Rightarrow \# \textit{Arrivals} = 8$

The 10 four-digit Random Numbers provided above would generate the following number of arrivals in each of the next 10 minutes:

Minute	Random Number	# Arrivals
1	9822	5
2	4931	2
3	7892	3
4	8765	4
5	7846	3
6	2013	1
7	3508	1
8	2929	1
9	5999	2
10	9633	5

Problem #2 (Simulation)

For the number of arrivals at the Payne Hospital emergency room generated in Problem 1, determine the following:

a) What is the expected number of arrivals in each minute?

The expected number of arrivals per minute is the Expected Value of the Poisson Random Variable with $\lambda = 2$ (i.e. two arrivals per minute).

b) What is the mean number of arrivals per minute for the sequence of arrivals at the emergency room determined for the 10 minutes?

There were a total of 27 (5 + 2 + 3 + • • • + 2 + 5) arrivals during the 10 minutes. The mean (average) number of arrivals per minute is 2.7.

c) Should one anticipate that the mean of the random arrivals will differ from the Expected Value of the Poisson Random Variable with $\lambda = 2$?

Due to randomness in the stream of uniformly distributed Random Numbers ranging from 0000 to 9999, it is not surprising that the sequence of arrivals in 10 minutes does not have exactly a mean of 2 arrivals per minute.

d) Assume that this simulation was being used to determine the size of the new emergency room to be built. Management would like the random arrival sequence to be close to what is expected (described by the Poisson probability distribution). What can be done to reduce the variance of the mean of the generated sequence of arrivals generated from its expected value?

Two methods for reducing variance of the mean of the arrival sequence from its expected value are as follows:

  1) Use a longer arrival sequence (a greater number of minutes). The Central Limit Theorem guarantees that in the long run the mean of the random arrival sequence will approach its expected value.

  2) Use sets of Random Numbers and their Anithetic Random Numbers. Due to the negative covariance between the outcomes generated by a set of Random Numbers and the outcomes generated by the corresponding set of antithetic Random Numbers, the variance of the mean (the sum of the number of arrivals divided by the number of minutes) from its expected value (two arrivals per minute) will be reduced.

Appendix A

Design of Experiments

"Quality control is achieved most efficiently, of course, not by the inspection operation itself, but by getting at causes."
Dodge and Romig

A | Design of Experiments

A.1 Introduction

There is an ongoing global competition to satisfy the need in the worldwide marketplace for a high quality product or service at an acceptable price. The American Society for Quality Control defines *quality* as the "totality of features and characteristics of a product or service that bear on its ability to satisfy stated or implied needs." Decision makers have become keenly interested in improving the *process* of producing a final product with:

1) Decreased variability and number of defects.

2) Lower production cost.

3) Shortened development cycle time.

Consider a typical production process (see Figure 1) for a product (or service). The production process transforms a number of input resources into a final output product. To improve the production process, it is imperative to understand the effect of each of the inputs on the final product quality.

Typical Production Process

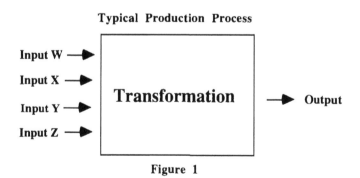

Figure 1

Design of Experiments (DOE) is a simple but powerful tool for characterizing how changes in inputs affect the output of a production process. DOE maximizes the information on the output of a production process using a minimum amount of input resources. DOE allows the decision maker to:

1) Determine which inputs have a significant effect on the output product and its variability and which inputs have no significant effect.

2) Determine the level (setting) of each input in order to optimize a desired feature (e.g. strength, speed, temperature, etc.) of the output product.

3) Build a *linear* mathematical model describing the effect of each input (and the interactions among inputs) on the output product.

Notes:

• Design of Experiments can be effectively applied to improve production processes already up and running. However, it is more efficient to apply Design of Experiments in the planning stage of a production process.

• It is cost-effective to gather a "brainstorming" group of process experts, implementers of the process, and customers of the product to discuss the inputs (what they are and what are the possible levels of each input) and the outputs (what are the features desired) for the production process. Design of Experiments can not be successfully applied unless the relevant inputs and desired output features of the production process have been identified.

• To the extent that the output of the production process is not a *linear* function of the inputs and their interactions, the utility of the Design of Experiments will decrease.

The concept of quality has evolved over the last 30 years. Consider the output of two production processes X and Y (see Figure 2) which produce an engine part with a target diameter T. Both Process X and Process Y produce a very high percentage of parts of *acceptable* quality (i.e. within the upper U and lower L specifications for the diameter of the part). However, Process Y and has the additional desirable quality of low variability in the diameter of its engine parts around the target diameter T.

Output for an Engine Part Production Process with Target Diameter T

Process X	Process Y

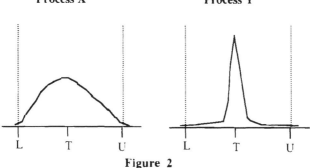

Figure 2

Design of Experiments can be used to transform a production process from one that produces output in a fashion similar to Process X into a one that produces output in a fashion similar to Process Y. This helps improve the performance, cost, and schedule for a production process:

> *Performance:* The feature of low variability in a production process has become increasingly important. Customers expect products which perform not just acceptably but perform *well* (e.g. everyone wants a car door which not only is within tolerance and closes but which is close to target specifications and has a tight seam). Merely acceptable performance for a product leads to increased customer complaints, increased requests for replacement products or rework under warranty, and eventually to disenchanted customers who switch to a more reliable product.

> *Cost:* Improved performance also leads to less defects and rework. Edward Deming in his book *Quality, Productivity, and Competitive Position* has stated that defective items are a burden to a company. The cost to produce and dispose of a defective item will exceed the cost to produce a good item. Furthermore, defective items do not produce revenue for the company. When a production process is well understood, one can focus attention on the critical inputs and not waste resources strictly controlling tolerances on non-critical inputs which do not have significant impacts on the quality of the output product.

> *Schedule:* Historically, managers would vary one input factor at a time to determine its effect on the quality of the output. For a large number of inputs each with many possible levels, the number of combinations of different possible levels for the inputs will become large. The schedule to accomplish this amount of testing and analysis of data does not promote short product development schedules. Product development lead times can be shortened by using efficient Design of Experiment test designs (discussed later in this Appendix) which use a much *smaller set of tests* to characterize the effect of changing individual input levels on the output product.

Design of Experiments attempts to: 1) determine the effects of changes in input levels on the output product, and 2) optimize the production process. Remember (see Chapter 6) that the effect of changes in the input levels on the change in the output of the production process can be determined statistically using regression. However, a regression will have difficulty in sorting out the effects of individual inputs on the output when the inputs are highly correlated (i.e. the multicollinearity problem). In Design of Experiments, high correlation between inputs is called <u>confounding</u>.

The fact that a change in the output is highly correlated to a change in an input to the process does not prove that the change in the output is caused by the change in the input. Causality between a change in a input and a change in the output must be determined logically and not statistically. Yet Design of Experiments does attempts to determine causality between changes in the inputs to the production process and its output. Design of Experiments relies on a well-considered experiment (test) design to aid in establishing causality:

> 1) Insure that the inputs controlled over the set of tests are uncorrelated to the maximum extent possible.

> 2) Randomize inputs not controlled over the set of tests. This process, called <u>randomization</u>, *averages out* the effects of inputs not controlled in the set of tests.

438

The criticality of randomization of inputs not controlled for in a test design is demonstrated in Example 1.

Example 1: (Randomization). Our Lady of Sorrows Hospital (OLSH) is considering whether to use reacting agent R1 or R2 to speed up its blood test turnaround times. OLSH decides to test the two reacting agents on blood samples which come in on over a period of three days. The following data on the turnaround time and the test day has been gathered:

Turnaround Time (R1)	Test Day	Turnaround Time (R2)	Test Day
5 hours	2	12 hours	3
6 hours	1	11 hours	3
7 hours	1	4 hours	2
7 hours	1	5 hours	1
10 hours	3	5 hours	2
7 hours	1	11 hours	3
Mean (R1) = 7 hours		**Mean (R2) = 9 hours**	

Determine which reacting agent (input) will optimize the turnaround time (output) of the blood tests (process) at OLSH.

Answer: Based on the turnaround time data collected at OLSH for the two reacting agents, R1 has a shorter mean turnaround time (7 hours) than R2 (9 hours).

However, OLSH neglected to randomize the day of the test (which was not controlled in the test design). On day 1, four blood samples were tested using R1 while only one was tested using R2. On day 2, only one blood sample was tested using R1 while two blood samples were tested using R2. On day 3, only one blood sample was tested using R1 while three blood samples were tested using R2.

A closer look at the data shows that all the tests on day 3 have significantly longer turnaround times than those on days 1 and 2. Perhaps some factor other than the reacting agents (such as short staffing) may have caused the longer turnaround times. When the tests taken on day 3 are disregarded, one can compute that R2 has a shorter mean turnaround time (4.66 hours) than R1 (6.4 hours).

If the tests using the two reacting agents had been randomized over the three days (so that the numbers of tests allocated to each day was approximately equal), the effect of long turnaround times on day 3 would have averaged out and the reacting agents R1 and R2 could have been fairly compared.

A.2 Design of Experiments Matrices

Design Matrix Representations

There are several equivalent representations of a given test design matrix possible for Design of Experiments. Consider a production process where there are three relevant inputs (W, X, and Y) each with two levels of interest (1 and 2, 0.25 and 0.50, and 10 and 15, respectively). It is decided (due to available resources such as time, manpower, cost, etc.) that the application of Design of Experiments to characterize the output response to the three inputs should consist of only four tests with the inputs at various high and low levels.

One representation of the four test design matrix uses the numerical values of the actual high and low levels of the inputs W, X, and Y (see Figure 3).

Four Test Design Matrix Using Actual Values

Test #	Inputs		
	W	X	Y
1	2	0.50	15
2	2	0.25	10
3	1	0.50	10
4	1	0.25	15

Figure 3

An equivalent representation of this four test design matrix uses a minus symbol (-) for the low level and a plus (+) symbol for the high level of each input (see Figure 4).

Four Test Design Matrix Using Plus and Minus Symbols

Test #	Inputs		
	W	X	Y
1	+	+	+
2	+	-	-
3	-	+	-
4	-	-	+

Figure 4

A third equivalent representation of this four test design matrix uses a negative one (-1) for the low level and a positive one (+1) for the high level of each input (see Figure 5).

Design Matrix Using Negative and Positive 1's

Test #	Inputs		
	W	X	Y
1	+1	+1	+1
2	+1	-1	-1
3	-1	+1	-1
4	-1	-1	+1

Figure 5

While all three representations of the design matrix are equivalent, the four test design matrix using negative and positive 1's (see Figure 5) is the most useful mathematically. This representation allows easy validation that a design matrix is balanced and orthogonal.

Notes:

• A design matrix is *balanced* if the sum of its entries for each column is zero. The four test design matrix in Figure 5 is balanced. That is, each column (input) has the same number of entries at the low setting (-1) as at the high setting (+1).

• A design matrix is *orthogonal* if the dot product of each pair of columns is zero. The four test design matrix in Figure 5 is orthogonal. That is, the *dot product* of any two columns (computed when the 4 entries in one column are multiplied by the corresponding 4 entries in the other column and the resulting 4 products are summed) always equals 0.

• An orthogonal test matrix allows one to test the effect of each input on the output *independent* of the effect of the other inputs.

Non-Orthogonal Test Design Matrices

Non-orthogonal design matrices for a test design have columns whose dot product with other columns does not equal zero. Some inputs to the production process are then correlated with other inputs. This results in some troublesome properties:

- Assume input W has a significant effect on the production process output while input Y does not. If W and Y are correlated (dependent), Y may also appear to have a significant effect.

- Assume input W and input Y have significant effects on the production process output. If W and Y are negatively correlated (one has a positive effect when the other has a negative effect), both may appear to have non-significant effects.

Consider the non-orthogonal ten test design matrix for inputs W, X, Y, and Z given in Figure 6. This design matrix demonstrates that a design matrix can be balanced (i.e. equal numbers of -1's and +1's in each column) and yet non-orthogonal (i.e. the dot products XY, XZ, and YZ do not equal 0).

Non-Orthogonal Balanced Test Design Matrix

Test #	Inputs			
	W	X	Y	Z
1	+1	-1	-1	+1
2	-1	+1	-1	+1
3	+1	-1	+1	-1
4	-1	-1	-1	+1
5	+1	+1	+1	-1
6	-1	+1	+1	-1
7	+1	-1	+1	-1
8	-1	+1	-1	+1
9	+1	+1	+1	-1
10	-1	-1	-1	+1

Figure 6

The correlation coefficient R (see Chapter 6) is used to determine the correlation between the four columns (inputs) in Figure 6. The pairwise correlation matrix for the inputs W, X, Y and Z is given in Figure 7.

**Correlation Coefficient Matrix
for Inputs W, X, Y and Z**

	W	X	Y	Z
W	1.0	-0.2	0.6	-0.6
X	-0.2	1.0	0.2	-0.2
Y	0.6	0.2	1.0	-1.0
Z	-0.6	-0.2	-1.0	1.0

Figure 7

The correlation coefficient matrix provides the following information on the dependency (i.e. the interaction) between a pair of inputs:

- An R of 0 (no correlation) indicates that the two inputs are independent (i.e. the dot product of the columns is 0).

- An R of 1 or -1 (perfect positive or negative correlation, respectively) indicates that the inputs are *aliases* (the columns are identical or opposite, respectively).

- As the absolute value of R increases from zero (no correlation) toward 1 (perfect correlation), the dependence (and the confounding) of the pair of inputs increases.

Figure 7 indicates that inputs W and X are weakly dependent (R=0.2), inputs W and Y are strongly dependent (R=0.6), inputs X and Y are weakly dependent (R=0.2), and that inputs Y and Z are aliases (R=-1.0, Y=-Z).

Three Level Test Design Matrices

Up to this point, test design matrices with only two levels (-1 and +1) for inputs have been examined. One may wish to include a third level (0) for each input. This approach requires more resources to perform these larger test design matrices but has two benefits:

- Granularity: A third setting provides greater choice for the level of each input to optimize the output response.

- Curvature Analysis: When plotting input vs. output, two (input,output) points always maps as a line. A third setting, however, provides another (input,output) point which can be use to check on the assumption by Design of Experiments of a *linear relationship* between each input and the output.

A balanced three level test design matrix (i.e. same number of -1's, 0's, and +1's in each column) is given in Figure 8.

442

Balanced Three Level Test Design Matrix

Test #	Inputs			
	W	X	Y	Z
1	+1	+1	+1	+1
2	+1	0	0	0
3	+1	-1	-1	-1
4	0	-1	0	+1
5	0	+1	-1	0
6	0	0	+1	-1
7	-1	0	-1	+1
8	-1	-1	+1	0
9	-1	+1	0	-1

Figure 8

Full Factorial Test Design Matrices

Despite the most careful upfront analysis of the relevant inputs to a production process, it is not always possible to determine which of the inputs might have a significant effect on the output and what levels of each input should be tested in order to characterize the effect on the output. One obvious strategy is to test each input at *all* its levels This approach is called the full factorial design in that it consists of all possible input test combinations.

Consider a production process where upfront analysis identifies four relevant inputs (W, X, Y, and Z) where each input has two levels given as low (-1) and high (+1). Then to test the four inputs at each of the two levels will require $2^4 = 16$ tests. A four input full factorial design matrix is described in Figure 9.

Full Factorial Test Design Matrix

Test #	Inputs			
	W	X	Y	Z
1	-1	-1	-1	-1
2	-1	-1	-1	+1
3	-1	-1	+1	-1
4	-1	-1	+1	+1
5	-1	+1	-1	-1
6	-1	+1	-1	+1
7	-1	+1	+1	-1
8	-1	+1	+1	+1
9	+1	-1	-1	-1
10	+1	-1	-1	+1
11	+1	-1	+1	-1
12	+1	-1	+1	+1
13	+1	+1	-1	-1
14	+1	+1	-1	+1
15	+1	+1	+1	-1
16	+1	+1	+1	+1

Figure 9

443

Besides the effect of the first-order (i.e. linear) interactions W, X, Y, and Z on the output, one might wish to test the effects of second-order interactions (e.g. WX, WY, XY, etc.), third-order interactions (e.g. WXY, XYZ, etc.) and fourth-order interactions (i.e. WXYZ) on the output of the production process.

Figure 10 provides the full listing of the possible interactions of the four inputs on the output which may be examined using Design of Experiments.

Possible Interactions For Inputs W, X, Y, and Z

First Order (Linear) (4)	Second-Order (Non-Linear) (6)	Third-Order (Non-Linear) (4)	Fourth-Order (Non-Linear) (1)
W	WX	WXY	WXYZ
X	WY	WYZ	
Y	WZ	XYZ	
Z	XY		
	XZ		
	YZ		

Figure 10

Notes:

• For k = 4 inputs with two levels, there are $2^4 - 1 = 15$ possible input (linear) and input interaction (non-linear) effects on the output of the production process. In general, for k inputs and 2 levels, there are $2^k - 1$ possible input and interaction effects.

• It is not common that input interactions of the third-order and above are significant.

• If it is decided to examine higher order interactions beyond the input (linear) effects W, X, Y, and Z, more tests will not be required. The full factorial design matrix provides enough information to analyze higher order interactions (see Example 2).

Example 2: (Full Factorial Design). A full factorial Design of Experiments is planned to characterize the effects of inputs W, X, Y, and Z on the output of a production process. Provide the test design matrix to test for the higher order interactions WX and XYZ as well as the first-order (linear) interactions.

Answer: The full factorial test design matrix (see Figure 9) is used as the basis of the test design matrix which will now include the second-order interaction WX and the third-order interaction XYZ.

The new test design matrix (see Figure 11) still contains the same 16 tests but two new higher-order interaction columns (WX and XYZ) are added to the full factorial test design matrix for four inputs and two levels given in Figure 9.

Higher Order
Full Factorial Test Design Matrix

Test #	Interactions					
	W	X	Y	Z	WX	XYZ
1	-1	-1	-1	-1	+1	-1
2	-1	-1	-1	+1	+1	+1
3	-1	-1	+1	-1	+1	+1
4	-1	-1	+1	+1	+1	-1
5	-1	+1	-1	-1	-1	+1
6	-1	+1	-1	+1	-1	-1
7	-1	+1	+1	-1	-1	-1
8	-1	+1	+1	+1	-1	+1
9	+1	-1	-1	-1	-1	-1
10	+1	-1	-1	+1	-1	+1
11	+1	-1	+1	-1	-1	+1
12	+1	-1	+1	+1	-1	-1
13	+1	+1	-1	-1	+1	+1
14	+1	+1	-1	+1	+1	-1
15	+1	+1	+1	-1	+1	-1
16	+1	+1	+1	+1	+1	+1

Figure 11

Notes:

- Each of the 16 row entries in the interaction column WX are determined by multiplying the corresponding row entries in input column W and input column X. Likewise, the entries in the interaction column XYZ are formed by multiplying the corresponding row entries in input columns X, Y, and Z.

- The columns WX and XYZ are also balanced (same number of 1's and -1's). The matrix remains orthogonal in that the dot product of any two columns is 0.

- The test matrix still consists of 16 tests. No new tests have been added to determine the effects of interactions WX and XYZ. However, we will see that some additional analysis must be done to determine the effects of the interactions WX and XYZ on the output.

Graphically, a full factorial test design matrix performs a test at each corner of the n-dimension response surface. Consider the full factorial test design matrix for 3 inputs W, X, and Y each with two levels (see Figure 12).

Full Factorial Design Matrix (3 Inputs)

Test #	Inputs		
	W	X	Y
1	-1	-1	-1
2	-1	-1	+1
3	-1	+1	-1
4	-1	+1	+1
5	+1	-1	-1
6	+1	-1	+1
7	+1	+1	-1
8	+1	+1	+1

Figure 12

The three-dimension cubic response surface specified by the full factorial test design matrix is given in Figure 13. The location of the 8 tests for the three inputs with two levels full factorial test design matrix is denoted by the dots at the corners of the cube. The full factorial design matrix tests the output response at the extremes of the input space in order to determine the best level setting for each input.

Graphical Three Dimension Full Factorial Response Surface

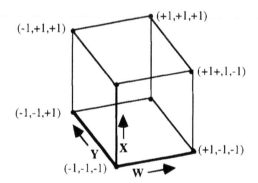

Figure 13

Fractional Factorial Test Design Matrices

While a full factorial approach provides the most information on the effect of relevant inputs on the production process, the full factorial approach is only practical (due to limitations of time, manpower, cost, etc.) for a small numbers of inputs and possible levels per input. It is more common to perform only a subset of the tests called for by the full factorial approach. Of course, this results is loss of *full information* on effects of all inputs and possible levels on the output of the production process.

The objective of a <u>fractional factorial</u> test design is to reduce the number of tests required but retain much of the information on the effects of changes in each input on the output of the production process. Assume there are k relevant inputs to a production process each with two levels of interest. A full factorial design matrix consists of 2^k tests. Based on the choice by the user of an integer j (where j < k), the fractional factorial design matrix will consist of 2^{k-j} tests. The integer j is chosen according to the need to reduce the total number of tests to match the test resources (e.g. cost, schedule, manpower, etc.) available.

Notes:

• Once the integer j is specified, the number of tests in a fractional factorial test design matrix is reduced by a factor 2^j tests from the number of tests in the full factorial test design matrix. For example, if k = 4 and j = 1, the fractional factorial test design matrix will consist of $2^3 = 8$ tests which is a reduction by $2^1 = 2$ in the number of tests in the full factorial test design matrix for k = 4 ($2^4 = 16$ tests).

• The *core* of the fractional factorial test design matrix is the full factorial test design matrix for k-j inputs. For k = 4 and j = 1, the core of the fractional factorial test design matrix is the k - j = 3 input full factorial test design matrix (see Figure 12) with $2^3 = 8$ tests.

• The fractional factorial test design matrix has 2^{k-j} rows (tests) and k columns. For k = 4 and j = 1, the fractional factorial test design matrix has $2^3 = 8$ rows (tests) and 4 columns.

• The first k-j columns consists of the first k-j inputs. The remaining j columns consist of higher order interactions which are assumed to have a *non-significant* effect on the output of the production process.

    - For inputs W, X, Y, and Z (k = 4), the first three columns of the fractional factorial design matrix are associated with the inputs W, X, and Y. The entries in these columns are the three input full factorial design matrix (see Figure 12).

    - The remaining column (j = 1) is associated with a higher order interaction such as the third-order interaction WXY. Remember that it is common that interactions of the third order and above are non- significant. The entries in the WXY column are determined by multiplying together the entries in the W, X, and Y columns.

A fractional factorial matrix for k = 4 inputs (two levels) where j is chosen to be 1 is given in Figure 14.

Fractional Factorial Test Design Matrix (k = 4 and j = 1)

Test #	Inputs			
	W	X	Y	WXY
1	-1	-1	-1	-1
2	-1	-1	+1	+1
3	-1	+1	-1	+1
4	-1	+1	+1	-1
5	+1	-1	-1	+1
6	+1	-1	+1	-1
7	+1	+1	-1	-1
8	+1	+1	+1	+1

Figure 14

Notes:

• The fractional factorial test design matrix is *balanced* (i.e. same number of 1's and -1's in each column).

• The fractional factorial test design matrix is *orthogonal* (i.e. the dot product of any two columns is 0).

• The multiplication of any column C by the transpose of itself (e.g. CC) results in a square matrix of all 0's except for 1's on the diagonal (called the *identity* matrix I). If C is defined to be the third-order interaction WXY, then CC = WXYC = I.

• Inputs and/or interactions which multiply to I (e.g. WXYC) are the same effect on the output and are called *aliases*. For example, the linear input W and the interaction XYC have the *same* effect on the output. Also, the interaction WY and the interaction XC are the same effect. Columns associated with aliases either have the same entries (perfect positive correlation) or opposite entries (perfect negative correlation).

Plackett-Burman Test Design Matrices

The number of tests associated with the fractional factorial test design matrices grows *geometrically* (i.e. 2^{k-j} = 2, 4, 8, 16, 32, 64, 128, etc. for k-j = 1, 2, 3, etc. inputs). It is often required to reduce further the number of tests to match available test resources. Plackett-Burman test design matrices have the desirable property of growing *arithmetically* (i.e. $4 \cdot k$ = 4, 8, 12, 16, 20, 24, 28, etc., for k = 1, 2, 3, etc., inputs).

Notes:

• The number of tests in a Plackett-Burman test design matrix is a factor of 4 . This arithmetic numerical sequence (i.e. 4, 8, 12, 16, 20, 24, 28, etc.) grows much more slowly than the geometric numerical sequence of tests required to implement the full factorial or fractional factorial test design matrices.

• Plackett-Burman test design matrices are balanced and orthogonal.

• Plackett-Burman design matrices of n = 4, 8, 12, etc. tests can evaluate up to n - 1 = 3, 7, 11, 15, etc. inputs and/or interactions. However, higher order interactions above the first order (linear) inputs are not typically included in Plackett-Burman test design matrices.

• Each Plackett-Burman test design matrix of n = 4, 8, 12, etc. tests (rows) is generated by a specific vector of n - 1 = 3, 7, 11, etc. entries. The specific 7-entry vector used to generate an 8 test (row) Plackett-Burman design matrix is (+1, +1, +1, -1, +1, -1, -1). Consider 7 relevant inputs to a production process given by T, U, V, W, X, Y, and Z.

- The first 7 row entries in the T column is the specific 7-entry vector provided.

- The first row entry in the U column is the 7th row entry in the T column (-1) followed by the first six row entries of the T column.

- The first row entry of the V column is the 7th entry of the U column (-1) followed by the first six row entries of the U column.

448

- The first row entry of the W column is the 7th row entry of the V column (+1) followed by the first six row entries of the V column.

- Any other columns (up to 7 columns) would be generated in a like manner.

- The last row (the eight test row for this matrix) is specified as all -1's.

The Plackett-Burman test design matrix for 8 tests (rows) and up to 7 inputs is given in Figure 15.

8 Test Plackett-Burman Test Design Matrix

Test #	Inputs						
	T	**U**	**V**	**W**	**X**	**Y**	**Z**
1	+1	-1	-1	+1	-1	+1	+1
2	+1	+1	-1	-1	+1	-1	+1
3	+1	+1	+1	-1	-1	+1	-1
4	-1	+1	+1	+1	-1	-1	+1
5	+1	-1	+1	+1	+1	-1	-1
6	-1	+1	-1	+1	+1	+1	-1
7	-1	-1	+1	-1	+1	+1	+1
8	-1	-1	-1	-1	-1	-1	-1

Figure 15

Taguchi Test Design Matrices

Taguchi identified a *small* number of minimal design matrices called Taguchi test design matrices which meet the objectives of Design of Experiments:

- Characterize the response of the quality of the output to changes in the inputs.

- Characterize the variability of the response of the output to changes in the inputs.

- Determine the input levels which optimize the quality of the output product and minimize the variability of the output process.

Notes: Taguchi test design matrices typically have the following properties:

• The columns associated with the inputs are orthogonal.

• The columns associated with higher order interactions are either completely correlated (aliased) or partially correlated (confounded) with the inputs.

These properties are not different than for many of the test design matrices discussed above. In fact, Taguchi test design matrices are a set of specific matrices chosen from the fractional factorial, Plackett-Burman, and other classes of test design matrices. Taguchi's contribution was performing analyses of quality using financial loss functions. Instead of implementing a loss only if the output of a process fell outside of specification limits,

Taguchi examined a financial loss function L which increases quadratically (i.e. $L \sim (O-T)^2$) as the output O deviates from the target value T.

Taguchi's analysis also demonstrated that this set of design matrices is *robust* in that they provide the desired optimum output response and low variability features even when all the assumptions implicit when using Design of Experiments (e.g. linearity of the response function, independence of the inputs, etc.) do not completely hold.

Taguchi test design matrices are typically designated by "Ln" where the integer n indicates the number of rows (tests) in the matrix. An Ln Taguchi test design matrix can accommodate up to n-1 inputs variables. Alternatively, if there are less than n-1 input variables, the remaining columns may be used for higher order interactions. Taguchi determined sets of test design matrices for inputs with 2 levels and with 3 levels.

An L4 Taguchi test design matrix for inputs with two levels is given in Figure 16. In this Taguchi matrix, two inputs and one higher order interaction (second-order) are used to fill out the four row by three column test design matrix.

Taguchi L4 Test Design Matrix (Two Levels)

Test #	W	X	-WX
1	-1	-1	-1
2	-1	+1	+1
3	+1	-1	+1
4	+1	+1	-1

Figure 16

An L8 Taguchi test design matrix for inputs with two levels is given in Figure 17. In this Taguchi matrix, three input variables and four higher order interactions (i.e. three second-order and one third-order) are used to fill out the eight row by 7 column matrix.

Taguchi L8 Test Design Matrix (Two Levels)

Test #	W	X	-WX	Y	-WY	-XY	WXY
1	-1	-1	-1	-1	-1	-1	-1
2	-1	-1	-1	+1	+1	+1	+1
3	-1	+1	+1	-1	-1	+1	+1
4	-1	+1	+1	+1	+1	-1	-1
5	+1	-1	+1	-1	+1	-1	+1
6	+1	-1	+1	+1	-1	+1	-1
7	+1	+1	-1	-1	+1	+1	-1
8	+1	+1	-1	+1	-1	-1	+1

Figure 17

Other popular Taguchi test design matrices for two level inputs are L12, L16, and L18. Taguchi also provided test design matrices for three level inputs. Popular three level input Taguchi test design matrices are L9, L18, and L27. These specific matrices can be found in references with fuller treatments of Design of Experiments.

A.3 Application of Design of Experiments

Users of Design of Experiments are encouraged to be creative and non-rigorous in brainstorming for the production process to determine the relevant inputs, the applicable levels of each input, and the response features desired. However, once the experiment (set of tests) is designed, it is extremely important that the tests be conducted with extreme discipline and that the results be carefully recorded.

Example 3 demonstrates how a fractional factorial test design (where the number of inputs k = 4 and the choice of j = 1) is used to determine which interactions have a significant effect on the output product and to determine optimal level for each input.

Example 3: (Fractional Factorial Design). A brainstorming session with a group of analysts, implementers, and potential customers of a cement production process determined that four inputs W (water), X (sand), Y (cement mix), and Z (mixing time) may be important to the quality of the cement produced and that the quality of the cement should be measured by its strength (Klb/in^2). The group also determined that the second-order interactions WX, WY, and XY may also be important to the quality of the cement and should be examined.

It was also determined that allowing two levels (a low level and a high level) for each input would be sufficient to test the sensitivity of the output (cement strength) to that input. The two levels for each of the four inputs are as follows:

Inputs (4)	Levels (2)
W: Water	1 gal., 2 gal.
X: Sand	0.25 cu. yd, 0.50 cu. yd
Y: Cement Mix	10 lbs., 15 lbs.
Z: Mixing Time	10 min., 20 min.

The group decided to choose j=1 which reduced the number of tests required by one half. The inputs W, X, and Y are chosen for the first three columns. The fourth column Z of the fractional factorial design matrix is defined as the third-order interaction WXY. Then any inputs and/or interactions which multiply to WXYZ = ZZ = I are aliases. Columns associated with the relevant interactions WX, WY, and XY are added to the matrix (no additional testing but some additional analysis is required).

Ten runs were performed for each of the 8 tests designated by the fractional factorial test design matrix (2^{4-1}=8 tests). The full test design matrix and the mean (μ) and standard deviation (σ) for the cement strength ($Klbs/in^2$) of the ten runs for each test are as follows:

Test #	W	X	Y	Z	WX	WY	XY	μ	σ
					Interactions				
1	-1	-1	-1	-1	+1	+1	+1	4.31	0.57
2	-1	-1	+1	+1	+1	-1	-1	5.05	0.66
3	-1	+1	-1	+1	-1	+1	-1	4.28	0.58
4	-1	+1	+1	-1	-1	-1	+1	5.07	0.76
5	+1	-1	-1	+1	-1	-1	+1	4.67	0.57
6	+1	-1	+1	-1	-1	+1	-1	3.85	0.77
7	+1	+1	-1	-1	+1	-1	-1	4.77	0.57
8	+1	+1	+1	+1	+1	+1	+1	3.71	0.70

Determine which inputs and interactions have a significant effect on cement strength.

Answer: For each input or interaction (column) identified in the above test design matrix, the average mean strength (response) using the -1 levels and the +1 levels is computed. For example, for the input (column) W, the mean cement strengths in Klbs/in^2 corresponding to the -1 levels are 4.31, 5.05, 4.28, and 5.07 while the mean cement strengths corresponding to the +1 levels are 4.67, 3.85, 4.77, and 3.71. Thus, for the input W, the average mean response for the -1 levels (called Ave-) and for the +1 levels (called Ave+) is given as follows:

$$Ave- = {}^{(4.31 + 5.05 + 4.28 + 5.07)}\!/_4 = 4.68 \; Klb/in^2$$

$$Ave+ = {}^{(4.67 + 3.85 + 4.77 + 3.71)}\!/_4 = 4.25 \; Klb/in^2$$

The Ave- and Ave+ for the remaining interaction columns are provided in Figure 18.

Mean Response For Low and High Levels of Each Input and Interaction

Test #	W	X	Y	Z	WX	WY	XY	μ	σ
			Interactions						
1	-1	-1	-1	-1	+1	+1	+1	4.31	0.57
2	-1	-1	+1	+1	+1	-1	-1	5.05	0.66
3	-1	+1	-1	+1	-1	+1	-1	4.28	0.58
4	-1	+1	+1	-1	-1	-1	+1	5.07	0.76
5	+1	-1	-1	+1	-1	-1	+1	4.67	0.57
6	+1	-1	+1	-1	-1	+1	-1	3.85	0.77
7	+1	+1	-1	-1	+1	-1	-1	4.77	0.57
8	+1	+1	+1	+1	+1	+1	+1	3.71	0.70
Ave-	4.68	4.47	4.51	4.50	4.47	4.89	4.49		
Ave+	4.25	4.46	4.42	4.43	4.46	4.04	4.44		

Figure 18

It is instructive to graph the Ave- and Ave+ mean output responses for each input and interaction and examine the slopes graphically (see Figure 19).

Average Mean Output Responses For Each Input and Interaction

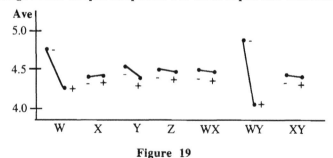

Figure 19

The steepness of the slopes in Figure 19 provide an indication importance of each input and interaction on the output. The most important effects on the output appear to be input W which provides the highest average cement strength at its low (-1) level and interaction WY which also provides the highest average strength at its low (-1) level. For the WY interaction to be at its low (-1) level while W is at its low (-1) level, the highest average cement strength is expected when input Y is at its high (+1) level.

Remember that interaction WY is an *alias* for interaction XZ (since WXYZ = I) for this fractional factorial design matrix. Hence, it is also desired that XZ be at its low (-1) level. To determine the best level for the individual X and Z inputs, one must plot the XZ interaction graph. Figure 20 provides the XZ interaction graph data (extracted from Figure 18).

XZ Interaction Graph Data

X	Z	Means	Ave
-1	-1	4.31,3.85	4.08
-1	+1	5.05,4.67	4.86
+1	-1	5.07,4.77	4.92
+1	+1	4.28,3.71	3.99

Figure 20

The XZ interaction graph can now be plotted (see Figure 21). X is held constant at its *low* (-1) level while the low (-1) and high (+1) levels of Z are plotted. Then X is held constant at its *high* (+1) level while the low (-1) and high (+1) levels of Z are plotted. The low (-1) levels and the high levels (+1) of Z are then connected to inspect the interactions.

XZ Interaction Graph

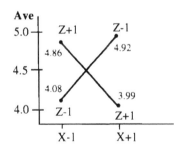

Figure 21

Figure 21 indicates that the highest average cement strength (4.92 Klb/in^2) is provided when the input X is at its high level (+1) and Z is at its low (-1) levels (i.e. XZ at -1). The cement strength is only slightly lower (4.86 Klb/in^2) when input X is at its low (-1) level and Z is at its high (+1) level (i.e. XZ at -1). Both these cement strengths indicate that the interaction XZ is desired to be at its low (-1) level.

Another desirable feature of this production process besides high cement strength is low variability for the output cement strength. Figure 22 provides a notional plot of the variability (defined by the standard deviation σ of the output responses) over the low (-1) and high (+1) levels of the inputs and interactions. The average standard deviation slopes are determined in a similar fashion (i.e. computing Ave- and Ave+) used to determine the average mean responses for the inputs and interactions (see Figure 18).

**Average Standard Deviation Output Responses
For Each Input and Interaction**

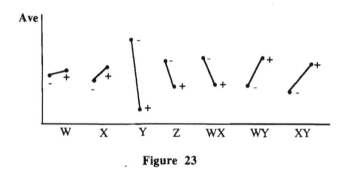

Figure 23

There will be cases for an input or interaction where the average mean output response graph (see Figure 18) does not strongly indicate whether a low (-1) or high (+1) level is more desirable. The average standard deviation output response graph should then be used to determine if a low (-1) or high (+1) level strongly affects the variability of the output. If so, the level which minimizes the output variability should be chosen.

The optimal levels for the inputs water (W), sand (X), cement mix (Y), and mixing time (Z) which maximize the output (cement strength in Klbs/in^2) and minimize the variability of the cement strength is given in Figure 24.

Optimal Input Levels

Factors	Levels
W: Water	Low (-1): 1 gal.
X: Sand	High (+1): 0.50 cu. yd
Y: Cement Mix	High (+1): 15 lbs.
Z: Mixing Time	Low (-1): 10 min.

Figure 24

Note: The event where the inputs (W, X, Y, Z) are at the levels (-1, +1, +1, -1) happens to correspond to test #4 in the fractional factorial test design matrix. However, the power of Design of Experiments is that the optimal levels for each of the inputs does *not* have to among those tested. Design of Experiments characterizes the response of the output to the inputs and derives the best level for each input even if that combination of input levels has not been tested.

Design of Experiments can also be used to determine a mathematical model defining the output response as a function of the significant inputs and interactions (see Example 4).

Example 4: (Mathematical Cement Strength Model). Using the data in the Fractional Factorial Design matrix of Example 3 (see Figure 18), determine a mathematical model which *predicts* the cement strength as a function of the significant inputs and interactions to the production process.

Answer: The derivation of a mathematical model relating inputs and interactions to the cement strength (output) requires that two quantities be added to the mean response matrix in Example 2 (see Figure 18). First, the Grand Mean (the mean of the means) $\bar{\mu}$ is computed. Second, for each input or interaction (i.e. each column) the difference Δ is computed as:

$$\Delta = (Ave+) - (Ave-)$$

These computations result in the prediction matrix given in Figure 25.

Cement Strength Prediction Matrix

Test #	\multicolumn{7}{c}{Inputs and Interactions}	μ	σ
	W	**X**	**Y**	**Z**	**WX**	**WY**	**XY**	μ	σ
1	-1	-1	-1	-1	+1	+1	+1	4.31	0.57
2	-1	-1	+1	+1	+1	-1	-1	5.05	0.66
3	-1	+1	-1	+1	-1	+1	-1	4.28	0.58
4	-1	+1	+1	-1	-1	-1	+1	5.07	0.76
5	+1	-1	-1	+1	-1	-1	+1	4.67	0.57
6	+1	-1	+1	-1	-1	+1	-1	3.85	0.77
7	+1	+1	-1	-1	+1	-1	-1	4.77	0.57
8	+1	+1	+1	+1	+1	+1	+1	3.71	0.70
Ave-	4.68	4.47	4.51	4.50	4.47	4.89	4.49		
Ave+	4.25	4.46	4.42	4.43	4.46	4.04	4.44	$\bar{\mu}=4.46$	
Δ	-0.43	-0.01	-0.09	-0.07	-0.01	-0.85	-0.05		

Figure 25

Then the equation to predict the cement strength (denoted by CS) as a function of the inputs and interactions provided in Example 3 is given by:

$$CS = \bar{\mu} + \frac{\Delta}{2} \cdot W + \frac{\Delta}{2} \cdot X + \frac{\Delta}{2} \cdot Y + \frac{\Delta}{2} \cdot Z + \frac{\Delta}{2} \cdot WX + \frac{\Delta}{2} \cdot WY + \frac{\Delta}{2} \cdot XY$$

Substituting $\bar{\mu}$ and the corresponding Δ's from Figure 25 yields the following cement strength prediction equation:

$$CS = 4.46 - 0.21 \cdot W - 0.005 \cdot X - 0.045 \cdot Y - 0.035 \cdot Z$$
$$- 0.005 \cdot WX - 0.425 \cdot WY - 0.025 \cdot XY$$

The optimum levels for the inputs (see Figure 24) are W = -1, X = +1, Y = +1, and Z = -1. The prediction equation computes the expected *maximum* cement strength (subject to variability about the mean) using these optimum input levels as follows:

$$CS = 4.46 - 0.21 \cdot (-1) - 0.005 \cdot (+1) - 0.045 \cdot (+1) - 0.035 \cdot (-1)$$
$$- 0.005 \cdot (-1) - 0.425 \cdot (-1) - 0.025 \cdot (+1)$$

$$= 5.06 \ Klbs/in^2$$

After examining the above prediction equation (i.e. the size of some coefficients versus the size of Grand Mean $\bar{\mu}$), it becomes apparent that some of the inputs and interactions have a negligible effect on the cement strength response. The prediction equation may be simplified (i.e. inputs and interactions with small effects on the output eliminated) to the following approximation:

$$CS = 4.46 - 0.21 \cdot W - 0.425 \cdot WY$$

This approximate equation relating cement strength CS to the significant input W and the significant second-order interaction WY is verified for test #1 ($W = -1$, $X = -1$, $Y = -1$, and $Z = -1$) as follows:

$$Actual \ CS = 4.31 \ Klbs/in^2$$

$$Predicted \ CS = 4.46 - 0.21 \cdot W - 0.425 \cdot WY$$

$$= 4.46 - 0.21 \cdot (-1) - 0.425 \cdot (-1 \cdot -1 = +1)$$

$$= 4.25 \ Klbs/in^2$$

The cement strength predictions versus the actual cement strengths (in Klbs/in^2) are given for each of the 8 fractional factorial tests in Figure 26:

Actual vs. Predicted Cement Strength CS using Design of Experiments

Test #	W	WY	Actual CS	Predicted CS
1	-1	+1	4.31	4.25
2	-1	-1	5.05	5.10
3	-1	+1	4.28	4.25
4	-1	-1	5.07	5.10
5	+1	-1	4.67	4.68
6	+1	+1	3.85	3.83
7	+1	-1	4.77	4.68
8	+1	+1	3.71	3.83

Figure 26

Notes:

• For this example, the test with the highest *actual* output (test #4 at 5.07 Klb/in^2) is also one of the tests with the highest *predicted* output (at 5.10 Klb/in^2). Due to randomness in the actual cement production process, on any given experiment the test with the highest actual output may differ from the test with the highest predicted output. However, over the long run (many runs), the test with the highest predicted output is expected to have the highest average actual output.

• The above method to determine the prediction equation (using the Grand Mean $\bar{\mu}$ and the half the difference between Ave+ and Ave-) is valid only when the inputs and interactions chosen are *orthogonal* (all dot products between columns equal 0).

Appendix B

Management of

Information Systems

"The genius of the future lies not in technology, but in our ability to manage it."
Anonymous

Appendix B | Management Of Information Systems

B.1 Introduction

There have been a number of "revolutions" in recent history which have changed the way people live and work in ways that could not have been predicted. Such revolutions would include:

- The industrial revolution (late 1800's/early 1900's)
- The computer revolution (post World War II)
- The information revolution (starting in the early 1980's)

The information revolution is still ongoing and its overall effects are still unfolding. However, it seems clear that the information revolution has already had a profound effect on the way people work and live. For example:

- People do not necessarily need to live near the organization which employs them (telecommuting has become increasingly common).

- People expect their jobs to be more interesting and empowering (the access to information and the power of the desktop computer has decentralized decision making in organizations).

- People in even the most remote locations have increasing access to identical information and entertainment choices provided in new and interesting formats (such as the World Wide Web, direct broadcast satellite systems, multimedia CDROMs, etc.).

This ability of new information technology to store, manage, and transfer large quantities of data has altered the constraints of location and time previously assumed in decision making. Information technology has made it easier to convert raw *data* into a useful format that provides *information* to a decision maker (see Figure 1). Information in the hands of cognitive human beings leads to *knowledge*. When knowledge is applied to the behavior of real world systems and comparisons of actual vs. forecasted outcomes are examined, *learning* occurs and better decisions are possible.

The Transformation of Data into Learning

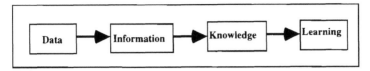

Figure 1

The purpose of this appendix is to prepare the decision maker to use information and information technology to the best advantage of his or her organization.

It is natural that the information revolution is viewed as a crisis to some who must adjust to a new paradigm under which decisions are being made. However, consider the fact that in the Chinese written language (obviously invented many centuries ago) the character for *crisis* is the combination of the character for *danger* and the character for *opportunity*. In parallel with this ancient insight into the nature of crises, this appendix on the management of information systems will examine:

> 1) The *opportunities* presented by the proper use of information technology: A general background on the fundamentals of information technology (hardware, software, and telecommunications) is provided. The role of information in achieving the strategic goals of an organization is examined. Also, a treatment of the structuring of the information systems department in a organization to exploit information opportunities is provided.

> 2) The *dangers* presented by the failure to use (or the improper use of) information technology: The potential issues associated with the application (and the misapplication) of information technology are provided throughout this appendix.

Note: The information revolution is rapidly unfolding in unpredictable directions. New information technology and new applications of existing information technology are being developed at an amazing rate. Even in the most nimble organizations, due to limited resources there will be a time lag before organizations implement the latest information technology. It is under this realistic scenario (where the available information technology may not be the best available) that a decision maker must operate.

In the information revolution, organizations have experienced a clear trend toward larger numbers of individuals at lower levels of management involved in decision making:

> - In the past, only a limited number of high level managers in an organization had access to the accounting, finance, contracts, etc. data required to derive and implement the decisions which benefited the organization. There was usually little interaction between the *centralized* data processing department and the other departments of the organization. Data processing was accomplished by the data processing department typically during off-hours (e.g. overnight).

> - With the advances in information technology (e.g. desktop computers, user-friendly software applications etc.), the processing of data and the resulting information has become *decentralized* and available to individuals at all levels of the

organization. Today individuals from multiple interfacing departments are working together daily as part of integrated teams to develop and implement decisions.

It is essential that a decision maker have a basic understanding of the fundamentals of information technology. Even if the future of the information revolution can not be reliably predicted, it is certain that a decision maker will have to operate comfortably in an environment which is rich in both new information technology and new applications of existing information technology.

B.2 The Fundamentals of Information Technology

Since the field of information technology is large and dynamic with new generations of information technology and applications continually on the near horizon, a comprehensive and current review of information technology can not be provided within this section. Instead the objective of this section is to provide the decision maker with the following:

> 1) A working knowledge of the language of information technology.

> 2) Enough technical information to serve as a basis for future learning of information technology during one's professional career.

Information technology includes an enormous array of information processing and transmission devices and peripherals. However, the two key information technology elements are: 1) the computer and 2) the attached telecommunication devices. In reality, the fields of computers and telecommunications have become increasingly intertwined over time. Many individuals in an organization now have a desktop computer system connected through telecommunication devices with other computers within and outside of the organization by means of networks internal and external to the organization.

The elements of a computer system include the following:

> • Hardware: The physical components (e.g. central processing unit (CPU), monitor, keyboard, printer, etc.).

> • Software: The programs which implement the operation of the computer system (e.g. DOS, Windows) and the programs which perform the required applications (e.g. Word, Lotus Notes, etc.).

> • Telecommunications: The electronic linking of a single computer to a peripheral device (e.g. a printer) and the electronic linking of a group of computers (e.g. local area networks (LANs), wide area networks (WANs), etc.).

Hardware

Personal computers used to support decision makers have common basic elements (see Figure 2) which include the following:

> • Input device(s) (e.g. a keyboard).

462

• Central processing unit (CPU).

• Memory (permanent read-only memory (ROM) and transient random access memory (RAM)).

• Display device(s) (e.g. a monitor).

• Output device(s) (e.g. a printer).

Basic Personal Computer (PC) Structure

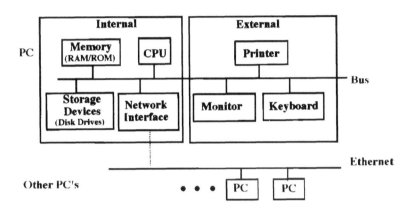

Figure 2

Other devices common to most computers include storage devices (including the hard drive and the floppy disk drive) and the CDROM (Compact Disc Read Only Memory). The floppy disk drive is also used as both an input device (e.g. reading a program into memory) and an output device (e.g. writing data or a program from memory onto a floppy disk).

The memory of a computer is a series of addressable cells (integrated circuits on a silicon chip) each of which holds a fixed quantity of data. When data is read into a particular memory cell, the previous data in the cell is destroyed. However, when data is read out of a particular cell, the data in that cell remains intact.

Notes:

• A single character of data is made of eight bits where a bit is a binary digit (i.e. either a zero or a one corresponding to the concept of off and on, respectively).

• A memory cell capable of storing a single character (8 bits) of data is called a byte.

• A memory cell capable of storing two or more characters of data is called a word.

The central processing unit (CPU) (which also consists of integrated circuits on a silicon chip) of a computer carries out the system operations required to control the computer as well as to perform both arithmetic functions (e.g. addition, multiplication, etc. of data), and logical functions (e.g. the comparison of whether one piece of data is greater than or equal to another piece of data). A collection of operations for the CPU to perform in order to accomplish a particular task is contained in a program which is also stored in memory.

Notes:

• The explicit set of instructions contained in a program is written in machine language.

• Each machine language instruction consists of two elements:

> - An *operation code* (e.g. M (multiply), P (print), etc.) which tells the CPU which operation to perform.

> - An *access code* (e.g. cells 7625 and 0891) which tells the CPU which cells are involved in the operation to be performed.

The memory and the central processing unit of a computer are often specified using the concept of bytes or bits. For example:

> - The size of the memory is given in terms of the total number of bytes (e.g. 16 Megabytes of RAM) available even if the data is actually stored in the memory cells in word format (i.e. two or more bytes in each memory cell).

> - The capability of the CPU may be given in terms of the amount of data handled by the CPU as a single unit (e.g. a 16 bit (two byte) machine, a 32 bit (four byte) machine, etc.).

Note: Another measure of the capability of a CPU is the number of instructions per second (ips) that can be executed. For example, a computer may have a 50 Mips (millions of instructions per second) capability. Computer instructions can be of varying length so this is an *average* measure of capability which is best used to compare the capability of *similar* computers (where the length of an instruction on the average is likely to be close).

The CPU of a computer executes an instruction much faster than it performs a data access from memory. For example, a given CPU may execute tens of thousands of arithmetic operations in the time it takes to access a single piece of data from memory. This speed mismatch can result in the CPU waiting for a data access to be completed before the execution of more arithmetic operations can begin. Methods to overcome this speed mismatch and improve the efficiency of the CPU include:

> - Use of a cache memory unit (separate from the main memory unit) where a block of data can be placed and high speed data accesses can occur. This method will improve the efficiency of the CPU when the data placed in cache memory is accessed repetitively (e.g. the same data is used over and over or the set of data used has nearby access codes).

> - Use of multiprogramming where several programs are run in the CPU *simultaneously*. Multiprogramming improves the efficiency of the CPU by

allowing the switching among several programs located in memory to minimize the time the CPU is idle waiting for a data access to occur.

Note: Multiprocessing is quite different from mutiprogramming. Multiprocessing occurs when the computer has two or more CPU's and is able to execute programs simultaneously (as opposed to the switching among several programs as performed during multiprogramming).

Software

Even though computers execute operations based on machine language instructions, programmers rarely write machine language code (instructions). Instead, higher order languages (HOL) (e.g. assembly, FORTRAN, C) are used to take advantage of the capability of the computer itself to write machine language. For example, say a programmer writes a program (called source code) in FORTRAN. A program resident in computer memory called a FORTRAN compiler is then used to convert the source code into machine language (called object code) which is executable by the computer.

Software is generally grouped into two major categories:

- Operating software: This is software (e.g. DOS, Windows, etc.) written to provide the background environment which supports the application software.

- Application software: This is software (e.g. WordPerfect, Excel, etc.) written to accomplish a particular task (e.g. word processing, spreadsheets, etc.).

A computer user will interface with the hardware and software through an operating system. An operating system will perform the following functions at high speed (much faster than human control will allow):

- Decide which task among a queue of tasks to run next.

- Store and retrieve data while keeping track of where all data is stored.

- Decide when to terminate a task (e.g. the task has completed or an error has occurred).

- Decide which task output to print next.

- Manage the software library including both operating and application software.

Time sharing is the mode of operation employed when there are a large number of users (often remotely located) tasking a computer simultaneously. Control of the operating system shifts back and forth among the users so each user has control of the computer for a fraction of the time. In a well-designed system, the computer works fast enough so that each user is given the perception of being the only person tasking the computer. However, if the number of simultaneous users exceed design specifications, users can experience substantial idle periods waiting for control of the computer to be switched over to them.

465

Application programs have become quite large in terms of the number of bytes of memory required for storage. Running several application programs simultaneously can easily exceed the amount of random access memory (RAM) available. This problem is eased through the use of virtual memory to *manage* the memory of a computer efficiently. Virtual memory takes advantage of the fact that only one segment of one application program is being executed at any given time while the rest of the application programs are idle. Thus only a small segment of each program will be kept in RAM at a time (and the operating system will switch in new segments of the application programs as required).

Two common approaches presently used for the construction of software programs are as follows:

- Structured programming: Where the software program is divided into blocks of code accomplishing a given *task*. For each block of code, there is only one point to enter and one point to exit the block. Structured programming allows users *other than the developer* to better understand the logic of a program and to better accomplish the maintenance of a program.

- Object oriented programming: Where the software program is divided into blocks of code which describe the *objects* (e.g. a radar) which make up the system (e.g. a satellite tracking system). Each object describes the *attributes* (e.g. site location, power, dish size, field of view, etc.) of the object and the *functions* (e.g. acquisition, tracking, handover to the next radar) carried out by the object. Hence the size of the block of code for an object varies depending on the complexity of the object. Object oriented programming (typically written in C language) is then concerned with the relationship among the objects identified.

There are two methods of processing the programs developed by users of a computer system:

- Batch processing: Where an organization accumulates a group of programs and processes them at the same time. This was common in the early days of the computer when magnetic tape was often used as the medium of storage. Such processing can be cost-effective but it is usually not timely enough to meet the information needs of an organization (e.g. transactions my be batch processed overnight resulting in accounts only being current first thing in the morning).

- On-line processing: Where each user in an organization interfaces directly with the computer in a real-time interactive mode. This type of processing is more timely (e.g. accounts can be made to be always up-to-date) and it is more expensive than batch processing. Whether on-line processing is cost-effective often depends on whether a system can be updated at the time the original data becomes available (e.g. inventory management at a supermarket driven by the real-time bar code entries of the on-line cash registers).

Notes: Other useful software definitions include:

• Database: A collection of logically related data.

• Database Management System (DBMS): Software which works with the operating system to store, modify and make the desired data accessible to users in a timely manner.

• Hypertext: Text, images, and databases which are linked together by buttons displayed on the monitor screen. A mouse is used to click the buttons that execute the links.

• Computer Aided Software Engineering (CASE) Tools: A set of software utilities helpful in the design, generation, and maintenance of software programs.

• Job Control Language: Instructions required to communicate to an operating system (e.g. DOS) the specifics of a desired operation (e.g. *copy a:FILENAME c:* might be keyed into a computer using the DOS operating system to instruct the computer to copy a file from the disk drive designated as *a* onto the hard drive designated as c). Job control language can vary significantly from one operating system to another. There has been a strong trend toward eliminating the need for a user to enter job control language to perform operations (e.g. the WINDOWS program designed with point and click operations).

Telecommunications

The technology explosion in the field of telecommunications has become an integral part of the information revolution. Today a large fraction of computers (including desktop computers) are connected via telecommunications devices directly both to nearby computers and to geographically dispersed computers by means of a variety of networks (e.g. Internet, America Online, etc.).

Networking within an organization allows certain individuals to share expensive or proprietary resources. A local area network (LAN) is often used to allow a number of individuals to share a high speed printer, a particular software program for which an organization has purchased a site license, or a closely held corporate data base (e.g. sales forecasts, pricing strategies, etc.).

Telecommunications involves the transmission of voice, text, and/or image signals over considerable distances via 1) an analog network or 2) a digital network.

1) An analog network (e.g. the public phone system) sends signals based on the change in a physical quantity (e.g. voltage) over time.

Notes:

• The computer (which deals with bit streams of zeros and ones) is incompatible with analog signals due to noise problems. The noise inherent in an analog network can easily cause zeros to be interpreted as ones and vice versa. The potential for garbling of messages is unacceptable.

• Computer systems typically use a modem (*m*odulator/*dem*odulator) to convert digital signals into analog signals (e.g. where there is a different voltage level transmitted for a zero than for a one) before the signals are sent over a transmission line to another computer. A second modem on the other end of the transmission then coverts the analog signals back to digital signals (which can be interpreted by the receiving computer).

2) A digital network (e.g. a local area network) sends signals based on unique patterns of 0's and 1's.

Notes:

• Digital networks typically have smaller transmission errors and allow for higher transmission speeds. Higher quality transmission (e.g. less noise on a digital telephone line) and higher quantity transmissions (e.g. movies, high resolution images, etc.) become possible. Modems are not required when computers are linked via a digital network.

• The speed of transmission over a network is measured in bits per second (e.g. 128 Kilobits per second). Equivalent terms for bits per second are hertz or baud.

There are three types of data transmission between two locations:

1) Simplex transmission: Where data can be transmitted only in a single direction (e.g. a public address system).

2) Half duplex transmission: Where data can travel in both directions but not simultaneously (e.g. a speaker phone).

3) Full duplex transmission: Where data can travel in both directions simultaneously (e.g. a telephone).

There are five physical (hardware) media most commonly used for the transmission of data:

1) Twisted pairs: Two insulated thin copper wires are twisted into a helix shape. The twisting reduces the interference from other nearby twisted pairs. This is the most common transmission media.

2) Coaxial cable ("coax"): A heavy copper wire is at the core of the cable. This is surrounded by a layer of insulating material. The next layer is a braided mesh conducting material. The outer layer is a protective plastic covering. Coaxial cable provides a relatively high transmission rate with low noise.

3) Fiber optic cable: Pulses of light (e.g. a light pulse signals a 1 while a lack of a light pulse signals a 0) are transmitted through a thin fiber of glass made of fused silica. There are three components of a fiber optics system: a light source (either a light emitting device (LED) or a laser diode), the fiber optic cable itself, and a light detector (a photodiode). When an electrical current is applied to the light source, a light pulse is generated; when the pulse of light traveling through the fiber hits the light detector, an electrical current is generated. The advantages of fiber optics are transmission speed, reliability, and security (difficult to pirate). The disadvantages of fiber optics are cost and the difficulty of working with thin fibers of glass.

4) Microwave: This is line-of-sight transmission (i.e. there must be an unobstructed path between the transmitter and the receiver) of radio signals. Due to the curvature of the earth, microwave towers must be spaced every 25 to 50 miles to support the relay of signals over long distances. This type of transmission is expensive but may be more cost-effective than burying underground long lines of coaxial or fiber optic cable to link sites.

5) <u>Satellite</u>: One or more transponders on the satellite will receive the signals, amplify them, and retransmit the signals. Satellites also transmit on a line-of-sight basis so that more than one satellite (possibly combined with a network of ground microwave towers) may be required to transmit data around the globe. Due to the large transmission distances, there may be noticeable delays between data transmission and data receipt.

Simple networks have one of the four basic topologies (arrangements) described below. Complex networks have various combinations of these four topologies:

1) <u>Bus</u>: In this linear topology (see Figure 3), a single cable (twisted pair, coaxial, or fiber optic) is shared by all nodes on the network. While the wiring for the topology is simple, a failure along the cable means that nodes on opposite sides of the point of failure can not communicate.

Basic Network Topologies

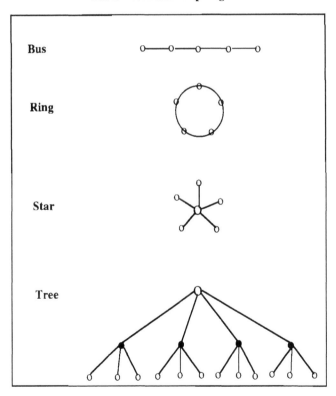

Figure 3

2) Ring: This topology is similar to the bus topology except that the two ends of the cable are connected (see Figure 3). This topology is not as susceptible to failure as the bus topology. For example, even after a single failure along the cable (in which case the ring topology degenerates in a bus topology), each node in the network can still communicate with every other node.

3) Star: This topology has a computer at its center with cables connecting the central computer to all other nodes in the network (see Figure 3). Local area networks often use the star topology. This topology offers ease in identification of the location of a failure (since each device has its own cable to the central computer), ease in installation (a single cable connects each node to the central computer), and low cost of cable when all the nodes are nearby (and the routes are short). However, the entire network fails when the central computer fails.

4) Tree: This topology is hierarchical. A central computer is at the top of the tree with several branches of star-like topologies (see Figure 3). Wide area networks (which are geographically dispersed) use this type of topology. Nodes can be connected at lower branches of the tree without long and expensive cable routing. However, the entire network fails when the central computer fails.

There are four categories of networks: computer networks, PBXs (private branch exchanges), LANs (Local Area Networks), and WANs (Wide Area Networks). The following provides a brief descriptions of each category:

1) Computer networks: Consist of a central computer and several other connected device nodes. All the other nodes operate as "slaves" to the central computer. Often a front-end processor is added to the central computer to ease the communications burden of driving the connected nodes.

2) PBXs: Consist of a digital switching system controlled by a dedicated computer. A PBX can simultaneously handle voice and data communications internal and external to the network. A PBX can serve as the central device in a star or tree network topology. The advantages of a PBX include its ability to use existing telephone lines and to carry voice and data over the same network. The disadvantages of a PBX include its cost and its throughput limitations on a single channel (too slow for large data transmissions).

3) LANs: Owned by a single organization and operated within a distance of a few miles or less. A LAN is a high speed *digital* data network capable of transmitting several millions of bits per second. A LAN differs from a computer network in that there is not a "master-slave" configuration (i.e. a LAN contains computers capable of doing their own data processing). A LAN differs from a PBX in that a LAN transmits only data and requires new wiring (i.e. it can not use the present phone system wiring). However, a LAN and a PBX share the goal of providing communication between a variety of devices for the purpose of sharing data and other resources.

4) WANs: Owned by several organizations and operated over long distances. A WAN is a long-haul network used to transmit both voice and data among geographically dispersed organizations. A WAN is an *analog* network with a slower data rate than a LAN. A WAN is highly dependent on equipment owned and operated by the public telephone company and other long distance carriers.

Both the number of people connected to wide area networks (e.g. Internet, Compuserve, America Online, etc.) and the number of services available on these global networks has been growing at an exponential rate. The methods used to set up such wide area networks include the following:

- Direct Distance Dialing (DDS): Available through the local telephone company and one of the several long distance carriers (e.g. MCI, SPRINT, etc.). The data transmission rate is low and the error rate (noise) is high. Such a wide area network is quite expensive to operate.

- Wide Area Telephone Service (WATS): Operates like a DDS but on the basis of a fixed monthly cost for unlimited long distance service. Hence the cost to operate such a wide area network is more acceptable than the cost of a DDS. However, the same issues of low data transmission rate and high error rate persist.

- T-1 line: The most common of the dedicated leased telephone line options. The cost of such links for a WAN is high. However, such dedicated lines may be cost-effective for an organization if there is extensive usage of the line and/or there is a requirement for a very high data transmission rate with a low error rate.

- Integrated Services Data Network (ISDN): Provides digital communication over the public telephone network. Modems are not required. However, special ISDN telephones which produce digital rather than analog signals are necessary to be a user on such a wide area network. Hardware (e.g. a digital PBX) and software at the sponsoring organization and the local telephone company allow a significant increase in transmission capacity while still using the twisted pair lines of the public telephone system.

Modern telecommunications systems are increasingly employing packet switching. Instead of dedicating a circuit over the entire length of a user's voice or data transmission session, packets of the user's data are interspersed with those of other users. Communications are sent in bundles of fixed length with control information on the front and back of the bundle. This allows the network to be used more efficiently (more users can be on the network simultaneously and shorter overall transmission delays will occur).

B.3 Information Systems Role Within an Organization

Many external forces inherent in the information revolution have caused even the most stable organizations to undergo significant changes in recent years. These forces include:

- Advances in desktop computers (lower cost and higher capability).

- Advances in commercially available software packages (more user-friendly and higher capability).

- Advances in telecommunications (such as E-mail, voice mail, networking, electronic bulletin boards, and video teleconferencing providing more capability for a team of workers in different locations to work together).

- Worldwide competition (reduced profit margins and reduced market shares).

- Customer demands (for new and diverse products at lower prices).

Adapting to the information revolution, customers have come to *expect* an ever increasing array of products and services (e.g. electronic transfer of funds between accounts, purchases from retail outlets by telephone from remote locations using any of a variety of credit vehicles, just-in-time inventory management, etc.) from organizations in the marketplace. Organizations are finding themselves investing and organizing to provide the information services required to meet customer demand. Individuals within organizations are finding that their ability to master information technology has a significant effect on their job performance and on their potential for career advancement.

The role of information systems within an organization must grow from the traditional role of reducing the cost of repetitive operations (e.g. weekly payroll) to that of supporting decision makers at all levels the organization in unique and unstructured decisions (e.g. developing the organization's market forecasts).

The information systems department should be viewed not only as an *operational* tool for internal cost reduction but also as a *strategic* tool for revenue enhancement through the introduction of: 1) new or improved products and services to external customers, and 2) a competitive advantage (e.g. lower cost, more customer convenience, etc.) for existing products. That is, the information systems department should be an integral part of the process of achieving the strategic goals of the organization.

Many organizations still maintain the view that the information systems department is merely a *utility* to support and enhance the activities of the strategic planning, finance, accounting, etc. departments. While it can indeed provide this role, the information systems department should also be viewed in the context of the external customer of the entire organization. The issue should be: how can diverse departments (including information systems) working on an equal footing as an integrated team best meet the demands of the customers of the organization?

An increasing number of organizations are now looking to the information systems department to support strategic objectives (e.g. reduce internal costs, increase market share, remain the technological leader in the marketplace, etc.) through the application of information technology. However, information technology is certainly not a panacea for all the issues within an organization. There are several important factors to be considered prior to the application of information technology:

• **Is this the right time to be applying information technology?** Classical cost-benefit calculations are often *not relevant* in this decision since there are often subtle benefits (which are hard to quantify) of greater access to information within an organization. Such subtle benefits might include an improved work environment, increased empowerment of employees, improved employee morale, etc. These benefits improve productivity but it is difficult to show a direct cause and effect relationship. Benefits may also accrue in the form of the competition having to spend money to match the new product or service. Thus *the application of information technology may be right even when it can not be strictly justified on a cost-benefit analysis* within the organization. On the other hand, an organization need not apply information technology to every issue just to appear to be on the cutting edge of technology. Such an expensive approach can be counter-productive to the strategic objectives of an organization.

• **Has the right application of information technology been identified?** The *identification* of the specific application of information technology within an organization can not be left solely to the information systems department to identify. The specific departments and/or external customers that will be the end-users should play an integral role in defining the requirements for information in order to assure that the right application for information technology has been identified.

• **Has the application of information technology been designed right?** The *design* of an application of information technology can not be done solely by the information systems department. Strong input from the end-user departments and/or the external customer is required. It is well documented in the hardware/software systems literature that it is much less expensive to change the specifications for a system during the design phase than during the prototype phase.

The key to successful applications of information technology (i.e. the development of the *right* design of the *right* application at the *right* time) is communications between the information systems department and the appropriate end-user departments and/or the external customers of the organization.

While communication to its interfacing departments is essential, the information systems department must also have its own internal strategic plan. Elements of this plan (which must be consistent with the organization's overall strategic plan) include:

- A vision statement of long range goals.

- A series of metrics to measure progress toward the vision.

- A short and focused list of near term initiatives.

B.4 Organizing to Exploit Information Opportunities

A system is generally defined as those assets (e.g. personnel, equipment, data, etc.) under the *immediate control* of the decision maker. However, a system is usually not an isolated entity. Systems typically have interfaces to other similar systems and are a part of a larger system (e.g. the strategic planning department interfaces with the finance department and both these departments are part of a larger system which is the organization). Thus a system is often made up of many smaller systems (i.e. a system is actually a system of systems). The objective of this section is to examine examples of organizational systems and how they can enhance the ability to exploit information opportunities.

Information is now viewed by observers of trends as a *resource* of large and growing importance to an organization's strategic goals. Such observers stress that the successful organizations of today and the future will be those that *manage and exploit information* as well as they manage and exploit other strategic assets (e.g. personnel, capital, etc.).

Notes:

• The use of information technology to accomplish strategic goals requires a continuity of purpose. A *one-time* investment in information technology to gain a competitive advantage through the development of a new product or a lower cost process can actually backfire in the marketplace. Competitors can typically duplicate a product or process at a cost significantly lower than the development cost. An organization in pursuit of its strategic goals must invest in information technology *on a sustained basis* in order to take advantage of opportunities for a competitive advantage.

• Strategic uses of information technology tend to be very visible to external customers. Poor implementation of information technology (e.g. which allows piracy of customer credit card numbers) due to human or technological problems can actually decrease customer goodwill and can cause significant public embarrassment to an organization.

A key factor in planning the organizational structure to exploit information opportunities in the pursuit of strategic goals is the promotion of *information sharing*. Information can only be exploited when an organization does the planning required to assure that necessary information is commonly available in a timely manner. This is a formidable goal for an organization and its implementation may require a "cultural change" among its personnel. Information is often viewed as power and there will be natural resistance to any system promising to implement information sharing processes.

The organization should be structured with the following in mind:

- Appropriate information has to be readily accessible to external customers.

- Information must readily flow to the location in the organization where a decision is being made.

- Information from diverse locations in the organization must readily come together in a common data base which is updated in a timely manner.

In a classical sense, information resides in an organization in one of the following models:

- *Centralized:* Each department accesses common data residing in a centralized data base. Departments do not have the opportunity to update the data base in real-time.

- *Decentralized:* Each department has its own data base operating independently of other departments in the organization.

- *Distributed:* Each department accesses common data residing in a centralized data base. Departments do have the opportunity to update the data base in real-time.

In reality, most organizations today use some form of the distributed information model featuring local access and centralized oversight. The issue becomes how to organize the information systems department to contribute best to the organization's goal of having the necessary information available at all levels in a timely manner. The information systems department must strike a delicate balance between supporting decision making within each department of the organization while avoiding the scenario where the departments become information islands onto themselves.

Note: Some observers of business trends believe that the organization of the future will not have an independent information system department. With advances in technology and increased understanding of the functions of information technology at all levels, the responsibilities of information systems will eventually be distributed throughout the organization. Such an evolution took place in the field of operations research where its functions (which once took place in a centralized group) are now distributed among the diverse departments of an organization.

Senior management has come to recognize that information technology (while it can not always be justified on a strict cost basis) has changed the way their organization functions. However, senior management often has the strong suspicion that the money allocated to information technology while necessary is not providing the best return on investment possible to the organization. To underscore its value-added to the organization, the information system department must perform the following functions:

- Work in an integrated team environment to exploit information technology to achieve a competitive advantage for the organization in the marketplace.

- Develop a vision for the organization's use of information technology and communicate this vision to all levels of the organization.

- Support the organization in the acquisition and implementation of the most appropriate and cost-effective new information technology.

- Maintain oversight on the proper use of information technology throughout the organization.

Senior management has also come to recognize the importance of appointing a chief information officer (CIO) whose responsibilities include the following:

- The CIO is not a management position but is a *staff* position whose responsibility it is to take a broad view of information needs of the organization.

- The CIO typically reports directly to the Chief Executive Officer (i.e. at a corporate level above that of the various departments of the organization among which information must be shared).

- The CIO is part of the team that makes the strategic decisions for the organization.

- The CIO is responsible for decentralizing data without the creation of islands of information throughout the company.

- The CIO is responsible for technical advice leading to the acquisition of the appropriate new information technology.

Note: It should be the director of the information systems department (and not the CIO) who manages the daily information systems activities within the organization.

Information systems departments within an organization are generally organized along one of three models:

    1) Centralized.

    2) Functional.

    3) Distributed.

1) In the *centralized* model (see Figure 4), the information systems department is monolithic and is focused on its technical proficiency in performing the mechanical tasks of information systems. The director of the information systems department is typically a very technically competent individual who concentrates on managing the group of information specialists rather than on achieving the strategic objectives of the organization. The director of information systems typically reports at the level of the director of administration (or human resources) or at some other relatively low level in the organization.

Centralized Information Systems Organization

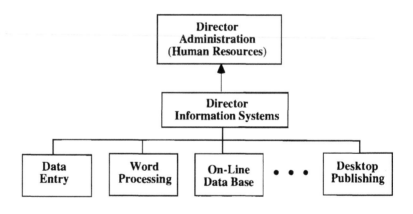

Figure 4

2) In the *functional* model (see Figure 5), the information systems department is segmented into groups each serving a particular functional department in the organization. Each group in the information systems department may have unique expertise appropriate to serve its target department. Such a segmented organization may have trouble focusing on the information needs of the organization as a whole. The director of information systems typically reports at a Vice President level (at a level below the Chief Executive Officer).

3) In the *distributed* model (see Figure 6), the information system department is organized along the major organizational elements of information systems: operations (e.g. data

entry), system development (e.g. software programming), technical services (e.g. software program maintenance), and administration (e.g. development of standards and procedures). The focus of the distributed information systems model is to develop an organizational vision and implementation plan for information systems and to assist the other departments to exploit information opportunities. The director of information systems (if he is not burdened with managing the operations of day to day activities) takes on the role of the Chief Information Officer (CIO) for the organization.

Functional Information Systems Organization

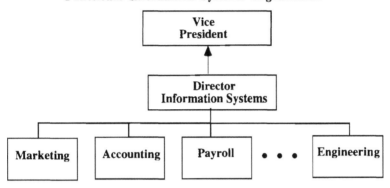

Figure 5

Distributed Information Systems Organization

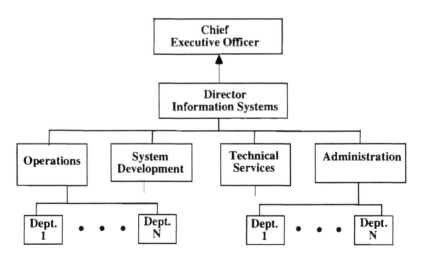

Figure 6

There is no "best" way to organize an information systems department. No single organizational model is a panacea for every organization to solve all their issues necessary to exploit their information opportunities and to achieve their strategic objectives. The appropriate model for an information systems department will depend on many circumstances including its vision, the corporate organization model, the strengths and weaknesses of key individual managers, and the internal politics of the organization. In fact, internal politics (based on the goals of key individuals rather than the goals of the organization) must be given all due consideration since it can sometimes overwhelm all logical considerations in the choice of an organizational model.

Historically, the focus of organizing the information systems department has shifted back and forth between the centralized and distributed models. At present, the distributed model for information systems (where all managers can expect to be given some responsibility for managing information systems tasks and perhaps even information systems personnel) is in fashion. At some point in the future, the popular wisdom will be that the problems associated with the current distributed model for information systems departments can only be solved with a dramatic move toward the centralized model. Due to such expected periodic reorganizations, it is critical that the director of information systems maintain a constant focus on the organization's vision for exploiting information opportunities and its implementation plan for information technology.

B.5 The Future of The Information Revolution

Early in the information revolution, executives were promised exciting returns on investment in information technology. The facts are that information technology has not met these lofty projections. This failure to achieve expected reductions in costs and personnel has increased the skepticism toward information systems. However, it must be realized that 1) firms can not function in today's marketplace without information technology and 2) information technology has played a significant role in the competitive advantages gained by organizations. Hence, even though its contribution is difficult to measure, information technology and its ability to exploit information opportunities are critical to the success of the organization of the future.

The information revolution has truly had a profound effect during the last two decades on the way people work and live. It is risky business trying to predict the future in relatively stable times. Trying to predict the future of the information revolution is extremely difficult. However, a few trends should be contemplated:

- The rate of change in information technology and its applications will only increase.

- People, with their ability to find uses for information technology that the inventors did not foresee, will remain the unknown parameter in the final scope of the information revolution.

- Information rather than traditional goods and services will be perceived as the main asset and product of an increasing number of organizations.

- Information technology implemented through the advances in telecommunications will transform the entire world into a "global village" and redefine once again our concepts of location and time.

478

- The evolution of the rules of society (legal, ethical, social, etc.) will continue to lag well behind the new scenarios created by the information revolution.

The amount of information available to individuals and organizations is already staggering. However, many potential users who could benefit by this information do not have mastery of the computer and telecommunications technology required to access this information. For the information revolution to maintain its momentum, there will have to be a focus on improving the ease by which the typical user can access desired information. Common standards must be agreed upon to clear up the confusion created by the array of platforms and operating systems which exist today.

The information revolution holds the potential to provide great benefits to society. There is also the potential for nightmare scenarios for the individual (e.g. loss of privacy, security, etc.). *Management of information systems* holds the key for decision makers trying to accentuate the positive effects and eliminate the negative effects of the information revolution.

Appendix C

Solutions to Exercises

Solutions To Exercise 1

Probability & Statistics

<u>Problem #1</u> (Simple Probability)

a) The sample space of possible outcomes of a fair die rolled twice is:

$$(1,1)\ (2,1)\ (3,1)\ \cdots\ (6,1)$$
$$(1,2)\ (2,2)\ (3,2)\ \cdots\ (6,2)$$
$$(1,3)\ (2,3)\ (3,3)\ \cdots\ (6,3)$$
$$(1,4)\ (2,4)\ (3,4)\ \cdots\ (6,4)$$
$$(1,5)\ (2,5)\ (3,5)\ \cdots\ (6,5)$$
$$(1,6)\ (2,6)\ (3,6)\ \cdots\ (6,6)$$

$$36\ possible\ \ outcomes = \underbrace{\frac{6}{1^{st}\ roll}} \cdot \underbrace{\frac{6}{2^{nd}\ roll}}$$

b) The outcomes $\{O_i\}$ for the Event E_1 = "doubles" = $\{(1,1), (2,2), (3,3), (4,4), (5,5), (6,6)\}$.

$$P(E_1) = P("doubles")\ = \Sigma\ P(O_i) \qquad \text{for all } O_i \in E_1$$

$$= 6\ (^1/_{36}) = ^1/_6$$

c) The outcomes $\{O_i\}$ for the event E_2 = "7" = $\{(1,6),(2,5), (3,4), (4,3), (5,2), (6,1)\}$.

$$P(E_2) = P("7") = \Sigma\ P(O_i) \qquad \text{for all } O_i \in E_2$$

$$= 6\ (^1/_{36}) = ^1/_6$$

d) The outcomes for Event E_1 = "doubles" = $\{(1,1),(2,2),(3,3),(4,4),(5,5),(6,6)\}$ while the outcomes for the Event E_2 = "7" = $\{(1,6),(2,5),(3,4),(4,3),(5,2),(6,1)\}$.

Then the Event $E_1 \cap E_2 = \phi$, so that

$$P(E_1 \cup E_2) = P(E_1) + P(E_2) - P(E_1 \cap E_2)$$

$$= P(E_1) + P(E_2) - 0$$

$$= ^1/_6 + ^1/_6$$

$$= ^1/_3$$

e) The outcomes for Event $E_1 = $ "doubles" $= \{(1,1),(2,2),(3,3),(4,4),(5,5),(6,6)\}$ while the outcomes for the Event $E_3 = $ "less than four" $= \{(1,1),(1,2),(2,1)\}$.

$P(E_3) = P(\text{"less than 4"}) = \Sigma\, P(O_i) \quad$ for all $O_i \in E_3$

$= \,^3\!/_{36}$

Then the Event $E_1 \cap E_3 = \{(1, 1)\}$, so that

$P(E_1 \cap E_3) = \Sigma\, P(O_i) \qquad\qquad$ for all $O_i \in E_1 \cap E_3$

$= \,^1\!/_{36}$

$P(E_1 \cup E_3) = P(E_1) + P(E_3) - P(E_1 \cap E_3)$

$\,^6\!/_{36} + \,^3\!/_{36} - \,^1\!/_{36}$

$= \,^8\!/_{36}$

$= \,^2\!/_9$

f) Define the Event $E_4 = $ "more than 3". Then the complement Event $\hat{E}_4 = $ "less than or equal to three" $= \{(1,1),(1,2),(2,1)\}$.

$P(E_4) = 1 - P(\hat{E}_4)$

$= 1 - \,^3\!/_{36}$

$= \,^{33}\!/_{36}$

Problem #2 (Expected Value)

a) Two possible outcomes of the game $= \{$Win \$30, Win \$0$\}$.

$P(Win\ \$30) = P(\leq 3) = 3/36$

$P(Win\ \$0) = P(> 3) = 33/36$

$Expected\ Value = \Sigma\, Outcome \cdot P(Outcome)$

$= \Sigma\, x_i \cdot P(x_i)$

$= \$30\ (^3\!/_{36}) + \$0\ (^{33}\!/_{36})$

$$= \$2.50$$

b) If one played this game a large number of times, the average winnings would be \$2.50 per game Of course, on any individual game, one would win \$30 or win \$0.

One should be willing to pay up to \$2.50 per game to play. At a cost to play of \$2.50, the game would be a *fair* game (Expected Winnings - Cost = \$0).

Note: Over a single game or a small number of games, a player could lose money in a game that was fair or even advantageous to the player.

c) There are two possible outcomes of the game = {Win \$30, Win \$0} with probabilities {3/36, 33/36}. The Expected Value of the game = \$2.50.

$$\sigma^2 = \Sigma \, (Outcome - Expected \; Value)^2 \cdot P(Outcome)$$

$$= \Sigma \, (x_i - EX)^2 \cdot P(x_i)$$

$$= (\$30 - \$2.50)^2 \cdot {}^{3}/_{36} + (\$0 - \$2.50)^2 \cdot {}^{33}/_{36}$$

$$= \$68.75$$

$$\sigma \;\; = \$8.30$$

Problem #3 (Normal Distribution)

a) X_1 is centered at 5 while X_2 is centered at 10. X_1 has a standard deviation of 2 while X_2 has a standard deviation of 3. Thus X_1 is taller at the mean value (less dispersion or variance around the mean value). The area under both curves is 1.

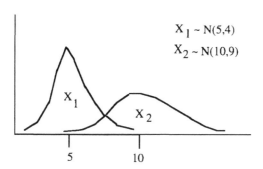

$$X_1 \sim N(5,4)$$
$$X_2 \sim N(10,9)$$

b) For $X_1 \sim N(5, 4)$,

$$P(X_1 < 10) \quad = P[(X_1 - 5)/2 < (10 - 5)/2]$$

$$= P(Z < 2.5) \qquad \text{where } Z = (X_1 - 5)/2 \sim N(0, 1)$$

$$= 0.994 \qquad \text{(Table 1a)}$$

c) For $X_2 \sim N(10, 9)$, find a and b such that

$$0.95 \quad = P[a < X_2 < b]$$

$$= P[Z_{.025} < Z < Z_{.975}] \qquad \text{where } Z = (X_2 - 10)/3 \sim N(0, 1)$$

$$= P[Z_{.025} < (X_2 - 10)/3 < Z_{.975}]$$

$$= P[10 + Z_{.025}(3) < X_2 < 10 + Z_{.975}(3)]$$

This implies that

$$a = 10 + Z_{.025}(3) = 10 - 1.96(3) = 4.12 \qquad \text{(Table 1b)}$$

$$b = 10 + Z_{.975}(3) = 10 + 1.96(3) = 15.88 \qquad \text{(Table 1b)}$$

$$0.95 = P(4.12 < X_2 < 15.88)$$

The symmetric interval around the mean 10 given by (4.12, 15.88) is the 95% confidence interval.

Problem #4 (Poisson Distribution)

a) Let X be a Poisson distributed Random Variable with mean $\lambda = 10$:

$$P(X = k) = \lambda^k e^{-\lambda} / k! \qquad\qquad k = 0, 1, 2, 3, \cdots$$

$$P(X = 8) = 10^8 e^{-10} / 8!$$

$$= 0.113 \qquad \text{(Use Calculator)}$$

b) For X a Poisson distributed Random Variable with mean $\lambda = 10$:

$$P(9 \le X \le 11) = P(X = 9) + P(X = 10) + P(X = 11)$$

$$= 10^9 e^{-10}/9! + 10^{10} e^{-10}/10! + 10^{11} e^{-10}/11!$$

$$= 0.125 + 0.125 + 0.114 \qquad \text{(Use Calculator)}$$

$$= 0.364$$

Problem #5 (Exponential Distribution)

a) For an Exponentially distributed Random Variable with its mean $1/\mu = 2$ (so that $\mu = 1/2$):

$$P(X \le x) = 1 - e^{-\mu x} \qquad\qquad x \ge 0$$

$$= 1 - e^{-x/2}$$

$$P(X \le 90 \ min) = P(X \le 1.5 \ hrs)$$

$$= 1 - e^{-1.5/2}$$

$$= 1 - e^{-0.75}$$

$$= 1 - 0.4724 \qquad\qquad (\text{Table 4})$$

$$= 0.5276$$

b) $P(30 \ min < X < 90 \ min) = P(X < 90 \ min) - P(X < 30 \ min)$

$$= P(X < 1.5 \ hr) - P(X < 0.5 \ hr)$$

$$= (1 - e^{-1.5/2}) - (1 - e^{-0.5/2})$$

$$= e^{-0.25} - e^{-0.75}$$

$$= 0.7788 - .4724 \qquad\qquad (\text{Table 5})$$

$$= 0.3064$$

Problem #6 (Statistics)

a) $\bar{X} = Mean = \Sigma x_i / N$ for all values x_i in the outcome stream for X

$$= (2 + 6 + 3 + 7 + 4 + 2 + 3 + 5) / 8$$

$$= 4$$

b) $s^2 = Variance = \Sigma(x_i - \bar{X})^2 / (N - 1)$

$$= [(2 - 4)^2 + (6 - 4)^2 + (3 - 4)^2 + \ldots + (5 - 4)^2] / 7$$

$$= 24 / 7$$

$$= 3.428$$

$s = Standard\ Deviation = (3.428)^{1/2} = 1.85$

Problem #7 (Chi-Square Test)

a) One must first pool data to get at least 5 *expected* outcomes per cell. Pool cells 1 and 2 and pool cells 4 and 5.

A_i	12	2	6	6	9
E_i	7	8	6	8	6

Note: The number of cells $M = 5$ and $k = 0$ since the number of parameters estimated to determine the expected outcomes in each cell is zero. The χ^2 statistic is given by:

$$\chi^2 = \Sigma\,(A_i - E_i)^2\,/\,E_i \qquad\qquad \text{for } i = 1 \text{ to } 5$$

$$= (12 - 7)^2\,/\,7 + (2 - 8)^2\,/\,8 + \cdots + (9 - 6)^2\,/\,6$$

$$= {}^{25}\!/_7 + {}^{36}\!/_8 + \cdots + {}^{9}\!/_6$$

$$= 10.07$$

b) χ^2 has $M - 1 = 5 - 1 = 4$ degrees of freedom (Case #1). From Table 3,

- $\chi^2_{4,\,0.10} = 7.78$

- $\chi^2_{4,\,0.05} = 9.49$

- $\chi^2_{4,\,0.025} = 11.14$

Since the χ^2 statistic of 10.07 is greater than $\chi^2_{4,\,0.05} = 9.49$, one is less than 5% confident that these streams were generated by the Exponential distribution.

Problem #8 (F-Distribution)

a) From Table 2a, $v_1 = 1,\, v_2 = 10$:

$$F_{1,\,10,\,.99} = 10.04$$

From Table 2a, $v_1 = 1,\, v_2 = 20$:

$F_{1, 20, .99} = 8.10$

b) From Table 2b, $v_1 = 1$, $v_2 = 10$:

$F_{1, 10, .95} = 4.96$

From Table 2b, $v_1 = 1$, $v_2 = 20$:

$F_{1, 10, .95} = 4.35$

Problem #9 (Conditional Probability)

Define Event E = "Heads on at least one flip" and Event F = "Heads on both flips". Then from the definition of conditional probability:

$$P(F|E) = \frac{P(E \cap F)}{P(E)}$$

For two flips of a fair coin, the outcomes in the sample space are given by:

$O_1 = \{H, H\}$
$O_2 = \{H, T\}$
$O_3 = \{T, H\}$
$O_4 = \{T, T\}$

Each outcome is equally likely and has a probability of $\frac{1}{4}$. The Event $E = \{O_1, O_2, O_3\}$ and the $P(E) = \frac{3}{4}$. The Event $F = \{O_1\}$ and the Event $E \cap F = \{O_1\}$. Hence, the $P(E \cap F) = \frac{1}{4}$.

Then the conditional probability of "Heads on both flips" given "Heads on at least one flip" is given by:

$$P(F|E) = \frac{P(E \cap F)}{P(E)}$$

$$= \frac{\frac{1}{4}}{\frac{3}{4}}$$

$$= \frac{1}{3}$$

Note: Another way to look at this problem of conditional probability is to use the property that probability is proportion in the long run. Suppose this process of flipping two coins is performed 100 times (see Figure below). On the first flip, one should expect 50 Heads (H_1) and 50 Tails (T_1). For the 50 Heads on the first flip (50 H_1), one should expect 25 Heads on the second flip (25 $H_1 \cap H_2$) and 25 Tails on the second flip (25 $H_1 \cap T_2$). For the 50 Tails on the first flip (50 T_1), one should expect 25 Heads on the second flip (25 $T_1 \cap H_2$) and 25 Tails on the second flip (25 $T_1 \cap T_2$). There are 75 trials with Heads on at least one flip (25 $H_1 \cap H_2$, 25 $H_1 \cap T_2$, 25 $T_1 \cap H_2$). The proportion that have Heads on the both flips ($H_1 \cap H_2$) is 25/75 or 0.333.

Conditional Probability Using Probability as a Proportion

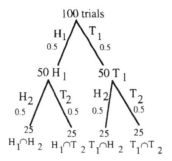

Problem #10 (Bayes' Theorem)

The initial (*a priori*) probabilities of a car being manufactured on a Monday (M), Tuesday (T), Wednesday (W), Thursday (Th), or Friday (F) are as follows:

$$P(M) = P(T) = P(W) = P(Th) = P(F) = 0.2$$

The effect that the day of the week the car is manufactured has on the event that the car is a lemon (L) is given in the problem statement:

$$P(L|M) = 0.04$$
$$P(L|T) = P(L|W) = P(L|Th) = 0.01$$
$$P(L|F) = 0.02$$

The event that the car is a lemon allows the initial probabilities P(M), P(T), P(W), P(Th), P(F) (where a car was equally likely to be manufactured on any day of the work week) to be updated.

Given that a given car turns out to be a lemon. it is now possible to compute the updated (*a posteriori*) probabilities P(M|L), P(T|L), P(W|L). P(Th|L), and P(F|L) that this car was manufactured on a certain day of the week. The probability that this car was manufactured on a Monday is given by:

$$P(M|L) = \frac{P(M \cap L)}{P(L)}$$

$$= \frac{P(L|M)P(M)}{P(L|M)P(M) + P(L|T)P(T) + P(L|W)P(W) + P(L|Th)P(Th) + P(L|F)P(F)}$$

$$= \frac{(0.04)(0.2)}{(0.04)(0.2)+(0.01)(0.2)+(0.01)(0.2)+(0.01)(0.2)+(0.02)(0.2)}$$

$$= 0.444$$

Without any other information, the probability that the car was built on Monday, P(M), equals 0.2. The information that this car turned out to be a lemon (L) allows the initial probability P(M) to be updated to P(M|L) equal to 0.444.

Note: Another way to look at this problem of conditional probability is to use the property that probability is proportion in the long run. Examine the production of 1000 cars (see Figure below). It is expected that 200 of the cars will be made on each of the days of the week (M, T, W, Th, and F). Of the 200 cars made on Monday (M), it is expected that 8 of the cars will be a lemon (M∩L) and 192 of the cars will be Good (M∩G). Of the 200 cars made on Tuesday (T), Wednesday (W), and Thursday (Th), it is expected that 2 of the cars will be a Lemon (L) (M∩L, T∩L, W∩L, Th∩L) and 198 of the cars are Good (G) (M∩G, T∩G, W∩G, Th∩G). Of the 200 cars made on Friday (F), 4 of the cars will be a Lemon (L) (198 F∩L) and 196 of the cars will be good (G) (196 F∩L). Then the probability of the car being made on Monday given the car is a lemon, P(M|L), is the P(M∩L) divided by the P(L). Of the 1000 cars built, there are 8 + 2+ 2 + 2 + 4 = 18 Lemons expected. Of the 18 Lemons, 8 were built on Monday. The probability of the car being built on Monday given the car is a Lemon is 8/18 = 0.4444.

Conditional Probability Using Probability as a Proportion

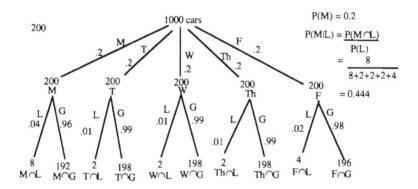

Solutions to Exercise 2

Decision Analysis

Problem #1 (Decision Matrix)

A decision matrix can be used since this is a single decision (how many passengers to overbook) and each action has the same set of outcomes. The decision matrix with actions, outcomes, costs, and probabilities is as follows:

Outcome Action	0 No Shows (0.50)	1 No Show (0.25)	2 No Shows (0.15)	3 No Shows (0.10)
Overbook 0	0	$600	$1200	$1800
Overbook 1	$1000	0	$600	$1200
Overbook 2	$2000	$1000	0	$600
Overbook 3	$3000	$2000	$1000	0

a) The expected cost for each action is computed as follows:

$$E(Overbook\ 0) = (0.5) \cdot (\$0) + (0.25) \cdot (\$600) + (0.15) \cdot (\$1200)$$
$$+ (0.10) \cdot (\$1800) = \$510$$

$$E(Overbook\ 1) = (0.5) \cdot (\$1000) + (0.25) \cdot (\$0) + (0.15) \cdot (\$600)$$
$$+ (0.10) \cdot (-\$1200) = \$710$$

$$E(Overbook\ 2) = (0.5) \cdot (\$2000) + (0.25) \cdot (\$1000) + (0.15) \cdot (\$0)$$
$$+ (0.10) \cdot (-\$600) = \$1310$$

$$E(Overbook\ 3) = (0.5) \cdot (\$3000) + (0.25) \cdot (\$2000) + (0.15) \cdot (\$1000)$$
$$+ (0.10) \cdot (\$0) = \$2150$$

The Expected Value (minimum expected cost) Decision is the action Overbook 0 passengers at $510.

b) The Minimax Decision calculations are as follows:

Maximum (worst case) cost (Overbook 0) $\Rightarrow$ $1800
Maximum (worst case) cost (Overbook 1) $\Rightarrow$ $1200
Maximum (worst case) cost (Overbook 2) $\Rightarrow$ $2000
Maximum (worst case) cost (Overbook 3) $\Rightarrow$ $3000

The Minimax Decision which minimizes the maximum (worst case cost) is the action Overbook 1. The value of the Minimax Decision is $1200.

c) The decision tree for this problem is as follows:

491

Decision Tree For Airline "No Show" Decision

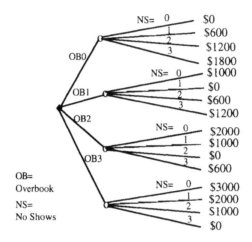

Problem #2 (Decision Tree)

For the decision tree in Problem #2, each of the outcome and action nodes are given a letter code and their expected payoffs are computed below:

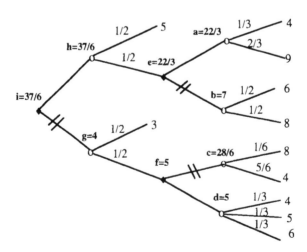

The values for the action nodes and outcome nodes are as follows:

The expected value of outcome node a = (4) · (1 / 3) +

$$(9) · (2 / 3) = 22 / 3$$

The expected value of outcome node b = (6) · (1 / 2) +

$$(8) · (1 / 2) = 7$$

The expected value of outcome node c = (8) · (1 / 6) +

$$(4) · (5 / 6) = 28 / 6$$

The expected value of outcome node d = (4) · (1 / 3) +

$$(5) · (1 / 3) + (6) · (1 / 3) = 5$$

The value of the action node e is Max (22 / 3, 7) = 22 / 3
Prune branch of value 7

The value of the action node f is Max (28 / 6, 5) = 5
Prune branch of value 28 / 6

The expected value of outcome node g = (5) · (1 / 2) +

$$(3) · (1 / 2) = 4$$

The expected value of outcome node h = (5) · (1 / 2) +

$$(22 / 3) · (1 / 2) = 37 / 6$$

The expected value of the initial action node i is Max (4, 37 / 6) = 37 / 6
Prune branch of value 4

The Expected Value Decision (maximum expected payoff) has a value of 37/6. The best actions are those which are not pruned.

If the scenario represented by this decision tree occurred many times and the actions which are not pruned are taken each time, the average payoff would be 37/6. Remember that on any single scenario, the Expected Value Decision could result in a payoff less or greater than 37/6.

Problem #3 (Decision Tree)

4a) The decision tree with actions, outcomes, probabilities and costs is as follows:

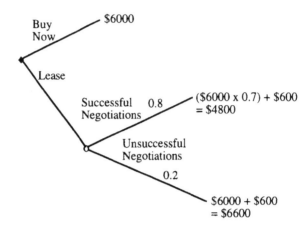

4b) The values for the action nodes and outcome nodes are as follows:

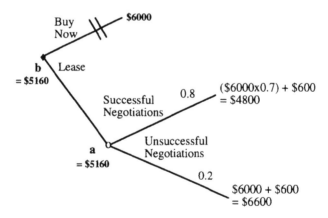

The expected cost of outcome node **a** = ($4800) · (0.8) +
($6600) · (0.2) = $5160

The cost of action node **b** = Min ($6000, $5160) = $5160
Prune action Buy Now

The Expected Value (minimum expected cost) Decision is the action Lease for two months and then buy the new computer (at the lower cost if the negotiations are successful). The expected cost of this action is $5160.

4c) The action Buy Now has an expected cost of $6000 - $5160 = $840 more than the action Lease. If the cost of Lease for two months increased by more than $840 to become more than $1440, the Expected Value Decision would then become the action Buy Now.

Problem #4 (Sequential Decision Tree)

3a) The decision tree for the game (complete with actions, outcomes, probabilities, and payoffs) is as follows:

FC = Face Card
NFC = Not Face Card

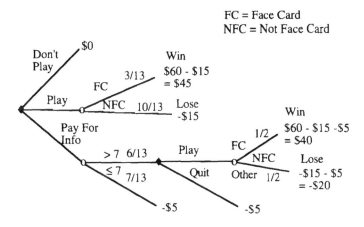

The probabilities are computed as follows:

   $P(Face\ Card) = 12/52 = 3/13$
   $P(Not\ Face\ Card) = 1 - 3/13 = 10/13$

   $P(> 7) = 24/52 = 6/13$
   $P(\leq 7) = 1 - 6/13 = 7/13$

   $P(Face\ Card \mid > 7) = P(Face\ Card\ and > 7) / P(> 7) = (12/52)/(24/52) = 1/2$
   $P(Not\ Face\ Card \mid > 7) = 1 - 1/2 = 1/2$

The conditional probabilities may be computed using the definition of probability as proportion in the long run.

Conditional Probability Using Probability as a Proportion

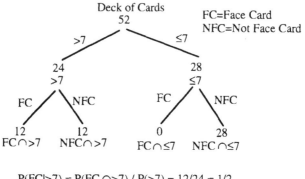

$P(FC|>7) = P(FC \cap >7) / P(>7) = 12/24 = 1/2$
$P(NFC|>7) = 1/2$

The solution to the decision tree is as follows:

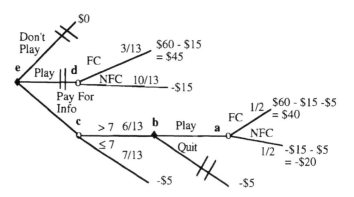

The values for the outcome and action nodes are as follows:

The expected value of outcome node a $= (\$40) \cdot (1/2) +$
$(- \$20) \cdot (1/2) = \$10$

The value of action node b $= Max (\$10, - \$5) = \$10$
Prune action of value $- \$5$

The expected value of outcome node c = ($10) · (6 / 13) +
$$(- \$5) \cdot (7 / 13) = \$25 / 13$$

The expected value of outcome node d = ($45) · (3 / 13) +
$$(- \$15) \cdot (10 / 13) = - \$15 / 13$$

The value of action node e = *Max*(25 / 13, – 15 / 13) = $25 / 13

Prune action of value – $15/13

The Expected Value Decision is Pay For Info at the start of the game. If the information is that the card chosen is >7, continue to Play; otherwise, Quit. If this game were played many times and this sequential decision were used in each game, the average (mean value) winnings would be 25/13 dollars (= $1.92) per game. For any individual game, the return to the card player would be $45, -$15, $40, -$20, or -$5 depending on the outcome of the game.

Solutions to Exercise 3

Break-Even Analysis

Problem #1 (Make vs. Buy Decision With Stepwise Fixed Costs)

1a) The fixed cost (FC), variable cost (VC), and total cost (TC) for Lake Woebegon General Hospital (LWGH) for the number of blood chemistry tests (output) performed up to the capacity of 50,000 tests/month are as follows:

• Option 1: Contract Out (Buy):

FC_{Buy} = $0
VC_{Buy} = $1.80 · #tests / month
TC_{Buy} = $1.80 · #tests / month

• Option 2: Establish Lab at LWGH (Make):

- # tests = 0 to 15,000 (1 Lab Technician)

FC_{Make} = $3000 + $2500·1 = $5500 / month
VC_{Make} = $1.50 · #tests / month
TC_{Make} = $5500 / month + $1.50 · #tests / month

- # tests = 15,000 to 30,000 (2 Lab Technicians)

FC_{Make} = $3000 + $2500·2 = $8000 / month
VC_{Make} = $1.50 · #tests / month
TC_{Make} = $8000 / month + $1.50 · #tests / month

- # tests = 30,000 to 40,000 (3 Lab Technicians)

FC_{Make} = $3000 + $2500·3 = $10,500 / month
VC_{Make} = $1.50 · #tests / month
TC_{Make} = $10,500 / month + $1.50 · #tests / month

- # tests = 40,000 to 50,000 (4 Lab Technicians)

FC_{Make} = $3000 + $2500·4 = $13,000 / month
VC_{Make} = $1.50 · #tests / month
TC_{Make} = $13,000 / month + $1.50 · #tests / month

For the Contract Out (Buy) Option, the total cost over the range of 0 to 50,000 tests/month can be plotted using the equation:

TC_{Buy} = $1.80 · #tests / month

498

For the Establish Lab at LWGH (Make) Option, the total cost over the range of 0 to 50,000 tests/month must be plotted in four line segments:

TC_{Make} = \$5500 / month + \$1.50 · #tests / month   (0 to 15K tests / month)

TC_{Make} = \$8000 / month + \$1.50 · #tests / month   (15K to 30K tests / month)

TC_{Make} = \$10,500 / month + \$1.50 · #tests / month   (30K to 40K tests / month)

TC_{Make} = \$13,000 / month + \$1.50 · #tests / month   (40K to 50K tests / month)

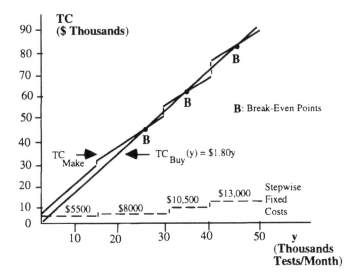

1b) The determination of the Break-Even Points is as follows:

Visually inspecting the total cost (TC) vs. tests/month (y) plot above, there seems to be no Break-Even Point in the range (step) 0 to 15,000 tests/month. Equating TC_{Buy} and TC_{Make} for this step will verify that there is no Break-Even Point in this step (range) for fixed cost:

$TC_{Buy}(y) = TC_{Make}(y)$

0 + \$1.80 y = \$5500 + \$1.50 y

\$5500 = \$0.30 y

y ≈ 18,333 tests / month

18,333 tests / month is out of the range 0 to 15,000

No Break – Even Point exists in this step

499

Visually inspecting the total cost (TC) vs. tests/month (y) plot above, there seems to be a Break-Even Point in the range (step) 15,000 to 30,000 tests/month. Equating TC_{Buy} and TC_{Make} for this step will verify that a Break-Even Point exists in this step:

$TC_{Buy}(y) = TC_{Make}(y)$

$0 + \$1.80\,y = \$8000 + \$1.50\,y$

$\$8000 = \$0.30\,y$

$y \approx 26{,}666\ tests\,/\,month$

$26{,}666\ tests\,/\,month$ is in the range 15,000 to 30,000

$26{,}666\ tests\,/\,month$ is a Break – Even Point

Visually inspecting the total cost (TC) vs. tests/month (y) plot above, there seems to be a Break-Even Point in the range (step) 30,000 to 40,000 tests/month. Equating TC_{Buy} and TC_{Make} for this step will verify that a Break-Even Point exists in this step:

$TC_{Buy}(y) = TC_{Make}(y)$

$0 + \$1.80\,y = \$10{,}5000 + \$1.50\,y$

$\$10{,}500 = \$0.30\,y$

$y \approx 35{,}000\ tests\,/\,month$

$35{,}000\ tests\,/\,month$ is in the range 30,000 to 40,000

$35{,}000\ tests\,/\,month$ is a Break – Even Point

Visually inspecting the total cost (TC) vs. tests/month (y) plot above, there seems to be a Break-Even Point in the range (step) 40,000 to 50,000 tests/month. Equating TC_{Buy} and TC_{Make} for this step will verify that a Break-Even Point exists in this step:

$TC_{Buy}(y) = TC_{Make}(y)$

$0 + \$1.80\,y = \$13{,}0000 + \$1.50\,y$

$\$13{,}000 = \$0.30\,y$

$y \approx 43{,}333\ tests\,/\,month$

$43{,}333\ tests\,/\,month$ is in the range 40,000 to 50,000

$43{,}333\ tests\,/\,month$ is a Break – Even Point

1c) The Make vs. Buy Decision Strategy for LWGH is as follows:

Below 26,666 tests/month, the Contract Out (Buy) Decision is less expensive for LWGH.

Above 26,666 tests/month, the total costs are very close as evidenced by the fact that the total cost plot displays more than one Break-Even Point. The decision for LWGH whether to Buy vs. Make will probably be made on other considerations besides total cost (e.g. acceptability of the 24 hr test results turnaround time for the Buy option).

1d) The fixed cost of space and leasing the analyzer is increased from $3000/month to $4500/month. The total cost for the Buy option (no significant fixed cost) is unchanged.

However the total cost associated with the Make option is increased by $1500 in each of the four line segments.

For the Establish Lab at LWGH (Make) Option, the total cost over the range of 0 to 50,000 tests/month is given by:

$$TC_{Make} = \$7000\,/\,month + \$1.50 \cdot \#tests\,/\,month \quad (0\ to\ 15K\ tests\,/\,month)$$
$$TC_{Make} = \$9500\,/\,month + \$1.50 \cdot \#tests\,/\,month \quad (15K\ to\ 30K\ tests\,/\,month)$$
$$TC_{Make} = \$12,000\,/\,month + \$1.50 \cdot \#tests\,/\,month \quad (30K\ to\ 40K\ tests\,/\,month)$$
$$TC_{Make} = \$14,500\,/\,month + \$1.50 \cdot \#tests\,/\,month \quad (40K\ to\ 50K\ tests\,/\,month)$$

For the Contract Out (Buy) Option, the total cost over the range of 0 to 50,000 tests/month remains:

$$TC_{Buy} = \$1.80 \cdot \#tests\,/\,month$$

The plot of the total cost (TC) vs. tests/months (y) when the fixed costs are increased is given below.

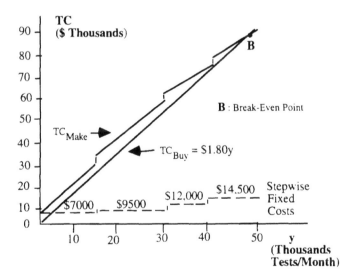

Visually inspecting the total cost (TC) vs. # tests/month (y) plot, the Break-Even Point is between 40,000 and 50,000 test/month. Equating TC_{Buy} and TC_{Make} for this range gives:

$$TC_{Buy}(y) = TC_{Make}(y)$$
$$0 + \$1.80\,y = \$14,5000 + \$1.50\,y$$
$$\$14,500 = \$0.30\,y$$
$$y \approx 48,333\ tests\,/\,month$$

501

48,333 *tests / month is in the range* 40,000 *to* 50,000

48,333 *tests / month is a Break – Even Point*

It can be verified that no Break-Even points exist in the steps (ranges) 0 to 15,000 tests/month, 15,000 to 30,000 tests/month, and 30,000 to 40,000 tests/month. For example, equating TC_{Buy} and TC_{Make} with for the range 0 to 15,000 tests/month gives:

$TC_{Buy}(y) = TC_{Make}(y)$

$0 + \$1.80\, y = \$7,000 + \$1.50\, y$

$\$7000 = \$0.30\, y$

$y \approx 23,333$ *tests / month*

23,333 *tests / month is not in the range* 0 *to* 15,000

No Break – Even Point exist in this step

Problem #2 (Compound Interest)

At the end of 10 years, the savings account will have accrued:

$\$1000 \cdot (1 + .06)^{10}$

$= \$1000 \cdot (\mathbf{1.7908})$ (*See Table* 7)

$= \$1790.80$

Problem #3 (Present Value)

The Present Value (*PV*) of a revenue of \$1 for the next 5 years at a current interest rate of 10% is \$3.791 (See Table 9). The Present Value of \$1000 revenue for each of next five years is then:

$PV = \$1000 \cdot (\mathbf{3.791})$ (*See Table* 9)

$= \$3791$

Investors interested in maximum return over the 5 year period would prefer the series of five yearly \$1000 revenues to \$3500 now. Other investors may wish access to the \$3500 now rather than wait the five years for the maximum return.

Problem #4 (Net Present Value)

The calculations for the Net Present Value (NPV_X) for Lathe X is as follows:

Year N	Revenue Flow	$[1/(1+.10)^N]$ (Table 8)	Discounted Revenue Flow
1	$2000	0.909	$1818
2	$2500	0.826	$2066
3	$2500	0.751	$1878
4	$2500	0.683	$1708
5	$2000	0.621	$1242
			Sum = $8712

$NPV_X = \$8712 - \$8000 = \$712$

The calculations for the Net Present Value (NPV_Y) for Lathe Y is as follows:

Year N	Revenue Flow	$[1/(1+.10)^N]$ (Table 8)	Discounted Revenue Flow
1	$2000	0.909	$1818
2	$3000	0.826	$2479
3	$3500	0.751	$2630
4	$3000	0.683	$2049
5	$2500	0.621	$1552
			Sum = $10528

$NPV_Y = \$10528 - \$10000 = \$528$

Both lathes have a positive NPV. Lathe X has a slightly higher NPV and should be preferred if only the financial aspect of the decision is considered.

Problem #5 (Internal Rate of Return)

The NPV for Lathe X is positive (equal to $712 for 10% yearly interest) so the Internal Rate of Return (IRR) must be greater than 10%.

If a 14% interest rate had been used, the calculation of the NPV would be as follows:

Year N	Revenue Flow	$[1/(1+.14)^N]$ (Table 8)	Discounted Revenue Flow
1	$2000	0.877	$1754
2	$2500	0.769	$1923
3	$2500	0.675	$1688
4	$2500	0.592	$1480
5	$2000	0.519	$1038
			Sum = $7884

$NPV_x = \$7884 - \$8000 = -\$116$

Hence, a 14% interest rate yields a negative Net Present Value. The two (Interest Rate, Net Present Value) points determined are:

(10%, + $712)
(14%, – $116)

Linear interpolation can be used to estimate the Internal Rate of Return (IRR) which provides a Net Present Value NPV of $0 as follows:

- The line segment connecting these two points will pass through NPV = 0 at the point (x, 0) where x is between 10% and 14%. The IRR is the ordinate of the point (x, 0) that lies on the line segment connecting (10%, +$712) and (14%, -$116).

- The NPV decreased by $712 - (-$116) = $828 as the interest rate increased by four percentage points. Thus for each $207 decrease in the NPV, the interest rate increased by one point.

- For the NPV to decrease by $712 to the value $0, the interest rate would have to increase by $712/$207 = 3.4 percentage points over 10%. The IRR for Lathe X is estimated to be 13.4%.

The Internal Rate of Return can also be determined graphically by plotting (10%, + $712) and (14%, – $116) as shown in the Figure below. An estimate for the Internal Rate of Return is found where the line segment connecting the two points crosses the x-axis (NPV = 0).

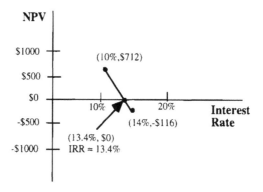

Solutions to Exercise 4

Inventory Decision Analysis

Problem #1 (Deterministic EOQ Formulation)

1a) Determine the EOQ for Ball Four novelty company:

k = $50/order
d = 13,000 mugs/year
h = $0.45/mug/year

$$EOQ = [2 d \cdot k / h]^{1/2}$$

$$= [2(50)(13000) / 0.45]^{1/2}$$

$$= 1699.76 \text{ mugs}$$

$$\approx 1700 \text{ mugs (round up!)}$$

1b) $TC(Q) = k \cdot d/Q + h \cdot Q/2$

$$= (50)(13000)/1700 + (0.45)(1700)/2$$

$$= \$382.35 + \$382.50$$

$$\approx \$765$$

Note: The Ordering Cost and Holding Cost are only approximately equal. They are not exactly equal since the value of EOQ was rounded up to the next higher integer value.

1c) Examine: $k_{-30\%}$ = $35/order (30% below $50)
$k_{+30\%}$ = $65/order (30% above $50)

• First, examine $k_{-30\%}$ = $35:

$$EOQ_{-30\%} = [2(35)(13000) / (0.45)]^{1/2}$$

$$= 1422.05$$

$$\approx 1423 \text{ mugs (round up!)}$$

For the case where the cost to order is actually $35, compare the *actual* Total Cost $TC(k_{-30\%}$ = $35, EOQ = 1700) using the non-optimal EOQ order size vs. the *optimal* Total Cost $TC(k_{-30\%}$ = $35, $EOQ_{-30\%}$ = 1423) using the optimal $EOQ_{-30\%}$ order size:

$$TC(k_{-30\%} = \$35, EOQ = 1700) = (\$35)(13000)/1700 + (\$0.45)(1700)/2$$

$$\approx \$268 + \$382$$

$$\approx \$650 \quad \text{(Actual Cost)}$$

Note: The Ordering Cost $268 and the Holding Cost $382 are not equal because EOQ = 1700 is not optimal when the Ordering Cost is $35 instead of $50.

$$TC(k_{-30\%} = \$35, EOQ_{-30\%} = 1423) = (\$35)(13000)/1423 + (\$0.45)(1423)/2$$

$$\approx \$320 + \$320$$

$$\approx \$640 \quad \text{(Optimal Cost for } k_{-30\%} = \$35)$$

Note: The Ordering Cost and Holding Cost are equal since $EOQ_{-30\%}$ is optimal when the Ordering Cost is $35 ($k$-30%).

% Error = $\dfrac{\text{Actual Cost - Optimal Cost}}{\text{Actual Cost}}$

$$= \frac{\$650 - \$640}{\$650}$$

$$\approx 1.5\%$$

• Next, examine $k_{+30\%} = \$65$:

$$EOQ_{+30\%} = [2(65)(13000) / (0.45)]^{1/2}$$

$$= 1937.92$$

$$\approx 1938 \text{ items (round up!)}$$

For the case where the Ordering Cost is actually $65, compare the *actual* Total Cost $TC(k_{+30\%} = \$65, EOQ = 1700)$ using the non-optimal EOQ value vs. the *optimal* Total Cost $TC(k_{+30\%} = \$65, EOQ_{+30\%} = 1938)$ using the optimal $EOQ_{+30\%}$ value:

$$TC(k_{+30\%} = \$65, EOQ = 1700) = (\$65)(13000)/1700 + (\$0.45)(1700)/2$$

$$\approx \$497 + \$382$$

$$\approx \$880 \text{ (Actual Cost)}$$

$$TC(k_{+30\%} = \$65, EOQ_{+30\%} = 1938) = (\$65)(13000)/1938 + (\$0.45)(1938)/2$$

$$\approx \$436 + \$436$$

$$\approx \$872 \text{ (Optimal Cost for } k_{+30\%} = \$65)$$

% Error = $\dfrac{\text{Actual Cost - Optimal Cost}}{\text{Actual Cost}}$

$$= \frac{\$880 - \$872}{\$880}$$

$$\approx 1\%$$

The sensitivity of total inventory cost to the Ordering Cost k is summarized as follows:

	k-30%	k	k+30%
Ordering Cost	$35	$50	$65
Optimal Order Size	1423	1700	1938
TC (Optimal Q)	$640	$765	$872
TC (EOQ = 1700)	$650	$765	$880
Per Cent Increase	1.5%	0%	1%

Note: As the Ordering Cost rises from $35 to $65, the size of the optimal order increases from 1423 mugs/order to 1938 mugs/order. As expected, larger (and fewer) orders are used to meet annual demand when the Ordering Cost rises.

1d) Inventory does not arrive simultaneously with placement of an order but takes 5 days to arrive. Demand per day is given by:

$$d = (13000 \text{ items/year}) / (365 \text{ days/year}) = 35.62 \text{ mugs/day}.$$

The number of mugs used in the 5 days Lead Time (LT) is the Reorder Point (ROP) quantity. The ROP is given by:

$$ROP = LT \cdot d$$

$$= (5 \text{ days}) \cdot (35.62 \text{ mugs/day})$$

$$\approx 178 \text{ mugs}.$$

Mugs are ordered when inventory level drops from the EOQ of 1700 mugs to 178 mugs. At the end of the 5 days Lead Time, the current inventory on hand will have dropped to zero just the new order of 1700 mugs arrive.

Problem #2 (Stochastic EOQ Formulation)

2a) The stochastic (variable) demand rate has an expected value of d. The calculation for the stochastic EOQ formulation is the same as for the deterministic EOQ formulation except that the *expected* demand rate d is substituted for the *constant* demand rate d.

$$d = 12,300 \text{ cases/year (expected value)}$$
$$k = \$200/\text{order}$$
$$h = (0.05)(\$19) = \$0.95/\text{case/year}$$

$$EOQ = [2d \cdot k / h]^{1/2}$$

$$= [2(12300)(200) / 0.95]^{1/2}$$

507

≈ 2276 cases of baby formula

2b) The Total Annual Cost TC(Q) is given by:

$$TC(EOQ) = k \cdot d/EOQ + h \cdot EOQ/2$$

$$TC(EOQ = 2276) = (\$200(12,300)/2276 + (\$0.95)(2276/2)$$

$$\approx \$1081 + \$1081$$

$$\approx \$2162$$

Note: The Safety Stock will be an additional one-time inventory cost (in inventory cycle 1) at $19 per case. Safety Stock once bought must be held over all the inventory cycles at a cost of $0.95 per case per year.

2c) The Deterministic Reorder Point is given by:

$$ROP_D = LT \cdot \text{Demand Rate d}$$

$$= 20 \text{ days} \cdot \frac{(12300 \text{ cases/year})}{(365 \text{ days/year})}$$

$$\approx 674 \text{ cases}$$

2d) The standard deviation of demand given in the description for Problem #2 for 1 day (OT) is 11 cases ($\sigma_{OT} = \sigma_1$). The standard deviation (σ_{LT}) of the desired Lead Time (LT) of 20 days is computed as follows:

$$\sigma_{LT} = (LT/OT)^{1/2} \cdot \sigma_{OT}$$

$$\sigma_{20} = (20/1)^{1/2} \cdot \sigma_1$$

$$\sigma_{20} = (20/1)^{1/2} \cdot 11$$

$$\approx 49.19 \text{ cases}$$

A 99% service level corresponds to a K multiplier of 2.33 (use the 99th percentile of the Normal N(0,1) distribution from Table 1b) for σ_{20}.

The Safety Stock required to meet Lead Time demand in 99% of the inventory cycles is given by:

$$\text{Safety Stock} = K \cdot \sigma_{20} = 2.33 \cdot \sigma_{20}$$

$$= (2.33) \cdot (49.19 \text{ cases})$$

$$\approx 115 \text{ cases}$$

Note: The one time inventory cost (in cycle 1) for the Safety Stock of 115 cases is $(115) \cdot (\$19) = \$2185$. The cost to hold the Safety Stock of 115 cases is $(115) \cdot (\$0.95) = \$109.25$ per year.

2d) The Stochastic Reorder Point is given by:

Stochastic Reorder Point $ROP_S = ROP_D + $ Safety Stock (SS)

$$= 674 \text{ cases } + 115 \text{ cases}$$

$$= 789 \text{ cases}$$

Problem #3 (Stochastic POQ Formulation)

3a) The stochastic (variable) demand rate has an expected value of d. The calculation for the stochastic EOQ formulation is the same as for the deterministic EOQ formulation except that the *expected* demand rate d is substituted for the *constant* demand rate d

$d = 36,800$ squirt guns/year
$r = 120$ squirt guns/day x 365 days/year = 43,800 squirt guns/year
$k = \$240$/order
$h = (0.05)(\$1) = \$0.05$/squirt gun/year

$$POQ = [2k \cdot d/h]^{1/2} \cdot [r/(r-d)]^{1/2}$$

$$= \{ [2(240)(36800) / (0.05)] [43800 / (43800\text{-}36800)] \}^{1/2}$$

$$\approx 47,017 \text{ squirt guns}$$

3b) The Total Annual Cost TC(Q) is:

$$TC(POQ) = k \cdot (d/POQ) + h \cdot (POQ/2) \cdot [(r-d)/r]$$

$$TC(19639) = (\$240) \cdot (36,800 / 47017)$$
$$+ (\$0.05)(47,017 / 2) \cdot [(43,800 - 36,800) / 43,800\}$$

$$= \$187.85 + \$187.85$$

$$\approx \$375.71$$

Note: The Safety Stock will be an additional one-time inventory cost (in inventory cycle 1) at a cost of $1 per squirt gun. Safety Stock once bought must be held over all the inventory cycles at a cost of $0.05 per squirt gun per year.

3c) The deterministic Lead time Demand is given by:

$$ROP_D = LT \cdot \text{Demand Rate d}$$

$$= 2 \text{ days} \cdot \frac{(36800 \text{ squirt guns/yr})}{(365 \text{ days/yr})}$$

$$\approx 202 \text{ squirt guns}$$

3d) The standard deviation of demand is given for 1 day (OT) as 100 squirt guns (σ_{OT}). The standard deviation $(\sigma_{OT} = \sigma_1)$ of the desired Lead Time (LT) of 2 days is computed as follows:

$$\sigma_{LT} = (LT/OT)^{1/2} \cdot \sigma_{OT}$$

$$\sigma_2 = (2/1)^{1/2} \cdot \sigma_1$$

$$\sigma_2 = (2/1)^{1/2} \times 100$$

$$\approx 141.42 \text{ squirt guns}$$

A 95% safety stock corresponds to a K multiplier of 1.645 (use the 95th percentile of the Normal distribution from Table 1b) for σ_2.

The Safety Stock required to meet Lead Time demand in 95% of the inventory cycles is given by:

$$\text{Safety Stock} \quad = \text{ K} \cdot \sigma_2 = 1.645 \cdot \sigma_2$$

$$= (1.645) \cdot (141.42 \text{ items})$$

$$\approx 233 \text{ squirt guns}$$

Note: The one time inventory cost (in cycle 1) for the Safety Stock of 233 squirt guns is $(233) \cdot (\$1) = \$233$. The cost to hold the Safety Stock of 233 squirt guns is $(233) \cdot (\$0.05) = \$11.65$ per year.

3e) The Stochastic Reorder Point is given by:

$$\text{Stochastic Reorder Point ROP}_S = \text{ROP}_D + \text{Safety Stock (SS)}$$

$$= 202 \text{ squirt guns} + 233 \text{ squirt guns}$$

$$= 435 \text{ squirt guns}$$

Solutions to Exercise 5

Project Management (PERT/CPM)

Problem 1 (Deterministic PERT)

a) The PERT network (precedences and completion times) for the house is given below. Note that a Start Task 0 (with zero completion time and no predecessors) has been added to the network. The Start Task 0 is connected to all tasks (Task 1 and Task 2) without predecessors.

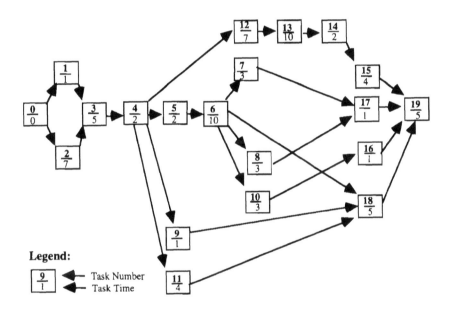

Legend:

$\boxed{\frac{9}{1}}$ ◀── Task Number
◀── Task Time

b) The earliest project completion time is determined by working forward through the network to find the individual task ECT's. The ECT for the final node (Task 19) is the earliest completion time (ECT) for the building the house.

Sample task ECT calculations for Tasks 3, 18 and 19 are as follows:

The ECT for Task 3 is 12 weeks. The derivation of the ECT is as follows:
    Task 3 has Tasks 1 and 2 as predecessors
    The ECT for Task 1 is 1 week
    The ECT for Task 2 is 7 weeks
    The Max (latest) ECT for any predecessor task is 7 weeks
    Task 3 itself takes 5 weeks to complete
    The ECT for Task 3 is 12 weeks

The ECT for Task 18 is 31 weeks. The derivation of the ECT is as follows:

511

Task 18 has Tasks 6, 9, and 11 as predecessors
The ECT for Task 6 is 26 weeks
The ECT for Task 9 is 15 weeks
The ECT for Task 11 is 18 weeks
The Max (latest) ECT for any predecessor is 26 weeks
Task 18 itself takes 5 weeks to complete
The ECT for Task 18 is 31 weeks

The ECT for Task 19 is 42 weeks. The derivation of the ECT is as follows:
Task 19 has Tasks 15, 17, 16, and 18 as predecessors
The ECT for Task 15 is 37 weeks
The ECT for Task 17 is 30 weeks
The ECT for Task 16 is 30 weeks
The ECT for Task 18 is 31 weeks
The Max(latest) ECT is 37 weeks
Task 19 itself takes 5 weeks to complete
The ECT for Task 19 is 42 weeks

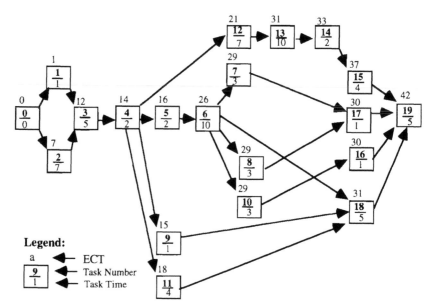

c) The critical path is determined by working backward through the network (final node to start node) to find the LCT for each task.

Sample task LCT calculations for Task 6 and Task 4 are as follows:

The LCT for Task 6 is 32 weeks. The derivation of the LCT is as follows:
Tasks 7, 8, 10, and 18 are the successor tasks for Task 6
The LCT for Task 7 is 36 weeks
Task 7 itself takes 3 weeks to complete
Task 6 must be complete by 33 weeks not to delay Task 7

The LCT for Task 8 is 36 weeks
Task 8 itself takes 3 weeks to complete
Task 6 must be complete by 33 weeks not to delay Task 8
The LCT for Task 10 is 36 weeks
Task 10 itself takes 3 weeks to complete
Task 6 must be complete by 33 weeks not to delay Task 10
The LCT for Task 18 is 37 weeks
Task 18 itself takes 5 weeks to complete
Task 6 must be complete by 32 weeks not to delay Task 7
The LCT for Task 6 is Min (36-3, 36-3, 36-3, 37-5) = 32 weeks
Task 6 must be complete by 32 weeks not to delay any successor Tasks

The LCT for Task 4 is 14 weeks. The derivation of the LCT is as follows:
Tasks 12, 5, 9, and 11 are successor Tasks to Task 4
The LCT for Task 12 is 21 weeks
Task 12 itself takes 7 weeks to complete
Task 4 must be complete by 14 weeks not to delay Task 12
The LCT for Task 5 is 22 weeks
Task 5 itself takes 2 weeks to complete
Task 4 must be complete by 20 weeks not to delay Task 5
The LCT for Task 9 is 32 weeks
Task 9 itself takes 1 weeks to complete
Task 4 must be complete by 31 weeks not to delay Task 9
The LCT for Task 11 is 32 weeks
Task 11 itself takes 4 weeks to complete
Task 4 must be complete by 28 weeks not to delay Task 11
Task 6 must be complete by time 32 weeks not to delay any subsequent Tasks
The LCT for Task 4 is Min (21-7, 22-2, 32-1, 32-4) = 14 weeks

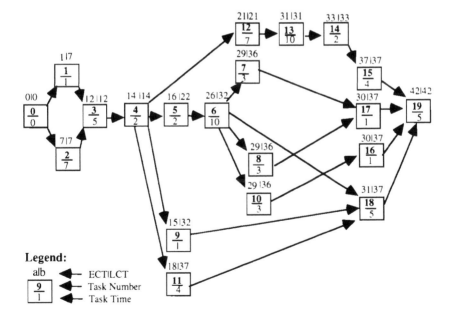

Legend:

a|b ⬅ ECTILCT

9/1 ⬅ Task Number

⬅ Task Time

513

The critical path for the Habitat house project consists of those tasks where Latest Completion Time equals the Earliest Completion Time (LCT = ECT):

Task 0→Task 2→Task 3→Task 4→Task 12→Task 13→Task 14→Task 15→Task 19

d) One would invest money in the task among the critical path tasks that is the least expensive to speed up one day. One would then recompute the critical path (there could be a new longest path through the network). One would continue to invest in the task on the current critical path which is the least expensive to speed up by one day until the overall project earliest completion time has been speeded up by three days.

Problem 2 (PERT With Stochastic Time Estimates)

a) The precedence relationships for the air conditioning project tasks are displayed in the network below. Note that a start node (Start) with zero completion time and no predecessors has been added to the network. The Start Task 0 is connected to all tasks (Task A, Task B, and Task D) without predecessors.

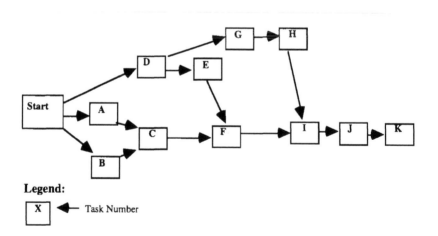

Legend:

[X] ← Task Number

The expected value of the task completion times $[= (o + 4m + p)/6]$ and the standard deviation of the task completion times $[= (p - 0)/6]$ are given as follows:

Solutions to Exercise 5 Project Management (PERT/CPM)

Task	Expected Value	Variance
Start	0.0000	0.0000
A	1.0833	0.0167
B	2.0000	0.1111
C	2.0000	0.1111
D	0.5000	0.0000
E	6.5000	1.3611
F	2.0000	0.1111
G	3.0000	0.1111
H	10.6667	1.7777
I	3.3333	0.4444
J	3.0000	0.1111
K	2.1667	0.2500

The PERT network with expected task completion times and precedences is given as follows:

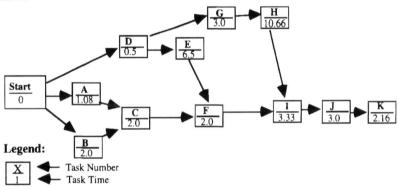

b) The PERT network with ECT|LCT's is given by:

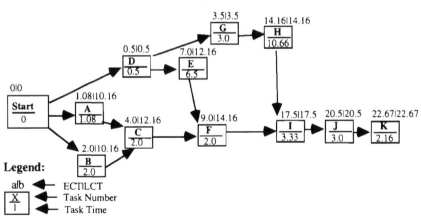

515

Sample ECT and LCT calculations are given as follows:

The Earliest Completion Time for Task C is 4.0 weeks. The derivation of the ECT is as follows:

Task C has Tasks A and B as predecessors
The ECT for Task A is 1.08 weeks
The ECT for Task B is 2.0 weeks
The Max (latest) ECT for any predecessor task is 2.0 weeks
Task C itself takes 2.0 weeks to complete
The ECT for Task C is 4.0 weeks

The Earliest Completion Time for Task F is 9.0 weeks. The derivation of the ECT is as follows:

Task F has Tasks C and E as predecessors
The ECT for Task C is 4.0 weeks
The ECT for Task E is 7.0 weeks
The Max (latest) ECT for any predecessor task is 7.0 weeks
Task F itself takes 2.0 weeks to complete
The ECT for Task F is 9.0 weeks

The Latest Completion Time for Task D is 0.5 weeks. The derivation of the LCT is as follows:

Tasks E and G are successor tasks for Task D
The LCT for Task E is 12.16 weeks
Task E itself takes 6.5 weeks
Task D must be completed by 5.66 weeks not to delay Task E
The LCT for Task G is 3.5 weeks
Task G itself takes 3.0 weeks
Task D must be completed by 0.5 weeks not to delay task G
The LCT for Task D is Min (5.66 weeks, 0.5 weeks) = 0.5 weeks
The LCT for Task D is 0.5 weeks

The Latest Completion Time for Task Start is 0.0 weeks. The derivation of the LCT is as follows:

Task A, B, and D are successor tasks to Task Start
The LCT for Task A is 10.16 weeks
Task A itself takes 1.08 weeks
Task Start must be completed by 10.16 - 1.08 = 9.08 weeks not to delay Task A
The LCT for Task B is 10.16 weeks
Task B itself takes 2.0 weeks
Task Start must be completed by 10.16 - 2.0 = 8.16 weeks not to delay Task B
The LCT for Task D is 0.5 weeks
Task D itself takes 0.5 week
Task Start must be complete by 0.5 - 0.5 = 0 weeks not to delay Task D
The LCT for Task Start is Min (9.08, 8.16, 0) = 0 weeks not to delay Task A
Task Start must be completed by 0.0 weeks not to delay the project

The critical path for the Fairbanks City Hall air conditioning installation project (where ECT = LCT) is given by:

Start→Task D→Task G→Task H→Task I→Task J→Task K (Final Node)

Solutions to Exercise 5 Project Management (PERT/CPM)

c) The expected project completion time $T_{exp,\,project}$ is given by:

$$T_{exp,\,project} \;=\; T_{exp}(Start) + T_{exp}(D) + T_{exp}(G) + T_{exp}(H) + T_{exp}(I) + T_{exp}(J) + T_{exp}(K)$$

$$= 0.0 + 0.5 + 3.0 + 10.6667 + 3.3333 + 3.0 + 2.1667$$

$$= 22.6667 \; weeks$$

The standard deviation of the project completion time is computed as follows:

$$\sigma^2_{project} = \sigma^2(Start) + \sigma^2(D) + \sigma^2(G) + \sigma^2(H) + \sigma^2(I) + \sigma^2(J) + \sigma^2(K)$$

$$= 0.0 + 0.0 + 0.1111 + 1.7777 + 0.4444 + 0.1111 + 0.25$$

$$= 2.70$$

Note: Do not attempt to determine the standard deviation of the project completion time by adding the standard deviations of the completion time for the tasks on the critical path. Statistically, it is valid only to add *variances* and not standard deviations. The standard deviation of the project completion time is the square root of the sum of the variances for the tasks on the critical path.

$$\sigma_{project} = (2.70)^{1/2}$$

$$= 1.6415$$

d) It is assumed that the overall project completion time $T_{project}$ is Normally distributed with expected value 22.667 and variance 2.70; that is, $T_{project} \sim N(22.67, 2.70)$.

The probability that the Fairbanks City Hall air conditioning project will be completed in 20 weeks or less is given below. Note that when the expected value $T_{exp,\,project}$ given by 22.67 weeks is subtracted from $T_{project}$ and the remainder is divided by the standard deviation $\sigma_{project}$ given by 1.6415 weeks, that quantity is distributed as a Normal Random Variable with expected value 0 and standard deviation 1. Table 1b for the Standard Normal [$N(0,1)$] distribution can be used to determine the probability the project will be complete in 20 weeks or less.

$$P(T_{project} \leq 20 \; weeks) \;= P[(T_{project} - 22.67)/1.6415 \leq (20 - 22.67)/1.6415]$$

$$= P(Z \leq -1.63) \quad Z = (T_{Project} - 22.67/1.6415 \sim N(0,1)$$

$$= 0.052 \qquad (Table\ 1b)$$

Solutions to Exercise 6

Simple Regression

Problem #1 (Method of Least Squares)

1a) The Least Squares best fit line is derived as follows:

Linear Regression Calculation Table

Time t (Year)	Demand D_t (# Patients/Day)	$D_t \cdot t$	t^2	D_t^2
1	20	20	1	400
2	22	44	4	484
3	25	75	9	625
4	29	116	16	841
5	30	150	25	900
Σt $=15$	ΣD_t $=126$	$\Sigma D_t \cdot t$ $=405$	Σt^2 $=55$	D_t^2 $=3250$

$$\bar{t} = \Sigma t / N = 15/5 = 3$$
$$\bar{D} = \Sigma D_t / N = 126/5 = 25.2$$

$$b = [N \Sigma D_t \cdot t - \Sigma t \Sigma D_t] / [N \Sigma t^2 - (\Sigma t^2)]$$

$$= [(5)(405) - (15)(126)] / [(5)(55) - (15)^2]$$

$$= 135 / 50$$

$$= 2.7$$

$$a = \bar{D} - b \cdot \bar{t}$$

$$= 25.2 - (2.7) \cdot (3)$$

$$= 17.1$$

Thus the Method of Least Squares line, the best line in terms of minimizing the sum of the squared deviations between the actual and estimated demand over the past 5 years, is given by:

$$D_t^* = 17.1 + 2.7 \cdot t$$

1b) The F distribution for the health clinic data has 1 and $N - 2 = 3$ degrees of freedom. The F-statistic is given by:

$$F = [e/1] / [u/3]$$

• Explained Deviation (e)

$$e = [\Sigma D_t \cdot t \ - \ (\Sigma t)(\Sigma D_t)/N]^2 / [\Sigma t^2 \ - \ (\Sigma t)^2/N]$$

$$= [405 \ - \ (15)(126)/5]^2 / [55 \ - \ (15)^2/5]$$

$$= 729 / 10$$

$$= 72.9$$

• Unexplained Deviation (u)

$$u = \Sigma D_t^2 \ - \ (\Sigma D_t)^2/N \ - \ e$$

$$= 3250 \ - \ (126)^2/5 \ - \ 72.9$$

$$= 1.9$$

• F-statistic

F = Explained Variance / Unexplained Variance

$$F = [72.9/1] / [1.9/3]$$

$$= 115.1$$

Since $F_{1, 3, 99} = 34.2$, we reject the null hypothesis of no trend in the data (b=0) at the 99% percentile (and accept the alternative hypothesis that the data represents a linear trend).

1c) The correlation coefficient R is given by:

$$R = [N(\Sigma D_t \cdot t) \ - \ (\Sigma t)(\Sigma D_t)] / \{[N(\Sigma t^2) \ - \ (\Sigma t)^2][N \Sigma D_t^2 \ - \ (\Sigma D_t)^2]\}^{1/2}$$

$$= [5(405) \ - \ (15)(126)] / \{[5(55) \ - \ (15)^2][5(3250) \ - \ (126)^2]\}^{1/2}$$

$$= 135 / 136.74$$

$$= 0.987$$

The correlation coefficient R has the same sign (positive) as the slope b (+ 2.7). The large positive correlation coefficient demonstrates a high positive linear correlation between the independent variable t and the dependent variable D_t. Since R is positive, as the independent variable time t increases, the dependent variable demand D_t generally increases. The increase in patient demand, however, can not be demonstrated by statistics to be *caused* by the increase in time. Causation must be established *logically*. In this example, time is a surrogate for other factors that have changed over the years (e.g. increased population, decrease in population health, increased awareness of the clinic, etc.).

The R^2-statistic is calculated as the square of the correlation coefficient R:

$$R^2 = 0.974$$

The Least Squares line $D_t^* = 17.1 + 2.7 \cdot t$ explains 97.4% of the total variation from the sample mean of the past demand data.

1d) Using the Least Squares line $D_t^* = 17.1 + 2.7 \cdot t$, the error term (residual) table is as follows:

t	D_t	D_t^*	$D_t - D_t^*$
1	20	19.8	0.2
2	22	22.5	-0.5
3	25	25.2	-0.2
4	29	27.9	1.1
5	30	30.6	-0.6
			$\Sigma(D_t - D_t^*)$ ≈ 0.0

The sample mean of the error terms (residuals) is given by:

$$\Sigma(D_t - D_t^*)/N \approx 0.0/5 \approx 0.0$$

The sample mean of the error terms (residuals) is quite close to zero. The following graph of the error terms (residuals) shows no discernible pattern over time which would preclude the assumption of randomness of the error terms. It may be assumed that the error terms (residuals) are *independent* (the magnitude or sign of a given error term (residual) does not depend on that of any other error term (residual)).

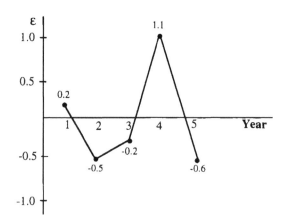

1e) The Least Squares best fit line $D_t^* = 17.1 + 2.7 \cdot t$ to the past demand data is used to forecast future demand:

$D_7^* = 17.1 + 2.7 \cdot (7) = 36$ patients/day

$D_{10}^* = 17.1 + 2.7 \cdot (10) = 44.1$ patients/day

$D_{100}^* = 17.1 + 2.7 \cdot (100) = 2717.1$ patients/day

One should have more faith in the demand forecast for year 7 than for year 10 (and much more than for the demand forecast for year 100 which would surely not be physically possible for a small walk-in health clinic to accommodate).

The model, however, assumes that the *past* linear trend will continue forever into the *future*. As one forecasts further into the future, one should be more circumspect about the assumption "all things remain the same" remaining valid.

1d) Since the number of degrees of freedom (N- 2 = 3) is well under 30, the Student t distribution with 3 degrees of freedom should be used to determine the confidence interval. As given in the problem statement, the symmetric 2.5th and 97.5th percentiles for 3 degrees of freedom are given by $t_{3,\ 025} = -3.182$ and $t_{3,\ 975} = 3.182$, respectively.

The expected value D_7^* for the confidence interval around year 7 is given by:

$D_7^* = 17.1 + 2.7 \cdot (7) = 36$ patients/day

The standard deviation s_7 for the confidence interval around year 7 is given by:

$$s_p = \{[(\Sigma D_t^2 - a\Sigma D_t - b\Sigma D_t \cdot t)/(N-2)] \cdot [(N+1/N) + ((p - \overline{7})^2/(\Sigma t^2 - N\overline{7}^2))]\}^{\frac{1}{2}}$$

$$s_7 = \{[(3250 - (17.1)(126) - (2.7)(405))/3] \cdot [6/5 + ((7-3)^2/(55 - 5(3)^2)]\}^{1/2}$$

$$= 1.33$$

The 95% confidence interval around year 7 is given by:

$[D_7^* + t_{3,\ 025} \cdot s_7 \ , \ D_7^* + t_{3,\ 975} \cdot s_7]$

$[36 - (3.182)(1.33) \ . \ 36 + (3.182)(1.33)]$

$[31.77 \ , \ 40.23]$

Note: $s_p = \{[0.63] \cdot [1.2 + (p - 3)^2/10]\}^{\frac{1}{2}}$. From this equation, it can be seen that $s_{10} > s_7$ for the health clinic. Thus the 95% confidence interval around year 10 will be larger than the confidence interval around year 7. The further one forecasts into the future (the further from the sample mean 3 of the independent variable time), the forecast is less certain and a given percentile confidence interval (e.g. 95%) will become larger.

Problem #2 (Linear Regression vs. Log-Linear Regression)

2a) The plot of Wheat Usage (Y) vs. Year (t) is given by:

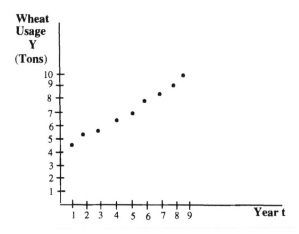

2b) The plot of the data seems to indicate a positive (upward) trend in the wheat usage data over time. Thus we expect the correlation coefficient R for both the linear and log-linear (exponential) models to be positive (i.e. wheat usage generally increases as time increases).

2c) There are N - 1 = 8 degrees of freedom in the data since the sample mean of the data has been calculated. There is only one independent variable (time) in the regression. Hence there are 8 degrees of freedom in the total deviation (from the sample mean), 1 degree of freedom in the explained deviation, and the remaining 7 degrees of freedom in the unexplained deviation.

2d) The Method of Least Squares line for the log-linear trend Wheat (Y) = $e^{1.4976 + 0.083458 \cdot \text{Year}(t)}$ is a better *statistical* fit to the demand data than the Least Squares Method line for the linear trend Wheat (Y) = 4.0944 + 0.57·Year(t). This is demonstrated statistically by the larger F-statistic (1565 vs. 1261) and the larger R^2-statistic (0.9955 vs. 0.9944).

If the error terms (residuals) were plotted, it would be seen that the error terms (residuals) for both the linear and log-linear trends are small with sample means close to zero.

2e) *Both* the log-linear (exponential) and linear trends are good statistical fits to the demand data. The decision as to which model should be used to predict future demand should be made by the Secretary of Agriculture based on *all available* information (factual and logical) about past and future demand for wheat for the nation.

Solutions to Exercise 7

Forecasting

Problem # 1 (Method of Moving Averages)

1a) The three and five month simple moving average forecasts for the Ann Guish Memorial Hospital are given as follows:

Month	Past Demand	3 Month Simple Moving Average	5 Month Simple Moving Average
1 (Jan)	2000	---	---
2 (Feb)	1350	---	---
3 (Mar)	1950	---	---
4 (Apr)	1975	1767	---
5 (May)	3100	1758	---
6 (Jun)	1750	2342	2075
7 (Jul)	1550	2275	2025
8 (Aug)	1300	2133	2065
9 (Sept)	2200	1533	1935
10 (Oct)	2770	1683	1980
11 (Nov)	2350	2092	1914
12 (Dec)	2550	2442	2035

Note: The first month in which a 3 month simple moving average forecast is available in month 4. The first month in which a five month simple moving average forecast is available is month 6.

Example simple moving average forecast calculations are given as follows:

Three Month Simple Moving Average Forecast For April $= (2000 + 1350 + 1950) / 3$

$$\approx 1767$$

Three Month Simple Moving Average Forecast For May $= (1350 + 1950 + 1975) / 3$

$$\approx 1758$$

Five Month Simple Moving Average Forecast For June $= (2000 + 1350 + 1950 + 1975 + 3100) / 5$

$$\approx 2075$$

1b) The past monthly demand for floral bouquets, the three month simple moving average forecasts, and the five month simple moving average forecasts are provided on the following graph:

523

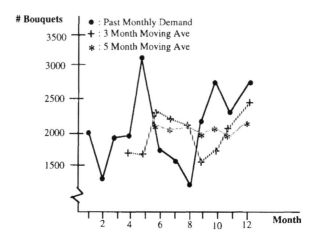

1c) The five month simple moving average forecasts (with the most # periods averaged) provides the better stability while the three month simple moving average forecasts (with the least # periods averaged) provides the better responsiveness to changes in the actual demand data.

Problem #2 (Method of Exponential Smoothing)

2a) Forecasts up to month 12 using the Method of Exponential Smoothing with $\alpha = 0.1$, 0.5 and 0.9 are given as follows:

Month	# Patients Observed	$\alpha = 0.1$	$\alpha = 0.5$	$\alpha = 0.9$
1 (Jan)	2000	2000	2000	2000
2 (Feb)	1350	2000	2000	2000
3 (Mar)	1950	1935	1675	1415
4 (Apr)	1975	1937	1813	1897
5 (May)	3100	1940	1894	1967
6 (Jun)	1750	2056	2497	2987
7 (Jul)	1550	2026	2123	1874
8 (Aug)	1300	1978	1837	1582
9 (Sept)	2200	1910	1568	1328
10 (Oct)	2770	1029	1884	2113
11 (Nov)	2350	2023	2330	2709
12 (Dec)	2550	2056	2340	2386

Note: Since there was no forecast available for month 1, the observed 2000 patients in month 1 is used as the forecast for month 1.

Example Method of Exponential Smoothing forecasts for March with $\alpha = 0.1$, 0.5 and 0.9 are computed as follows:

$$Y_3^* = Y_2^* + \alpha \cdot (Y_2 - Y_2^*) \qquad\qquad \alpha = 0.1$$

$$= 2000 + 0.1 \cdot (1350 - 2000)$$

$$= 1935$$

$$Y_3^* = Y_2^* + \alpha \cdot (Y_2 - Y_2^*) \qquad\qquad \alpha = 0.5$$

$$= 2000 + 0.5 \cdot (1350 - 2000)$$

$$= 1675$$

$$Y_3^* = Y_2^* + \alpha \cdot (Y_2 - Y_2^*) \qquad\qquad \alpha = 0.9$$

$$= 2000 + 0.9 \cdot (1350 - 2000)$$

$$= 1415$$

2b) The observed (actual) past monthly demand and the Method of Exponential Smoothing forecasts with $\alpha = 0.1$ and 0.9 are given on the following graph:

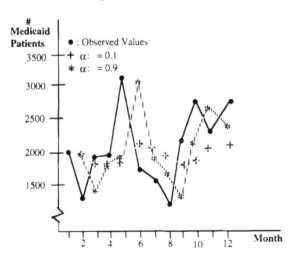

2c) The smaller the value of α, the greater the stability of the Method of Exponential Smoothing forecasts. The larger the value of α, the greater the responsiveness the Method of Exponential Smoothing forecasts to changes in the observed data.

Note: A small α for the Method of Exponential Smoothing forecasts corresponds in smoothing capability to a large N for the Method of Moving Averages forecasts. In fact, the Method of Exponential Smoothing with $\alpha = 2/(N+1)$ provides the same smoothing performance by an N period moving average.

Problem #3 (Method of Seasonal Indices)

3a) The quarterly means and the grand mean for the # visits to the Home Care Program at Payne Hospital over the past five years is given as follows:

Quarter Q Year t	1 Qtr 1	2 Qtr 2	3 Qtr 3	4 Qtr 4
1 (1989)	2	15	4	61
2 (1990)	63	71	99	138
3 (1991)	125	125	164	216
4 (1992)	186	180	228	294
5 (1993)	248	254	293	372
Quarterly Mean $(\bar{Y}_Q)$	124.8	125.0	163.6	216.2

Grand Mean $\bar{\bar{Y}} = \Sigma \bar{Y}_Q/4 = 629/4 = 157.4$

The quartery indices I_Q (Q = 1, 2, 3, 4) are given by:

$$I_Q = \bar{Y}_Q / \bar{\bar{Y}} \qquad\qquad Q = 1, 2, 3, 4$$

$$I_1 = \bar{Y}_1/ \bar{\bar{Y}} = 124.8/157.4 \qquad = 0.793$$
$$I_2 = \bar{Y}_2/ \bar{\bar{Y}} = 125.0/157.4 \qquad = 0.794$$
$$I_3 = \bar{Y}_3/ \bar{\bar{Y}} = 163.6/157.4 \qquad = 1.037$$
$$I_4 = \bar{Y}_4/ \bar{\bar{Y}} = 216.2/157.4 \qquad = \underline{1.374}$$
$$\qquad\qquad\qquad\qquad\qquad 4.000 \quad (= \Sigma I_Q)$$

Note: $\Sigma I_Q /4 = 1$. That is, the mean of the quarterly indices is equal to 1.

3b) The Method of Least Squares has been used to forecast a yearly demand Y^*_{1995} of 15000 visits in 1995. The forecast for quarter 2 and quarter 4 of 1995 will use the appropriate quarterly index derived above applied to the Least Squares forecast of 15000 in 1995.

• Year 1995 Quarter 2

Method of Least Squares forecast $Y^*_{1995} = 15000$ visits
Average quarterly demand in 1995 = 15000/4 = 3750 visits
Method of Seasonal Indices forecast $= (3750).I_2$
$= (3750) \cdot (0.794)$
$= 2977.5$ visits

526

• Year 1995 Quarter 4

Method of Least Squares forecast $Y^*_{1995} = 15000$ visits
Average quarterly demand in 1995 = $15000/4 = 3750$ visits
Method of Seasonal Indices forecast $= (3750) \cdot I_4$
$\qquad\qquad\qquad\qquad\qquad\qquad = (3750) \cdot (1.374)$
$\qquad\qquad\qquad\qquad\qquad\qquad = 5152.5$ visits

Problem 4 (Times Series vs. Forecasting Methods)

4a) The monthly data seems to show a negative linear trend over time. The Method of Least Squares can be used to determine how well the monthly data fits a linear trend. There also seems to be a seasonal or cyclical pattern to the demand data. The Method of Indices (using monthly indices) can be applied to aggregate (e.g. yearly) forecasts using the Method of Least Squares to determine if the monthly forecasts can be improved (lower mean squared error).

4b) The monthly data does not seem to show a trend over time. The sample mean appears to be the best forecast for each month. When the Method of Least Squares is applied, the F-statistic can be used to verify whether the sample mean of the data provides the best fit to the monthly demand data. There is a strong indication of a cyclical or seasonal component to the demand data. The Method of Seasonal Indices (using monthly indices) can be applied to the sample mean to determine whether the monthly forecasts can be improved (lower mean squared error).

4c) The monthly data seems to show a slow-varying positive trend. The Method of Exponential Smoothing can be applied to the data to provide monthly forecasts for demand. There does not seem to be a cyclical or seasonal pattern in the demand data. However, the Method of Seasonal Indices can be applied to the Method of Exponential Smoothing monthly forecasts to determine if the monthly forecasts can be improved (lower mean squared error).

Solutions to Exercise 8

Multiple Regression

Note: The regressions used to answer the questions for this exercise were performed using STORM Personal Version 3.0 © 1992 by STORM Software, Inc.

Problem #1 (Method of Least Squares)

a) The coefficients for the Designer Pet regression model (1) were determined to be:

$$PROFIT = 1.996349 + 35.56463 \cdot PRICE + 10.85774 \cdot ADV$$

b) The coefficients for the Designer Pet regression model (2) were determined to be:

$$PROFIT = 1.996349 + 35.56463 \cdot PRICE + 0.01085 \cdot ADV$$

c) When the monthly advertising budget (ADV) was given in units of thousands of dollars (the advertising data in regression (1) was multiplied by 1000), the coefficient of the independent variable ADV in the Designer Pet regression (2) became 0.01085 (10.85 divided by 1000).

It is not correct to say that the effect of the monthly advertising budget (ADV) has a larger effect on the monthly profit (PROFIT) in regression (1) than in regression (2). As we have seen, the value of the coefficient of ADV is a function of the units in which the monthly advertising budget is expressed.

The correct interpretation of the coefficient of the independent variable ADV in regressions (1) and (2) is that if all other independent variables remaining constant, the effect of an increase in the monthly advertising budget of $1M and $1K is an increased monthly PROFIT of $10.85 M and $0.01085 M ($10.85 K), respectively. As expected, the effect of the independent variable ADV on the dependent variable PROFIT is the same regardless of the units chosen for the independent variable ADV.

d) The goodness of fit statistics for the Designer Pet regression (1) were determined to be:

R^2-statistic = 0.946498
Adjusted R^2-statistic = 0.938855
Standard Error of the Estimate = 4.121409
F-statistic = 123.84

Notes:

• Regression (1) and regression (2) are equivalent except that the independent variable ADV is expressed in units of millions and thousands of dollars, respectively. While the coefficient of the independent variable ADV changed in regression (2), all the goodness of fit statistics remain the same. This is to be expected since the two regressions are using the same data.

• All the above statistics are dimensionless except for the Standard Error of the Estimate which is a function of the units chosen for the dependent variable PROFIT. Since the units chosen is $M in both regressions (1) and (2), the Standard Error of the Estimate will remain the same. Remember the Standard Error of the Estimate provides a measure of the scatter of the data points around the Least Squares best fit line. Since the Least Squares lines for regressions (1) and (2) will plot as the same line, the scatter of the data around the line will be the same for both regressions.

e) The correlation matrix of the Designer Pet regression (1) was determined to be:

	PRICE	ADV	PROFIT
PRICE	1.0000	0.5316	0.6887
ADV	0.5316	1.0000	0.9481
PROFIT	0.6887	0.9481	1.0000

The correlation matrix can be computed using the data prior to any regressions being performed. If no variables are added to the regression, the correlation matrix will remain constant regardless of the independent variables one chooses to include in the regression.

Notes:

• The correlation between two variables is a measure of the tendency of the two variables to move together rather than independently.

• A correlation matrix is symmetric. The correlation of variable ADV with variable PROFIT is the same as the correlation of variable PROFIT with variable ADV.

Problem #2 (Method of Least Squares)

a) The coefficients of the Pet Designer regression (3) were determined to be:

$$PROFIT = 2.473257 + 35.23896 \cdot PRICE + 10.244813 \cdot ADV + 0.157486 \cdot MONTH$$

b) The goodness of fit statistics for the Designer Pet regression (3) were determined to be:

R^2-statistic = 0.946552
Adjusted R^2-statistic = 0.934218
Standard Error of the Estimate = 4.274832
F-statistic = 76.743

c) The R^2-statistic for regression (3) increased as is expected when the additional independent variable MONTH is added to the regression. However, the adjusted R^2-statistic for regression (3) decreased. The loss of the degree of freedom in the data (the coefficient for the independent variable MONTH had to be estimated) was not compensated by the increase in the amount of explained deviation when the independent variable MONTH is included in the regression. Since the adjusted R^2-statistic decreased when MONTH was added to the regression, we expect to find that the t-statistic for MONTH will be less than 1.

The value of the Standard Error of the Estimate increased. There is more scatter in the data points around the Least Squares best fit line determined for regression (3) than the Least Squares line for regression (1)..

The F-statistic for regression (3) decreased. One can not reject the null hypothesis of no trend in the data (and accept the alternative hypothesis of a linear trend) to as high a level of significance. Thus regression (3) does not fit a linear trend as well as regression (1). However, both have large F-statistics and are significant to the 99th percentile ($F_{3,13,.99}$ = 5.74).

d) The sample standard deviation (error) of the dependent variable PROFIT in the Pet Designer data is reported to be 16.6674. The Standard Error of the Estimate for the dependent variable PROFIT using regression (3) is reported to be 4.274832. The regression (3) which decreased the error in forecasting the dependent variable PROFIT by a factor of approximately 4 does provide explanatory power for the dependent variable PROFIT.

e) The correlation matrix for the Designer Pet data including the additional independent variable MONTH was determined to be:

	PRICE	ADV	MONTH	PROFIT
PRICE	1.0000	0.5316	0.5602	0.6887
ADV	0.5316	1.0000	0.9872	0.9481
MONTH	0.5602	0.9872	1.0000	0.9462
PROFIT	0.6887	0.9481	0.9462	1.0000

Note: The values of all correlations between variables not involving the new variable MONTH are the same as for the Designer Pet data in Problem #1.

The independent variables ADV and MONTH have a very large correlation coefficient (0.9872) which is indicative of strong linear correlation between the two independent variables. Based on the correlation matrix, the independent variable ADV is more highly correlated with the dependent variable PROFIT than is the independent variable MONTH. Based on the statistical information provided by the correlation matrix, the independent variable MONTH should be dropped from the regression (this decision would have to be reviewed from a logical point of view).

Problem #3 (Calculation of the Adjusted R^2-Statistic)

There are 17 observations and 3 independent variables in the regression. The adjusted R^2-statistic may be computed as a function of the R^2-statistic (reported for regression (3) to be 0.946552) as follows:

$$Adjusted\ R^2 = 1 - \left[\frac{(N-1)}{(N-1)-k} \cdot (1-R^2) \right]$$

$$Adjusted\ R^2 = 1 - \left[\frac{(17-1)}{(17-1)-3} \cdot (1-0.946552) \right]$$

Adjusted $R^2 = 0.934218$

Problem #4 (Calculation of the F-Statistic)

An easier method to compute the F-statistic makes use of the R^2-statistic which has been reported by the regression software package to be 0.876541. The F-statistic may then be computed as a function of the R^2-statistic, the # observations N = 8, and the number of independent variables k = 1 estimated as follows:

$$F = \frac{R^2/_k}{(1-R^2)/_{(N-1-k)}}$$

$$= \frac{0.876541/_1}{(1-0.876541)/_{(8-1-1)}}$$

$$= 42.6$$

Problem #5 (Calculation of the t-Statistic in Simple Regression)

a) The F-statistic for the Round Table Furniture Company data is reported by the regression software package to be 42.6.

b) The t-statistic for YEAR is reported to be 6.52679.

b) The t-statistic for the single independent variable YEAR in the simple regression is indeed the square root of the F-statistic. Remember that the sign of the t-statistic is the same as the sign of the coefficient in the regression. The t-statistic is positive here because the sign (slope) of the coefficient for YEAR in the simple regression is also positive (55.95).

Problem #6 (Method of Least Squares)

a) The t-statistics for the Designer Pet Regression (3) were determined to be:

Variable	Coefficient	Standard Error	t-Statistic
Constant	2.47326	6.81694	0.36281
PRICE	35.23896	10.83905	3.25111
ADV	10.24413	5.45093	1.87933
MONTH	0.15749	1.37464	0.11457

b) There are 17 observations and 3 independent variables in the regression. The number of degrees of freedom in the data is 17 - 1 - 3 = 13. Since there are less than 30 degrees of freedom in the data, the Student t distribution will be used to determine significance of the coefficients of regression (3).

c) The 95th percentile of the Student t distribution with 13 degrees of freedom is 1.771. The independent variables PRICE and ADV for regression (3) are significant (non-zero). The independent variable MONTH and the constant are not statistically significant.

Note: Not significant statistically does not imply not important in explaining the dependent variable. The data has randomness in it. Another set of observations might indicate that MONTH is significant. However, the small t-statistic (0.11457) does indicate that the likelihood of the independent variable MONTH will be significant in another set of observations is not high.

d) The overall regression is significant (the F-statistic is high) and the R^2-statistic is also high indicating good explanatory power in the regression. And yet there is a coefficient in the regression which is not significant. One should suspect multicollinearity among the independent variables.

Recall that the correlation coefficient between the independent variable ADV and MONTH is 0.9872. This indicates a strong linear relationship between these two independent variables. The correlation coefficient between ADV and PRICE is 0.0.5316 while the correlation coefficient between MONTH and PRICE is 0.5602. Both of these correlation coefficients are moderate (below approximately 0.70) and not likely to cause large multicollinearity problems in the regression.

e) It is never recommended that the constant term be dropped in regression analysis unless there is a strong reason to do so. Dropping the constant term forces the regression through the origin. If there are relevant independent variables omitted from the regression, the coefficients for the independent variables will be biased (since there is no constant term to attribute the expected value of the omitted independent variables). The constant term is a "garbage-can term" which is helpful in regression when estimating coefficients, computing goodness of fit statistics, and in comparing models. The constant term may be dropped from the final regression (all remaining independent variables significant) without any statistical consequences.

f) The estimates for PRICE of $0.75 and for ADV of $4.5 Million are substituted into regression (3) to forecast the PROFIT in month 18:

$$PROFIT = 2.473257 + 35.23896 \cdot PRICE + 10.244813 \cdot ADV + 0.157486 \cdot MONTH$$

$$PROFIT = 2.473257 + 35.23896 \cdot (\$0.75) + 10.244813 \cdot (\$4.5) + 0.157486 \cdot (18)$$

$$PROFIT = \$77.83 \, M$$

The forecast is in the neighborhood of the data used to determine the regression. Hence, there is statistic confidence in the forecast. Remember that the statistical confidence decreases (the confidence interval around the forecast increases) the further from the estimates of the independent variables are from their sample mean.

Problem #7 (Durbin-Watson Statistic)

a) The regression of the percentage (the dependent variable PERCENT) of the U. S. population which lives on a farm as a function of the year (the independent variable TIME) is determined to be:

532

$PERCENT = 15.3527 - 0.394808 \cdot TIME$

b) There is only 1 independent variable (TIME) in this regression. There are 37 observations in the data. The appropriate Durbin-Watson values are as follows:

$d_L = 1.40$
$d_U = 1.53$

c) The Durbin-Watson statistic is reported to be 0.081829. This value is in the region $(0, d_L)$ which indicates positive autocorrelation.

d) Positive autocorrelation is indicated when one error term is positive (the data point is above the regression line) and error terms in its vicinity are likely to be positive as well. Likewise, when one error term is negative (below the line), error terms in the vicinity are then likely to be negative.

Note: Positive autocorrelation for time series data is not uncommon and should always be suspected.

Problem #8 (Dependent vs. Independent Variable)

a) Logically, COWS are a function of the # ACRES on the farm and thus COWS should be designated as the dependent variable. The McDonald County regression using Y (COWS) as the dependent variable and X (ACRES) as the independent variable is determined to be:

(1) $COWS = 0.032727 + 0.734545 \cdot ACRES$

b) The McDonald County regression using X (ACRES) as the dependent variable and Y (COWS) as the independent variable is determined to be:

(2') $ACRES = 4.245283 + 0.95283 \cdot COWS$

c) Both regressions (1) and (2) must provide the expression for COWS as a function of ACRES in order to be compared directly. Regression (2) is solved for COWS as a function of ACRES and written as (2'):

(1) $COWS = 0.032727 + 0.734545 \cdot ACRES$

(2) $COWS = -4.46 + 1.05 \cdot ACRES$

The R^2-statistic and the adjusted R^2-statistic are the same for both regressions. The same percentage of the deviation in the model is explained. Also, the F-statistics are identical indicating that the ratio of the explained to the unexplained deviation is identical. However, the values of the explained deviation, unexplained deviation, and total deviation are different for the two regressions. They are larger when the deviation is ACRES is minimized since the data values for ACRES are generally larger than for COWS.

The standard error of the estimate (which indicates how widely the data points are scattered around the regression line) is smaller when COWS is the dependent variable (2.82) and the error terms along the COW (Y) axis are minimized than when ACRES is the dependent variable (3.21) and the error terms along the ACRES (X) are minimized. Again the magnitude of the data values for ACRES are generally larger than for COWS.

The t-statistic for the independent variable in both regressions are identical. However, the t-statistic for the constant is different in the two regressions. In fact, in regression (1), the constant is not significant and zero may be used for the content without any statistical difference.

Solutions to Exercise 9

Linear Programming (LP)

Note: The LP module, Assignment Problem module, and Transportation module from STORM Personal Version 3.0 © 1992 by STORM Software, Inc. were used to perform the computations required by this exercise.

Problem #1 (Linear Programming)

1a) Define the decision variables as A = # adults and C = # children. The objective function which describes the revenue per week is given by $1000·A + $200·C. There is a limit of 60 photo session hours per week and 360 development procedure hours per week. The LP formulation to maximize the revenue per week for the Glitz Photography Studio is given by:

$$Max \quad 1000 \cdot A + 200 \cdot C$$
$$1.5 \cdot A + \quad 3 \cdot C \leq 60 \; (Photo \; Session \; Hours)$$
$$30 \cdot A + \quad 10 \cdot C \leq 360 \; (Development \; Hours)$$
$$A \geq 0$$
$$C \geq 0$$

1b) Graph the constraints 1.5·A + 0.3·C = 60 and 30·A + 10·C = 360 in the upper right hand quadrant given by A ≥ 0, C ≥ 0. The feasible region for the LP with corner points labeled is given by the area below both constraint lines and above the A axis and C axis. The hatched area in the following figure denotes the feasible region for this LP.

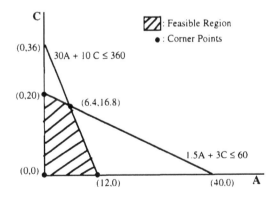

<u>Corner Point Evaluation</u> (objective function where revenue = $1000·A + $200·C)

(0,20)   ⟹ Revenue = $4000

(0,0)   ⟹ Revenue = $0

(12,0)   ⟹ **Revenue = $12000**

(6.4,16.8)   ⟹ Revenue = $9760

The corner point A = 12 and C = 0 (12 adults and 0 children) maximizes the revenue per week and is the optimal solution for the above LP formulation.

1c) The Simplex Method will start at the initial feasible corner point (0,0). The adjacent corner point which brings the most improvement in the objective function $1000·A + $200·C is the corner point (12,0). Since no adjacent corner point to (12,0) will improve the value of the objective function $1000·A + $200·C, the corner point (12,0) is optimal.

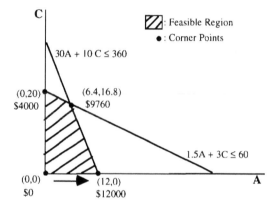

1d) The decision variables are still A = # adults and C = # children. The objective function which maximizes clients per week is given by A + C. There is a still a limit of 60 photo session hours per week and 360 development hours per week. The LP formulation to maximize the # clients per week at the Glitz Photography Studio is given by:

$$
\begin{aligned}
Max \quad & A + \quad C \\
& 1.5 \cdot A + \quad 3 \cdot C \leq 60 \ (Photo\ Session\ Hours) \\
& 30 \cdot A + \quad 10 \cdot C \leq 360 \ (Development\ Hours) \\
& A \geq 0 \\
& C \geq 0
\end{aligned}
$$

1e) The same feasible region given for the maximize revenue LP formulation applies to the maximize # clients LP formulation since the constraints remain the same (only the objective function changed):

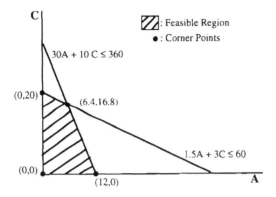

Corner Point Evaluation (objective function where # clients = A + C)

(0,20)	⇒ # Clients = 20
(0,0)	⇒ # Clients = 0
(12,0)	⇒ # Clients = 12
(6.4,16.8)	⇒ **# Clients = 23.2**

The corner point A = 6.4 and C =16.8 (approximately 6 adults and 17 children) maximizes the # clients objective function and is the optimal solution for the this LP formulation.

1f) The Simplex Method will start at the initial feasible corner point (0,0). The adjacent corner point which brings the most improvement in the objective function A + C is the corner point (0,20). The Simplex Method will move from corner point (0,0) to corner point (0,20). The adjacent corner point to (0,20) which brings the most improvement in the objective function A + C is the corner point (6.4,16.8) which is the only option available. The Simplex Method will move from corner point (0,20) to corner point (6.4,16.8). Since no adjacent corner point to (6.4,16.8) will improve the value of the objective function A + C. the corner point (6.4, 16.8) is optimal.

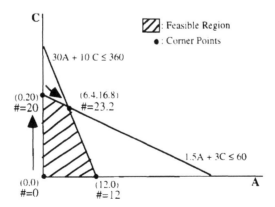

1g) In general, it is best to use the optimal solution from the previously solved LP as the initial solution to the new LP. This reduces the number of pivots (movements) of the Simplex Method. Starting at the corner point (12,0), the Simplex Method would move to the optimal corner point (6.4,16.8) in only one pivot rather than the two used when the initial corner point used is (0,0).

Problem #2 (Linear Programming)

2a) The decision variables are P = # post-surgical patients, H = # hip fracture patients, and S = # stroke patients. The objective function which describes the yearly revenue to the Green Pastures nursing home is given by $3000·P + $15300·H + $26600·S. There is a limit of 36500 bed days/year (BD), 75100 nursing hours/year (NH), and 39000 hours/year (PT). The LP formulation to find the number of admissions per year of each patient type which maximizes total yearly revenue at the nursing home is given by:

$$Max \quad \$3000 \cdot P + \$15300 \cdot H + \$26600 \cdot S$$

$$20 \cdot P + \quad 80 \cdot H + \quad 140 \cdot S \leq 36500 \; (BD)$$

$$35 \cdot P + \quad 170 \cdot H + \quad 360 \cdot S \leq 75100 \; (NH)$$

$$4 \cdot P + \quad 100 \cdot H + \quad 100 \cdot S \leq 39000 \; (PT)$$

$$P \geq 0$$

$$H \geq 0$$

$$S \geq 0$$

2b) The optimal number of post-surgical patients P, hip fracture patients H, and stroke patients S is 312.06, 377.52, and 0, respectively. The fractional number of each patient type would need to be rounded off to the nearest integer value. The maximum yearly revenue possible with the limited resources available is approximately $3000·(312.06) + $15,300·(377.52) + $26600·(0) = $6,712,191.

The basic (non-zero) variables are post-surgical patients P, hip fracture patients H, and the slack variable bed days BD = 312.06, 377.52, and 57.45, respectively. The non-basic (zero) variables are stroke patients S, nurse hours NH, and physical therapy hours PT.

The resource bed days BD is not used up at the optimal solution. Hence, the shadow price (the additional yearly revenue if another bed day could be added) is zero. The shadow prices for the resources nurse hours NH and physical therapy PT hours (all used up at the optimal solution) are $84.68 and $9.04, respectively.

2c) The optimal number of post-surgical patients P, hip fracture patients H, and stroke patients S is 239.33, 359.11, and 21.32, respectively. The fractional number of each patient type would need to be rounded off to the nearest integer value. The maximum yearly revenue possible with the extra nursing resources available is approximately $3000·(239.33) + $15,300·(359.11) + $26600·(21.32) = $6,779,444.

If a nurse is hired at $32,000 per year and is able to provide an additional 2000 nursing hours, the increased yearly revenue is $6,779,444 minus $6,712,191 = $67,253. Hiring the nurse seems to be a worthwhile investment.

2d) The comparison of the LP solutions for the maximum yearly revenue and the maximum yearly net revenue is as follows:

		Max Yearly Revenue LP	Max Yearly Net Revenue LP
Decision Variables	Post-surgical Patients P	312	1142
	Hip Patients H	378	0
	Stroke Patients S	0	98
Slack Variables	Bed Days BD	57	0
	Nurse Hours NH	0	0
	Physical Therapy PT	0	24672

The optimal mix of patient types changed. Stroke Patients was not a basic variable for the optimal solution to the maximum yearly revenue LP while Hip Patients was not a basic variable for the optimal solution to the maximum yearly net revenue LP.

The binding constraints also changed. The constraint associated with the resource Bed Days was non-binding on the optimal solution to the maximum yearly revenue LP while the constraint associated with the constraint Physical Therapy hours was non-binding on the optimal solution to the maximum yearly revenue LP.

Note that the total number of basic variables remained constant at three (equal to the number of constraints).

Problem 3 (Simplex Method)

3a) The basic (non-zero) variables correspond to the identity columns (all zeros except for a single 1). The basic variables are $X1$, $X2$, and $S2$ = 12, 15, and 20, respectively. The non-basic (zero) variables correspond to the non-identity columns associated with the slack variables $S1$ and $S3$.

3b) The non-binding constraints at any corner point will have a positive (non-zero) value for its slack variable. For this final Simplex Method tableau corresponding to the optimal corner point, the slack variable $S2$ is positive (20 units remain at the optimal solution) denoting that the constraint on the resources $S2$ is non-binding. The constraints corresponding to the resources $S1$ and $S3$ are binding since their slack variables are zero (resources $S1$ and $S3$ are all used up at the optimal solution).

3d) Since not all of the resource $S2$ is used up at the optimal solution, its shadow price (the marginal utility of one extra unit of the resource) is zero. The shadow price for a resource is found in the objective row of the Simplex Method tableau for the optimal solution. From the objective row of the Simplex Method tableau for the optimal solution, the shadow price for resource $S2$ is found to be 0. Likewise, the shadow prices for the resources $S1$ and $S3$ (binding constraints) are found to be $25/unit and $40/unit, respectively.

Problem #4 (Assignment Problem)

4a) The solution to the Assignment Problem using the student ranking of hospitals is as follows:

Student	Ranking
Al	2
Betty	1
Carol	1
Dave	2
Eph	1
Fran	1
Gert	1
Hal	1
Irma	1
Jack	1
Total Sum of Rankings = 12	

4b) The solution to the Assignment Problem using the hospital ranking of the students is as follows:

Student	Ranking
Al	1
Betty	2
Carol	1
Dave	2
Eph	1
Fran	2
Gert	3
Hal	1
Irma	4
Jack	1
Total Sum of Rankings = 18	

4c) The university is better able to match the students to the hospital of their first choice than to match the hospitals to the student of their first choice. This is because many of the hospitals made the same students (namely Al and Carol) their first choice. On the hand, only two hospitals (namely A and I) were chosen by more than 1 student as their first choice.

Problem #5 (Transportation Problem)

5a) The optimal (minimum cost) routing of the 19300 blood tests to the 3 labs are given as follows:

	Lab #1	Lab #2	Lab #3	TOTAL
Agony		2500		2500
Bland			5400	5400
Costly	1900			1900
Dow	3100		700	3800
Eager		2500	1900	4400
Farr	1300			1300
TOTAL	6300	5000	8000	19300

The cost of this routing is 2500·($0.60) +   5400·($0.20) +   1900·($0.40) + 3100·($0.60) + 700·($0.40) + 2500·($0.50) + 1900·($0.50) + 1300·($0.20) = $7940.

5b) When the maximum volume for Lab #2 is increased by 2500 tests/week while the volume at Lab #1 is decreased by the same amount, the optimal (minimum cost) routing of the 19300 blood tests to the 3 labs are given as follows:

	Lab #1	Lab #2	Lab #3	TOTAL
Agony		2500		2500
Bland		600	4800	5400
Costly	1900			1900
Dow	600		3200	3800
Eager		4400		4400
Farr	1300			1300
TOTAL	3800	7500	8000	19300

The cost of this routing is 2500·($0.60) + 600·($0.40) + 4800·($0.20) + 1900·($0.40) + 600·($0.60) + 3200·($0.40) + 4400·($0.50) + 1300·($0.20) = $7560.

This relocation of resources saves $7940 - $7560 = $380 per week. However, since the weekly cost to MegaMedical of this relocation is $1500, this move will probably not be considered.

Solutions to Exercise 10

Queueing

Note: The Queueing Module from STORM Personal Version 3.0 © 1992 by STORM Software, Inc. was used to provide the quantitative answers for this exercise.

Problems #1 (Yet Another Frozen Yogurt Shoppe)

1a) Define queuing parameters and assumptions:

• Queueing Parameters:

- Arrival process distribution

- Service process distribution

- # servers of frozen yogurt - decide on range of interest

- Queue discipline - who is the next person served frozen yogurt

- Annual cost of maintaining each employee

- Annual cost of each minute of expected waiting line time prior to frozen yogurt order being taken by the employee

- Annual cost of each customer turnaway due to the expected waiting line length at the frozen yogurt business upon arrival

• Queueing Assumptions:

- Markovian arrival process (constant average arrival rate) - perhaps Poisson distributed

- Markovian service distribution (constant average service completion time) - perhaps Exponential distributed

- Constant number of servers over the day

- First In - First Out queue discipline

- Average arrival rate < average service completion rate

- The "steady-state" characteristics of the queue are of interest rather than characteristics at particular times like just after startup or during peak times.

1b) Appropriate data to collect or estimate to quantify the queueing problem:

The arrival distribution could be characterized by examination of frozen yogurt businesses in similar shopping malls. Or the business could survey patrons of the shopping mall to determine how many (and how often) patrons would frequent the frozen yogurt business

542

once it is open for business. This data can be refined once the business opens and actual arrival data can be gathered.

The service distribution for the frozen yogurt business could be established from gathering data from other similar frozen yogurt businesses. This data can be refined once the business opens and actual service data can be gathered.

Cost of maintaining a staff of employees can be determined by accountants based on salaries, benefits, training required, etc.

The costs of customer delay (loss of goodwill) and customer turnaway (due to the line being unacceptably long to a customer) are subjective. These would have to be established by expert opinion (by those in similar service industries) or by surveying potential customers prior to opening the business and actual customers after opening the business.

Problem #2 and #3 (Copy Spot M/M/1 Queueing Process)

For this example:

λ = 10 customers per hour (arrival rate)

$1/\mu$ = 4 minutes per customer (service completion time)
  = 0.066667 hour per customer

μ = 15 customers per hour (service completion rate)

Since λ = 10 customers per hour (arrival rate) < μ = 15 customers per hour (service completion rate), the queue will not grow without bound and will attain a steady-state mode (i.e. an analytical queueing solution exists).

2a) The steady-state expected waiting line length L is given as follows:

$$L = \lambda^2 / (\mu)(\mu-\lambda)$$

$$= (10)^2 / (15)(15 - 10)$$

$$= 100 / 75$$

$$= 1.333 \text{ customers}$$

2b) The steady-state expected waiting line time W in the queue before service is given as follows:

$$W = L / \lambda$$

$$= (1.333 \text{ customers}) / (10 \text{ customers/hour})$$

$$= 0.1333 \text{ hours}$$

$$\approx 8 \text{ minutes}$$

2c) The probability that the waiting line time (prior to receiving service) is less that 6 minutes is given by the following Exponential ($1/\mu = W = 8$ minutes) cumulative distribution function:

$$F(t) = 1 - e^{-\mu t}$$

$$= 1 - e^{-t/W}$$

$$= 1 - e^{-t/(8 \, minutes)}$$

$$F \, (6 \text{ minutes}) = 1 - e^{-(6 \, minutes/(8 \, minutes))}$$

$$= 1 - e^{-0.75}$$

$$= 0.5276$$

2d) The probabilities of exactly 1 and 2 customers at Copy Spot (in the waiting line or being served) are given by:

$$P \text{ (exactly N in the queueing system)} = (1 - \lambda/\mu)(\lambda/\mu)^N$$

$$P \text{ (exactly 1 customer at Copy Spot)} = (1 - 10/15)(10/15)^1$$

$$= (1 - 2/3)(2/3)$$

$$= 2/9$$

$$= 0.2222$$

$$P \text{ (exactly 2 customers at Copy Spot)} = (1 - 10/15)(10/15)^2$$

$$= (1 - 2/3)(2/3)^2$$

$$= 4/27$$

$$= 0.1481$$

2e) The expected time a customer spends in the queueing system includes the expected waiting line time W and the expected service completion time $1/\mu$. The expected time at the copying kiosk is then the expected waiting line time W (1.33 hours) and the expected service completion time $1/\mu$ (4 minutes = 1/15 hours = 0.067 hours).

$$E(Time \; in \; queueing \; system) = E(Waiting \; line \; time) + E(Service \; time)$$

$$= W + 1/\mu$$

$$= 0.1333 \text{ hrs} + 0.067 \text{ hrs}$$

$$= 0.20 \text{ hrs}$$

2f) The probability of at least 1 customer in the queueing system is given by:

P(At least 1 customer) = 1 - P(no customers in the queueing system)

$$= 1 - (1 - \lambda/\mu)(\lambda/\mu)^0$$

$$= 1 - (1 - \lambda/\mu)$$

$$= \lambda/\mu$$

$$= 10/15$$

$$= 2/3$$

$$= 0.66$$

Another way to compute the probability of at least one customer in the queueing system is to compute the server utilization rate given by:

Server utilization rate $= \rho$

$$= \lambda/\mu$$

$$= 10/15$$

$$= 2/3$$

Problem #4 (M/M/c Queue)

4a) The probability that a given line is being utilized (the server utilization rate ρ) for each of the c = 7 servers (switchboard operators) is given as a function of the call arrival rate λ = 60 calls per hour and the service completion rate μ = 12 calls per hour:

Server utilization rate $= \lambda/c\mu$

$$= \frac{60 \; calls \; per \; hour}{(7 \; servers) \cdot (12 \; calls \; per \; hour/server)}$$

$$= \frac{60}{84}$$

$$= 71.42\%$$

4b) The expected number of customers on hold (the length of the waiting line) is given by the expected waiting line length L(7). This is reported by the queueing software package to be 0.8104.

4c) The expected number of customers in the switchboard system (either on hold or being answered) is reported by the queueing software package to be 5.8103.

Note: The expected number of calls being answered at any given time is equal to the number of operators times the probability each one is busy = $(7) \cdot (0.7142) = 4.99$. Then the expected number of calls in the switchboard system (5.8103) consists of the expected waiting line length $L(7)$ equal to 0.8104 plus the expected number of calls being answered equal to 4.99.

4d) The probability that a customer calling will find all lines busy and will be put on hold is 0.324. That is, with probability 0.676, a customer will have his or her call answered immediately and will not be put on hold.

4e) The expected waiting time a customer will spend on hold (in the waiting line) is given by the expected waiting line time $W(7)$. This is reported by the queueing software package to be 0.0135 hours (0.81 minutes).

4f) The expected time a call will spend in the switchboard system (either on hold or being answered) is 0.0968 hours (5.81 minutes).

Note: The expected time a call will spend in the switchboard system (5.81 minutes) consists of the expected waiting line time $W(7)$ equal to 0.81 minutes plus the expected service completion time for a call equal to 5 minutes.

Solutions to Exercise 11

Simulation

Problem #1 (Seed Numbers)

Consider a bank performance simulation which requires that the number of customer arrivals to the bank during an eight hour day be generated. Independent trials (with independent sets of eight hours of customer arrival data) are required in order to compute statistics and perform hypothesis testing on the bank performance data (e.g. average waiting time) provided by the simulation.

If the Random Number streams for each individual trial are chosen beginning in the same location in a Random Number table or the same starting point for a generated series of Random Numbers, the Random Number streams will not be independent and, in fact, will be identical. Then each trial will have the exact same eight hour customer arrival stream.

Seed numbers provided by the user point to different (and thus independent) locations in a Random Number table or a generated sequence of Random Numbers. Different seed numbers for each trial ensure that each trial (eight hours of customer arrival data) is independent and that the statistics computed are valid.

Problem #2 (Generating Random Values)

A four-digit Random Number (between 0000 and 9999) can be used the following manner to generate random arrivals for the Payne Hospital emergency room:

$$0000 \leq RN \leq 1999 \Rightarrow \# Arrivals = 0$$
$$2000 \leq RN \leq 5999 \Rightarrow \# Arrivals = 1$$
$$6000 \leq RN \leq 8999 \Rightarrow \# Arrivals = 2$$
$$9000 \leq RN \leq 9999 \Rightarrow \# Arrivals = 3$$

The 10 four-digit Random Numbers would generate the following number of arrivals at the emergency room in the first 10 hours:

3	1	2	2	2
1	1	1	1	0

Problem #3 (Antithetic Values)

The antithetic Random Number to the four-digit Random Number 2389 is determined by subtracting 2389 from the highest possible value (9999) for a four-digit Random Number. Thus 7610 is the antithetic Random Number to the Random Number 2389.

The Random Number 2389 and its antithetic Random Number 7610 are considered as the decimal values (equally likely to be between 0 and 1) 0.2389 and 0.7610, respectively. Table 1b is used to determine the values $z_{.2389}$ and $z_{.7610}$ for the Standard Normal $N(0,1)$ cumulative distribution $F(z)$ such that:

$$F(z_{.2389}) = 0.2389$$
$$F(z_{.7610}) = 0.7610$$

Table 1b indicates that the values $z_{.2389} \approx -0.71$ and $z_{.7610} \approx +0.71$. Then the outcomes O1 and O2 for the Random Variable $X \sim N(10,9)$ associated with the Random Number RN1 = 2389 and the antithetic Random Number RN2 = 7610, respectively, are determined as follows:

$$RN1 = 2389$$
$$RN2 = 7610$$

$$\mu + z_{.2389} \cdot \sigma$$
$$\mu + z_{.7610} \cdot \sigma$$

$$10 + (-0.71 \cdot 3)$$
$$10 + (+0.71 \cdot 3)$$

$$O1 = 7.87$$
$$O2 = 12.13$$

Problem #4 (Simulation of a Fair Coin)

a) The flip of a fair coin has a probability of 0.50 of an outcome "Heads" and a probability of 0.50 of an outcome "Tails".

The table of four-digit Random Numbers may be used as follows to determine outcomes of the flip of a fair coin:

$$RN < 5000 \Rightarrow \text{"Heads"}$$
$$RN \geq 5000 \Rightarrow \text{"Tails"}$$

The outcomes of the 10 flips of a fair coin based on the table of 10 four-digit Random Numbers are:

"Tails"	"Heads"	"Tails"	"Tails"	"Tails"
"Heads"	"Heads"	"Heads"	"Tails"	"Tails"

b) No, due to the randomness in a sequence of Random Numbers, there will typically be some *variance* from the expected number of outcomes of each type. For example, the above sequence of 10 Random Numbers provided 4 "Heads" and 6 "Tails" rather than the expected 5 "Heads" and 5 "Tails".

However, for any long sequence of Random Numbers), the probability (proportion in the long run) of the outcomes "Heads" and "Tails" will approach 0.50.

c) The 20 outcomes based on the table of 10 four-digit Random Numbers and their 10 antithetic Random Numbers are:

"Tails"	"Heads"	"Tails"	"Tails"	"Tails"
"Heads"	"Heads"	"Heads"	"Tails"	"Tails"

"Heads"	"Tails"	"Heads"	"Heads"	"Heads"
"Tails"	"Tails"	"Tails"	"Heads"	"Heads"

d) Yes, the outcomes of 20 flips using 10 Random Numbers and their 10 antithetic Random Numbers will always produce the expected number of "Heads" (10) and "Tails" (10). For example, the 20 flips based on the above sequence of four-digit Random Numbers (the first two rows) and their antithetic Random Numbers (the last two rows) produced 10 "Heads" and 10 "Tails".

e) In general, the *variance* of the mean (total number of "Heads" divided by 20 in this problem) of simulation outcomes using a *fixed* number of trials N (= 20 in this problem) can be *reduced* by using input data (e.g. number of arrivals, service completion times, etc.) determined by N/2 (=10 in this problem) Random Numbers and their N/2 (=10 in this problem) antithetic Random Numbers.

Appendix D

References

References

Dilworth, James, B. Production and Operations Management (Second Edition), Random House Business Division, 1983.

Emmons, Hamiltion, Flowers, A. Dale, Khot, Chandrashekar M., and Mathur, Kamlesh, STORM Personal Version 3.0: Quantitative Modeling for Decision Support, Prentice Hall/Allyn & Bacon, 1992.

Fishman, George S., Concepts and Methods in Discrete Event Digital Simulation, Wiley, 1973.

Johnson, Aaron C., Jr., Johnson, Marvin B., and Buse, Rueben C., Econometrics Basic and Applied, Macmillan Publishing Company, 1987.

Makridakis, Spyros, and Wheelwright, Steven C., Forecasting Methods and Applications, John Wiley & Sons, 1978.

Martin, E. Wainright, DeHayes, Daniel W., Hoffer, Jeffrey A., and Perkins, William C., Managing Information Technology: What Managers Need to Know, Macmillan Publishing Company, 1991.

Ramanathan, Ramu, Introductory Econometrics with Applications (Second Edition), Harcourt, Brace, and Jovanovich, Inc., 1992.

Schmidt, Stephen R. and Launsby, Robert G., Understanding Industrial Designed Experiments, 2nd Edition, CQC Ltd, 1989.

Warner, D.M., Holloway D. C., and Grazier, K.L., Decision Making and Control for Health Administration (Second Edition), Health Administration Press, 1984.

Appendix E

Tables

Table 1a
Area Under The Standard Normal Distribution Curve (From 0 to z)

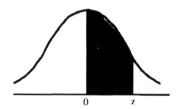

z	0.00	0.01	0.02	0.03	0.04	0.05	0.06	0.07	0.08	0.09
0.0	0.0000	0.0040	0.0080	0.0120	0.0160	0.0199	0.0239	0.0279	0.0319	0.0359
0.1	0.0398	0.0438	0.0478	0.017	0.0557	0.0596	0.0636	0.0675	0.0714	0.0753
0.2	0.0793	0.0832	0.0871	0.0910	0.0948	0.0987	0.1026	0.1064	0.1103	0.1141
0.3	0.1179	0.1217	0.1255	0.1293	0.1331	0.1368	0.1406	0.1443	0.1480	0.1517
0.4	0.1554	0.1591	0.1628	0.1664	0.1700	0.1736	0.1772	0.1808	0.1844	0.1879
0.5	0.1915	0.1950	0.1985	0.2019	0.2054	0.2088	0.2123	0.2157	0.2190	0.2224
0.6	0.2257	0.2291	0.2324	0.2357	0.2389	0.2422	0.2454	0.2486	0.2518	0.2549
0.7	0.2580	0.612	0.2642	0.2673	0.2704	0.2734	0.2764	0.2794	0.2823	0.2852
0.8	0.2881	0.2910	0.2939	0.2967	0.2995	0.3023	0.3051	0.3078	0.3106	0.3133
0.9	0.3159	0.3186	0.3212	0.3238	0.3264	0.3289	0.3315	0.3340	0.3365	0.3389
1.0	0.3413	0.3438	0.3461	0.3485	0.3508	0.3531	0.3554	0.3577	0.3599	0.3621
1.1	0.3643	0.3665	0.3686	0.3708	0.3729	0.3749	0.3770	0.3790	0.3810	0.3830
1.2	0.3849	0.3869	0.3888	0.3907	0.3925	0.3944	0.3962	0.3980	0.3997	0.4015
1.3	0.4032	0.4049	0.4066	0.4082	0.4099	0.4115	0.4131	0.4147	0.4162	0.4177
1.4	0.4192	0.4207	0.4222	0.4236	0.4251	0.4265	0.4279	0.4292	0.4306	0.4319
1.5	0.4332	0.4345	0.4357	0.4370	0.4382	0.4394	0.4406	0.4418	0.4429	0.4441
1.6	0.4452	0.4463	0.4474	0.4484	0.4495	0.4505	0.4515	0.4525	0.4535	0.4545
1.7	0.4554	0.4564	0.4573	0.4582	0.4591	0.4599	0.4608	0.4616	0.4625	0.4633
1.8	0.4641	0.4649	0.4656	0.4664	0.4671	0.4678	0.4686	0.4693	0.4699	0.4706
1.9	0.4713	0.4719	0.4726	0.4732	0.4738	0.4744	0.4750	0.4756	0.4761	0.4767
2.0	0.4772	0.4778	0.4783	0.4788	0.4793	0.4798	0.4803	0.4808	0.4812	0.4817
2.1	0.4821	0.4826	0.4830	0.4834	0.4838	0.4842	0.4846	0.4850	0.4854	0.4857
2.2	0.4861	0.4864	0.4868	0.4871	0.4875	0.4878	0.4881	0.4884	0.4887	0.4890
2.3	0.4893	0.4896	0.4898	0.4901	0.4904	0.4906	0.4909	0.4911	0.4913	0.4916
2.4	0.4918	0.4920	0.4922	0.4925	0.4927	0.4929	0.4931	0.4932	0.4934	0.4936
2.5	0.4938	0.4940	0.4941	0.4943	0.4945	0.4946	0.4948	0.4949	0.4951	0.4952
2.6	0.4953	0.4955	0.4956	0.4957	0.4959	0.4960	0.4961	0.4962	0.4963	0.4964
2.7	0.4965	0.4966	0.4967	0.4968	0.4969	0.4970	0.4971	0.4972	0.4973	0.4974
2.8	0.4974	0.4975	0.4976	0.4977	0.4977	0.4978	0.4979	0.4979	0.4980	0.4981
2.9	0.4981	0.4982	0.4982	0.4983	0.4984	0.4984	0.4985	0.4985	0.4986	0.4986
3.0	0.4986	0.4987	0.4987	0.4988	0.4988	0.4989	0.4989	0.4989	0.4990	0.4990
3.1	0.4990	0.4991	0.4991	0.4991	0.4992	0.4992	0.4992	0.4992	0.4993	0.4993
3.2	0.4993	0.4993	0.4994	0.4994	0.4994	0.4994	0.4994	0.4995	0.4995	0.4995
3.3	0.4995	0.4995	0.4995	0.4996	0.4996	0.4996	0.4996	0.4996	0.4996	0.4997
3.4	0.4997	0.4997	0.4997	0.4997	0.4997	0.4997	0.4997	0.4997	0.4998	0.4998
3.5	0.4998	0.4998	0.4998	0.4998	0.4998	0.4998	0.4998	0.4998	0.4998	0.4998
3.6	0.4998	0.4998	0.4999	0.4999	0.4999	0.4999	0.4999	0.4999	0.4999	0.4999
3.7	0.4999	0.4999	0.4999	0.4999	0.4999	0.4999	0.4999	0.4999	0.4999	0.4999
3.8	0.4999	0.4999	0.4999	0.4999	0.4999	0.4999	0.4999	0.5000	0.5000	0.5000
3.9	0.5000	0.5000	0.5000	0.5000	0.5000	0.5000	0.5000	0.5000	0.5000	0.5000

Table 1b
Area Under The Standard Normal Distribution Curve (-∞ to z)

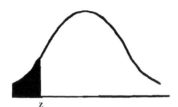

z	0	1	2	3	4	5	6	7	8	9
-3.0	.0013	.0010	.0007	.0005	.0003	.0002	.0002	.0001	.0001	.0000
-2.9	.0019	.0018	.0017	.0017	.0016	.0016	.0015	.0015	.0014	.0014
-2.8	.0026	.0025	.0024	.0023	.0023	.0022	.0021	.0021	.0020	.0019
-2.7	.0035	.0034	.0033	.0032	.0031	.0030	.0029	.0028	.0027	.0026
-2.6	.0047	.0045	.0044	.0043	.0041	.0040	.0039	.0038	.0037	.0036
-2.5	.0062	.0060	.0059	.0057	.0055	.0054	.0052	.0051	.0049	.0048
-2.4	.0082	.0080	.0078	.0075	.0073	.0071	.0069	.0068	.0066	.0064
-2.3	.0107	.0104	.0102	.0099	.0096	.0094	.0091	.0089	.0087	.0084
-2.2	.0139	.0136	.0132	.0129	.0126	.0122	.0119	.0116	.0113	.0110
-2.1	.0179	.0174	.0170	.0166	.0162	.0158	.0154	.0150	.0146	.0143
-2.0	.0228	.0222	.0217	.0212	.0207	.0202	.0197	.0192	.0188	.0183
-1.9	.0287	.0281	.0274	.0268	.0262	.0256	.0250	.0244	.0238	.0233
-1.8	.0359	.0352	.0344	.0336	.0329	.0322	.0314	.0307	.0300	.0294
-1.7	.0446	.0436	.0427	.0418	.0409	.0401	.0392	.0384	.0375	.0367
-1.6	.0548	.0537	.0526	.0516	.0505	.0495	.0485	.0475	.0465	.0455
-1.5	.0668	.0655	.0643	.0630	.0618	.0606	.0594	.0582	.0570	.0559
-1.4	.0808	.0793	.0778	.0764	.0749	.0735	.0722	.0708	.0694	.0681
-1.3	.0968	.0951	.0934	.0918	.0901	.0885	.0869	.0853	.0838	.0823
-1.2	.1151	.1131	.1112	.1093	.1075	.1056	.1038	.1020	.1003	.0985
-1.1	.1357	.1335	.1314	.1292	.1271	.1251	.1230	.1210	.1190	.1170
-1.0	.1587	.1562	.1539	.1515	.1492	.1469	.1446	.1423	.1401	.1379
-.9	.1841	.1814	.1788	.1762	.1736	.1711	.1685	.1660	.1635	.1611
-.8	.2119	.2090	.2061	.2033	.2005	.1977	.1949	.1922	.1894	.1867
-.7	.2420	.2389	.2358	.2327	.2297	.2266	.2236	.2206	.2177	.2148
-.6	.2743	.2709	.2676	.2643	.2611	.2578	.2546	.2514	.2483	.2451
-.5	.3085	.3050	.3015	.2981	.2946	.2912	.2877	.284	.2810	.2776
-.4	.3446	.3409	.3372	.3336	.3300	.3264	.3228	.3192	.3156	.3121
-.3	.3821	.3783	.3745	.3707	.3669	.3632	.3594	.3557	.3520	.3483
-.2	.4207	.4168	.4129	.4090	.4052	.4013	.3074	.3936	.3897	.3859
-.1	.4602	.4562	.4522	.4483	.4443	.4404	.4364	.4325	.4286	.4247
-.0	.5000	.4960	.4920	.4880	.4840	.4801	.4761	.4721	.4681	.4641

Table 1b (Continued)
Area Under The Standard Normal Distribution Curve (-∞ to z)

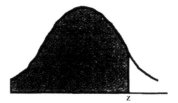

z	0	1	2	3	4	5	6	7	8	9
.0	.5000	.5040	.5080	.5120	.5160	.5199	.5239	.5279	.5379	.5359
.1	.5398	.5438	.5478	.5517	.5557	.5596	.5636	.5675	.5714	.5753
.2	.5793	.5832	.5871	.5910	.5948	.5987	.6026	.6064	.6103	.6141
.3	.6179	.6217	.6255	.6293	.631	.6368	.6406	.6443	.6480	.6517
.4	.6554	.6591	.6628	.6664	.6700	.6736	.6772	.6808	.6844	.6879
.5	.6915	.6950	.6985	.7019	.7054	.7088	.7123	.7157	.7190	.7224
.6	.7257	.7291	.7324	.7357	.7389	.7422	.7454	.7486	.7517	.7549
.7	.7580	.7611	.7642	.7673	.7703	.7734	.7764	.7794	.7823	.7852
.8	.7881	.7910	.7939	.7967	.7995	.8023	.8051	.8078	.8106	.8133
.9	.8159	.8186	.8212	.8238	.8264	.8289	.8315	.8340	.8365	.8389
1.0	.8413	.8438	.8461	.8485	.8501	.8531	.8554	.8577	.8599	.8621
1.1	.8643	.8665	.8686	.8708	.8729	.8749	.8770	.8790	.8810	.8830
1.2	.8849	.8869	.8888	.8907	.8925	.8944	.8962	.8980	.8997	.9015
1.3	.9032	.9049	.9066	.9082	.9099	.9115	.9131	.9147	.9162	.9177
1.4	.9192	.9207	.9222	.9236	.9251	.9265	.9278	.9292	.9306	.9319
1.5	.9332	.9345	.9357	.9370	.9382	.9394	.9406	.9418	.9430	.9441
1.6	.9452	.9463	.9474	.9484	.9495	.9505	.9515	.9525	.9535	.9545
1.7	.9554	.9564	.9573	.9582	.9591	.9599	.9608	.9616	.9625	.9633
1.8	.9641	.9648	.9656	.9664	.9671	.9678	.9686	.9693	.9700	.9706
1.9	.9713	.9719	.9726	.9732	.9738	.9744	.9750	.9756	.9762	.9767
2.0	.9772	.9778	.9783	.9788	.9793	.9798	.9803	.9808	.9812	.9817
2.1	.9821	.9826	.9830	.9834	.9838	.9842	.9846	.9850	.9854	.9857
2.2	.9861	.9864	.9868	.9871	.9874	.9878	.9881	.9884	.9887	.9890
2.3	.9893	.9896	.9898	.9901	.9904	.9906	.9909	.9911	.9913	.9916
2.4	.9918	.9920	.9922	.9925	.9927	.9929	.9931	.9932	.9934	.9936
2.5	.9938	.9940	.9941	.9943	.9945	.9946	.9948	.9949	.9951	.9952
2.6	.9953	.9955	.9956	.9957	.9959	.9960	.9961	.9962	.9963	.9964
2.7	.9965	.9966	.9967	.9968	.9969	.9970	.9971	.9972	.9973	.9974
2.8	.9974	.9975	.9976	.9977	.9977	.9978	.9979	.9979	.9980	.9981
2.9	.9981	.9982	.9982	.9983	.9984	.9984	.9985	.9985	.9986	.9986
3.0	.9987	.9990	.9993	.9995	.9997	.9998	.9998	.9999	.9999	1.0000

Table 2a

The F Distribution with v_1 and v_2 Degrees of Freedom

(95th Percentile)

$s\ F_{v_1, v_2, .95}$

	2	3	4	5	6	7	8	9	10	12	15	20	24	30	40	60	120	∞
1.4	199.5	215.7	224.6	230.2	234.0	236.8	238.9	240.5	241.9	243.9	245.9	248.0	249.1	250.1	251.1	252.2	253.3	254.3
51	19.00	19.16	19.25	19.30	19.33	19.35	19.37	19.38	19.40	19.41	19.43	19.45	19.45	19.46	19.47	19.48	19.49	19.50
13	9.55	9.28	9.12	9.01	8.94	8.89	8.85	8.81	8.79	8.74	8.70	8.66	8.64	8.62	8.59	8.57	8.55	8.53
71	6.94	6.59	6.39	6.26	6.16	6.09	6.04	6.00	5.96	5.91	5.86	5.80	5.77	5.75	5.72	5.69	5.66	5.63
61	5.79	5.41	5.19	5.05	4.95	4.88	4.82	4.77	4.74	4.68	4.62	4.56	4.53	4.50	4.46	4.43	4.40	4.36
99	5.14	4.76	4.53	4.39	4.28	4.21	4.15	4.10	4.06	4.00	3.94	3.87	3.84	3.81	3.77	3.74	3.70	3.67
59	4.74	4.35	4.12	3.97	3.87	3.79	3.73	3.68	3.64	3.57	3.51	3.44	3.41	3.38	3.34	3.30	3.27	3.23
32	4.46	4.07	3.84	3.69	3.58	3.50	3.44	3.39	3.35	3.28	3.22	3.15	3.12	3.08	3.04	3.01	2.97	2.93
12	4.26	3.86	3.63	3.48	3.37	3.29	3.23	3.18	3.14	3.07	3.01	2.94	2.90	2.86	2.83	2.79	2.75	2.71
96	4.10	3.71	3.48	3.33	3.22	3.14	3.07	3.02	2.98	2.91	2.85	2.77	2.74	2.70	2.66	2.62	2.58	2.54
84	3.98	3.59	3.36	3.20	3.09	3.01	2.95	2.90	2.85	2.79	2.72	2.65	2.61	2.57	2.53	2.49	2.45	2.40
75	3.89	3.49	3.26	3.11	3.00	2.91	2.85	2.80	2.75	2.69	2.62	2.54	2.51	2.47	2.43	2.38	2.34	2.30
67	3.81	3.41	3.18	3.03	2.92	2.83	2.77	2.71	2.67	2.60	2.53	2.46	2.42	2.38	2.34	2.30	2.25	2.21
60	3.74	3.34	3.11	2.96	2.85	2.76	2.70	2.65	2.60	2.53	2.46	2.39	2.35	2.31	2.27	2.22	2.18	2.13
54	3.68	3.29	3.06	2.90	2.79	2.71	2.64	2.59	2.54	2.48	2.40	2.33	2.29	2.25	2.20	2.16	2.11	2.07
49	3.63	3.24	3.01	2.85	2.74	2.66	2.59	2.54	2.49	2.42	2.35	2.28	2.24	2.19	2.15	2.11	2.06	2.01
45	3.59	3.20	2.96	2.81	2.70	2.61	2.55	2.49	2.45	2.38	2.31	2.23	2.19	2.15	2.10	2.06	2.01	1.96
41	3.55	3.16	2.93	2.77	2.66	2.58	2.51	2.46	2.41	2.34	2.27	2.19	2.15	2.11	2.06	2.02	1.97	1.92
38	3.52	3.13	2.90	2.74	2.63	2.54	2.48	2.42	2.38	2.31	2.23	2.16	2.11	2.07	2.03	1.98	1.93	1.88
35	3.49	3.10	2.87	2.71	2.60	2.51	2.45	2.39	2.35	2.28	2.20	2.12	2.08	2.04	1.99	1.95	1.90	1.84
32	3.47	3.07	2.84	2.68	2.57	2.49	2.42	2.37	2.32	2.25	2.18	2.10	2.05	2.01	1.96	1.92	1.87	1.81
30	3.44	3.05	2.82	2.66	2.55	2.46	2.40	2.34	2.30	2.23	2.15	2.07	2.03	1.98	1.94	1.89	1.84	1.78
28	3.42	3.03	2.80	2.64	2.53	2.44	2.37	2.32	2.27	2.20	2.13	2.05	2.01	1.96	1.91	1.86	1.81	1.76
26	3.40	3.01	2.78	2.62	2.51	2.42	2.36	2.30	2.25	2.18	2.11	2.03	1.98	1.94	1.89	1.84	1.79	1.73
24	3.39	2.99	2.76	2.60	2.49	2.40	2.34	2.28	2.24	2.16	2.09	2.01	1.96	1.92	1.87	1.82	1.77	1.71
23	3.37	2.98	2.74	2.59	2.47	2.39	2.32	2.27	2.22	2.15	2.07	1.99	1.95	1.90	1.85	1.80	1.75	1.69
21	3.35	2.96	2.73	2.57	2.46	2.37	2.31	2.25	2.20	2.13	2.06	1.97	1.93	1.88	1.84	1.79	1.73	1.67
20	3.34	2.95	2.71	2.56	2.45	2.36	2.29	2.24	2.19	2.12	2.04	1.96	1.91	1.87	1.82	1.77	1.71	1.65
18	3.33	2.93	2.70	2.55	2.43	2.35	2.28	2.22	2.18	2.10	2.03	1.94	1.90	1.85	1.81	1.75	1.70	1.64
17	3.32	2.92	2.69	2.53	2.42	2.33	2.27	2.21	2.16	2.09	2.01	1.93	1.89	1.84	1.79	1.74	1.68	1.62
08	3.23	2.84	2.61	2.45	2.34	2.25	2.18	2.12	2.08	2.00	1.92	1.84	1.79	1.74	1.69	1.64	1.58	1.51
00	3.15	2.76	2.53	2.37	2.25	2.17	2.10	2.04	1.99	1.92	1.84	1.75	1.70	1.65	1.59	1.53	1.47	1.39
92	3.07	2.68	2.45	2.29	2.17	2.09	2.02	1.96	1.91	1.83	1.75	1.66	1.61	1.55	1.50	1.43	1.35	1.25
84	3.00	2.60	2.37	2.21	2.10	2.01	1.94	1.88	1.83	1.75	1.67	1.57	1.52	1.46	1.39	1.32	1.22	1.00

Table 2b

The F Distribution with v_1 and v_2 Degrees of Freedom

(99th Percentile)

Entry is $F_{v_1, v_2, .99}$

v_2 \ v_1	1	2	3	4	5	6	7	8	9	10	12	15	20	24	30	40	60
1	4052	4999.5	5403	5625	5764	5859	5928	5982	6022	6056	6106	6157	6209	6235	6261	6287	6313
2	98.50	99.00	99.17	99.25	99.30	99.33	99.36	99.37	99.39	99.40	99.42	99.43	99.45	99.46	99.47	99.47	99.48
3	34.12	30.82	29.46	28.71	28.24	27.91	27.67	27.49	27.35	27.23	27.05	26.87	26.69	26.60	26.50	26.41	26.32
4	21.20	18.00	16.69	15.98	15.52	15.21	14.98	14.80	14.66	14.55	14.37	14.20	14.02	13.93	13.84	13.75	13.65
5	16.26	13.27	12.06	11.39	10.97	10.67	10.46	10.29	10.16	10.05	9.89	9.72	9.55	9.47	9.38	9.29	9.20
6	13.75	10.92	9.78	9.15	8.75	8.47	8.26	8.10	7.98	7.87	7.72	7.56	7.40	7.31	7.23	7.14	7.06
7	12.25	9.55	8.45	7.85	7.46	7.19	6.99	6.84	6.72	6.62	6.47	6.31	6.16	6.07	5.99	5.91	5.82
8	11.26	8.65	7.59	7.01	6.63	6.37	6.18	6.03	5.91	5.81	5.67	5.52	5.36	5.28	5.20	5.12	5.03
9	10.56	8.02	6.99	6.42	6.06	5.80	5.61	5.47	5.35	5.26	5.11	4.96	4.81	4.73	4.65	4.57	4.48
10	10.04	7.56	6.55	5.99	5.64	5.39	5.20	5.06	4.94	4.85	4.71	4.56	4.41	4.33	4.25	4.17	4.08
11	9.65	7.21	6.22	5.67	5.32	5.07	4.89	4.74	4.63	4.54	4.40	4.25	4.10	4.02	3.94	3.86	3.78
12	9.33	6.93	5.95	5.41	5.06	4.82	4.64	4.50	4.39	4.30	4.16	4.01	3.86	3.78	3.70	3.62	3.54
13	9.07	6.70	5.74	5.21	4.86	4.62	4.44	4.30	4.19	4.10	3.96	3.82	3.66	3.59	3.51	3.43	3.34
14	8.86	6.51	5.56	5.04	4.69	4.46	4.28	4.14	4.03	3.94	3.80	3.66	3.51	3.43	3.35	3.27	3.18
15	8.68	6.36	5.42	4.89	4.56	4.32	4.14	4.00	3.89	3.80	3.67	3.52	3.37	3.29	3.21	3.13	3.05
16	8.53	6.23	5.29	4.77	4.44	4.20	4.03	3.89	3.78	3.69	3.55	3.41	3.26	3.18	3.10	3.02	2.93
17	8.40	6.11	5.18	4.67	4.34	4.10	3.93	3.79	3.68	3.59	3.46	3.31	3.16	3.08	3.00	2.92	2.83
18	8.29	6.01	5.09	4.58	4.25	4.01	3.84	3.71	3.60	3.51	3.37	3.23	3.08	3.00	2.92	2.84	2.75
19	8.18	5.93	5.01	4.50	4.17	3.94	3.77	3.63	3.52	3.43	3.30	3.15	3.00	2.92	2.84	2.76	2.67
20	8.10	5.85	4.94	4.43	4.10	3.87	3.70	3.56	3.46	3.37	3.23	3.09	2.94	2.86	2.78	2.69	2.61
21	8.02	5.78	4.87	4.37	4.04	3.81	3.64	3.51	3.40	3.31	3.17	3.03	2.88	2.80	2.72	2.64	2.55
22	7.95	5.72	4.82	4.31	3.99	3.76	3.59	3.45	3.35	3.26	3.12	2.98	2.83	2.75	2.67	2.58	2.50
23	7.88	5.66	4.76	4.26	3.94	3.71	3.54	3.41	3.30	3.21	3.07	2.93	2.78	2.70	2.62	2.54	2.45
24	7.82	5.61	4.72	4.22	3.90	3.67	3.50	3.36	3.26	3.17	3.03	2.89	2.74	2.66	2.58	2.49	2.40
25	7.77	5.57	4.68	4.18	3.85	3.63	3.46	3.32	3.22	3.13	2.99	2.85	2.70	2.62	2.54	2.45	2.36
26	7.72	5.53	4.64	4.14	3.82	3.59	3.42	3.29	3.18	3.09	2.96	2.81	2.66	2.58	2.50	2.42	2.33
27	7.68	5.49	4.60	4.11	3.78	3.56	3.39	3.26	3.15	3.06	2.93	2.78	2.63	2.55	2.47	2.38	2.29
28	7.64	5.45	4.57	4.07	3.75	3.53	3.36	3.23	3.12	3.03	2.90	2.75	2.60	2.52	2.44	2.35	2.26
29	7.60	5.42	4.54	4.04	3.73	3.50	3.33	3.20	3.09	3.00	2.87	2.73	2.57	2.49	2.41	2.33	2.23
30	7.56	5.39	4.51	4.02	3.70	3.47	3.30	3.17	3.07	2.98	2.84	2.70	2.55	2.47	2.39	2.30	2.21
40	7.31	5.18	4.31	3.83	3.51	3.29	3.12	2.99	2.89	2.80	2.66	2.52	2.37	2.29	2.20	2.11	2.02
60	7.08	4.98	4.13	3.65	3.34	3.12	2.95	2.82	2.72	2.63	2.50	2.35	2.20	2.12	2.03	1.94	1.84
120	6.85	4.79	3.95	3.48	3.17	2.96	2.79	2.66	2.56	2.47	2.34	2.19	2.03	1.95	1.86	1.76	1.66
∞	6.63	4.61	3.78	3.32	3.02	2.80	2.64	2.51	2.41	2.32	2.18	2.04	1.88	1.79	1.70	1.59	1.47

Table 3

The Chi-Square Distribution with v Degrees of Freedom

Q is the area under the Chi-Square curve to the right of a vertical line at the table entry. For example, for 15 degrees of freedom, the area to the right of 7.26 is 0.95.

Q / v	0.995	0.990	0.975	0.950	0.900	0.750	0.500
1	$3.92 \cdot 10^{-10}$	$1.57 \cdot 10^{-10}$	$9.821 \cdot 10^{-9}$	$3.932 \cdot 10^{-8}$	0.0157908	0.1015308	0.454937
2	0.0100251	0.0201007	0.0506356	0.102587	0.210720	0.575364	1.38629
3	0.0717212	0.114832	0.215795	0.351846	0.584375	1.212534	2.36597
4	0.206990	0.297110	0.484419	0.710721	1.063623	1.92255	3.35670
5	0.411740	0.554300	0.831211	1.145476	1.61031	2.67460	4.35146
6	0.675727	0.872085	1.237347	1.63539	2.20413	3.45460	5.34812
7	0.989265	1.239043	1.68987	2.16735	2.83311	4.25485	6.34581
8	1.344419	1.646482	2.17973	2.73264	3.48954	5.07064	7.34412
9	1.734926	2.087912	2.70039	3.32511	4.16816	5.89883	8.34283
10	2.15585	2.55821	3.24697	3.94030	4.86518	6.73720	9.34182
11	2.60321	3.05347	3.81575	4.57481	5.57779	7.58412	10.3410
12	3.07382	3.57056	4.40379	5.22603	6.30380	8.43842	11.3403
13	3.56503	4.10691	5.00874	5.89186	7.04150	9.29906	12.3398
14	4.07468	4.66043	5.62872	6.57063	7.78953	10.1653	13.3393
15	4.60094	5.22935	6.26214	7.26094	8.54675	11.0365	14.3389
16	5.14224	5.81221	6.90766	7.96164	9.31223	11.9122	15.3385
17	5.69724	6.40776	7.56418	8.67176	10.0852	12.7912	16.3381
18	6.26481	7.01491	8.23075	9.39046	10.8649	13.6753	17.3379
19	6.84398	7.63273	8,90655	10.1170	11.6509	14.5620	18.3376
20	7.43386	8.26040	9.59083	10.8508	12.4426	15.4518	19.3374
21	8.03366	8.89720	10.28293	11.5913	13.2396	16.3444	20.3372
22	8.64272	9.54249	10.9823	12.3380	14.0415	17.2396	21.3370
23	9.26042	10.19567	11.6885	13.0905	14.8479	18.1373	22.3369
24	9.88623	10.8564	12.4011	13.8484	15.6587	19.0372	23.3367
25	10.5197	11.5240	13.1197	14.6114	16.4734	19.9393	24.3366
26	11.1603	12.1981	13.8439	15.3791	17.2919	20.8434	25.3364
27	11.8076	12.8786	14.5733	16.1513	18.1138	21.7494	26.3363
28	12.4613	13.5648	15.3079	16.9279	18.9392	22.6572	27.3363
29	13.1211	14.2565	16.0471	17.7083	19.7677	23.5666	28.3362
30	13.7867	14.9535	16.7908	18.4926	20.5992	24.4776	29.3360
40	20.7065	22.1643	24.4331	26.5093	29.0505	33.6603	29.3354
50	27.9907	29.7067	32.3574	34.7642	37.6886	42.9421	49.3349
60	35.5346	37.4848	40.4817	43.1879	46.4589	52.2938	59.3347
70	43.2752	45.4418	48.7576	51.7393	55.3290	61.6983	69.3344
80	51.1720	53.5400	57.1532	60.3915	64.2778	71.1445	79.3343
90	59.1963	61.7541	65.6466	69.1260	73.2912	80.6247	89.3342
100	67.3276	70.0648	74.2219	77.9295	82.3581	90.1332	99.3341

Table 3 (Continued)

The Chi-Square Distribution with v Degrees of Freedom

Q \ v	0.250	0.100	0.050	0.025	0.010	0.005	0.001
1	1.3230	2.70554	3.84146	5.02389	6.63490	7.87944	10.828
2	2.77259	4.60517	5.99147	7.37776	9.21034	10.5966	13.816
3	4.10835	6.25139	7.81473	9.34840	11.3449	12.8381	16.266
4	5.38527	7.77944	9.48773	11.1433	13.2767	14.8602	18.467
5	6.62568	9.23635	11.0705	12.8325	15.0863	16.7496	20.515
6	7.84080	10.6446	12.5916	14.4494	16.8119	18.5476	22.458
7	9.03715	12.0170	14.0671	16.0128	18.4753	20.2777	24.322
8	10.2188	13.3616	15.5073	17.5346	20.0902	21.9550	26.125
9	11.3887	14.6837	16.9190	19.0228	21.6660	23.5893	27.877
10	12.5489	15.9871	18.3070	20.4831	23.2093	25.1882	29.588
11	13.7007	17.2750	19.6751	21.9200	24.7250	26.7569	31.264
12	14.8454	18.5494	21.0261	23.3367	26.2170	28.2995	32.909
13	15.9839	19.8119	22.3621	24.7356	27.6883	29.8194	34.528
14	17.1170	21.0642	23.6848	26.1190	29.1413	31.3193	36.123
15	18.2451	22.3072	24.9958	27.4884	30.5779	32.8013	37.697
16	19.3688	23.5418	26.2962	28.8454	31.9999	34.2672	39.252
17	20.4887	24.7690	27.5871	30.1910	33.4087	35.7185	40.790
18	21.6049	25.9894	28.8693	31.5264	34.8053	37.1564	42.312
19	22.7178	27.2036	30.1435	32.8523	36.1908	38.5822	43.820
20	23.8277	28.4120	31.4104	34.1696	37.5662	39.9968	43.315
21	24.9348	29.6151	32.6705	35.4789	38.9321	41.4010	46.797
22	26.0393	30.8133	33.9244	36.7807	40.2894	42.7956	48.268
23	27.1413	32.0069	35.1725	38.0757	41.6384	44.1813	49.728
24	28.2412	33.1963	36.4151	39.3641	42.9798	45.5585	51.179
25	29.3389	34.3816	37.6525	40.6465	44.3141	46.9278	52.620
26	30.4345	35.5631	38.8852	41.9232	45.6417	48.2899	54.052
27	31.5284	36.7412	40.1133	43.1944	46.9630	49.6449	55.476
28	32.6205	37.9159	41.3372	44.4607	48.2782	50.9933	56.892
29	33.7109	39.0875	42.5569	45.7222	49.5879	52.3356	58.302
30	34.7998	40.2560	43.7729	46.9792	50.8922	53.6720	59.703
40	45.6160	51.8050	55.7585	59.3417	63.6907	66.7659	73.402
50	56.3336	63.1671	67.5048	71.4202	76.1539	79.4900	86.661
60	66.9814	74.3970	79.0819	83.2976	88.3794	91.9517	99.607
70	77.5766	85.5271	90.5312	95.0321	100.425	104.315	112.317
80	88.1303	96.5782	101.879	106.629	112.329	116.321	124.839
90	98.6499	107.565	113.145	118.136	124.116	128.299	137.208
100	109.141	118.498	124.342	129.561	135.807	140.169	140.449

Table 4

The Poisson Distribution

Entries are the probability that a Poisson process with expected value λ will take on the non-negative integer value N.

N/λ	0.1	0.2	0.3	0.4	0.5	0.6	0.7	0.8	0.9	1.0
0	0.905	0.819	0.741	0.670	0.607	0.549	0.497	0.449	0.407	0.368
1	0.090	0.164	0.222	0.268	0.303	0.329	0.348	0.359	0.366	0.368
2	0.005	0.016	0.033	0.054	0.076	0.099	0.122	0.144	0.165	0.184
3	0.000	0.001	0.003	0.007	0.013	0.020	0.028	0.038	0.049	0.061
4	0.000	0.000	0.000	0.001	0.002	0.003	0.005	0.008	0.011	0.015
5	0.000	0.000	0.000	0.000	0.000	0.000	0.001	0.001	0.002	0.003
6	0.000	0.000	0.000	0.000	0.000	0.000	0.000	0.000	0.000	0.001
7	0.000	0.000	0.000	0.000	0.000	0.000	0.000	0.000	0.000	0.000
8	0.000	0.000	0.000	0.000	0.000	0.000	0.000	0.000	0.000	0.000

N/λ	1.1	1.2	1.3	1.4	1.5	1.6	1.7	1.8	1.9	2.0
0	0.333	0.301	0.273	0.247	0.223	0.202	0.183	0.165	0.150	0.135
1	0.366	0.361	0.354	0.345	0.335	0.323	0.311	0.298	0.284	0.271
2	0.201	0.217	0.230	0.242	0.251	0.258	0.264	0.268	0.270	0.271
3	0.074	0.087	0.100	0.113	0.126	0.138	0.150	0.161	0.171	0.180
4	0.020	0.026	0.032	0.039	0.047	0.055	0.064	0.072	0.081	0.090
5	0.004	0.006	0.008	0.011	0.014	0.018	0.022	0.026	0.031	0.036
6	0.001	0.001	0.002	0.003	0.004	0.005	0.006	0.008	0.010	0.012
7	0.000	0.000	0.000	0.001	0.001	0.001	0.001	0.002	0.003	0.003
8	0.000	0.000	0.000	0.000	0.000	0.000	0.000	0.000	0.001	0.001

N/λ	2.2	2.4	2.6	2.8	3.0	3.2	3.4	3.6	3.8	4.0
0	0.111	0.091	0.074	0.061	0.050	0.041	0.033	0.027	0.022	0.018
1	0.244	0.218	0.193	0.170	0.149	0.130	0.113	0.098	0.085	0.073
2	0.268	0.261	0.251	0.238	0.224	0.209	0.193	0.177	0.162	0.147
3	0.197	0.209	0.218	0.222	0.224	0.223	0.219	0.212	0.205	0.195
4	0.108	0.125	0.141	0.156	0.168	0.178	0.186	0.191	0.194	0.195
5	0.048	0.060	0.074	0.087	0.101	0.114	0.126	0.138	0.148	0.156
6	0.017	0.024	0.032	0.041	0.050	0.061	0.072	0.083	0.094	0.104
7	0.005	0.008	0.012	0.016	0.022	0.028	0.035	0.042	0.051	0.060
8	0.002	0.002	0.004	0.006	0.008	0.011	0.015	0.019	0.024	0.030
9	0.000	0.001	0.001	0.002	0.003	0.004	0.006	0.008	0.010	0.013
10	0.000	0.000	0.000	0.000	0.001	0.001	0.002	0.003	0.004	0.005
11	0.000	0.000	0.000	0.000	0.000	0.000	0.001	0.001	0.001	0.002
12	0.000	0.000	0.000	0.000	0.000	0.000	0.000	0.000	0.000	0.001
13	0.000	0.000	0.000	0.000	0.000	0.000	0.000	0.000	0.000	0.000
14	0.000	0.000	0.000	0.000	0.000	0.000	0.000	0.000	0.000	0.000
15	0.000	0.000	0.000	0.000	0.000	0.000	0.000	0.000	0.000	0.000

Table 4 (Continued)

The Poisson Distribution

Entries are the probability that a Poisson process with expected value λ will take on the non-negative integer value N.

N/λ	4.2	4.4	4.6	4.8	5.0	5.2	5.4	5.6	5.8	6.0
0	0.015	0.012	0.010	0.008	0.007	0.006	0.005	0.004	0.003	0.002
1	0.063	0.054	0.046	0.040	0.034	0.029	0.024	0.021	0.018	0.015
2	0.132	0.119	0.106	0.095	0.084	0.075	0.066	0.058	0.051	0.045
3	0.185	0.174	0.163	0.152	0.140	0.129	0.119	0.108	0.098	0.089
4	0.194	0.192	0.188	0.182	0.175	0.168	0.160	0.152	0.143	0.134
5	0.163	0.169	0.173	0.175	0.175	0.175	0.173	0.170	0.166	0.161
6	0.114	0.124	0.132	0.140	0.146	0.151	0.156	0.158	0.160	0.161
7	0.069	0.078	0.087	0.096	0.104	0.113	0.120	0.127	0.133	0.138
8	0.036	0.043	0.050	0.058	0.065	0.073	0.081	0.089	0.096	0.103
9	0.017	0.021	0.026	0.031	0.036	0.042	0.049	0.055	0.062	0.069
10	0.007	0.009	0.012	0.015	0.018	0.022	0.026	0.031	0.036	0.041
11	0.003	0.004	0.005	0.006	0.008	0.010	0.013	0.016	0.019	0.023
12	0.001	0.001	0.002	0.003	0.003	0.005	0.006	0.007	0.009	0.011
13	0.000	0.000	0.001	0.001	0.001	0.002	0.002	0.003	0.004	0.005
14	0.000	0.000	0.000	0.000	0.000	0.001	0.001	0.001	0.002	0.002
15	0.000	0.000	0.000	0.000	0.000	0.000	0.000	0.000	0.001	0.001

Table 5

The Exponential Distribution

(Values of e^{-x})

x	e^{-x}	x	e^{-x}	x	e^{-x}	x	e^{-x}	x	e^{-x}
0.00	1.0000	2.00	0.1353	4.00	0.0183	6.00	0.0025	8.00	0.0003
0.05	0.9512	2.05	0.1287	4.05	0.0174	6.05	0.0024	8.05	0.0003
0.10	0.9048	2.10	0.1225	4.10	0.0166	6.10	0.0022	8.10	0.0003
0.15	0.8607	2.15	0.1165	4.15	0.0158	6.15	0.0021	8.15	0.0003
0.20	0.8187	2.20	0.1108	4.20	0.0150	6.20	0.0020	8.20	0.0003
0.25	0.7788	2.25	0.1054	4.25	0.0143	6.25	0.0019	8.25	0.0003
0.30	0.7408	2.30	0.1003	4.30	0.0136	6.30	0.0018	8.30	0.0002
0.35	0.7407	2.35	0.0954	4.35	0.0129	6.35	0.0017	8.35	0.0002
0.40	0.6703	2.40	0.0907	4.40	0.0123	6.40	0.0017	8.40	0.0002
0.45	0.6376	2.45	0.0863	4.45	0.0117	6.45	0.0016	8.45	0.0002
0.50	0.6065	2.50	0.0821	4.50	0.0111	6.50	0.0015	8.50	0.0002
0.55	0.5770	2.55	0.0781	4.55	0.0106	6.55	0.0014	8.55	0.0002
0.60	0.5488	2.60	0.0743	4.60	0.0101	6.60	0.0014	8.60	0.0002
0.65	0.5220	2.65	0.0707	4.65	0.0096	6.65	0.0013	8.65	0.0002
0.70	0.4966	2.70	0.0672	4.70	0.0091	6.70	0.0012	8.70	0.0002
0.75	0.4724	2.75	0.0639	4.75	0.0087	6.75	0.0012	8.75	0.0002
0.80	0.4493	2.80	0.0608	4.80	0.0082	6.80	0.0011	8.80	0.0002
0.85	0.4274	2.85	0.0578	4.85	0.0078	6.85	0.0011	8.85	0.0001
0.90	0.4066	2.90	0.0550	4.90	0.0074	6.90	0.0010	8.90	0.0001
0.95	0.3867	2.95	0.0523	4.95	0.0071	6.95	0.0010	8.95	0.0001
1.00	0.3679	3.00	0.0498	5.00	0.0067	7.00	0.0009	9.00	0.0001
1.05	0.3499	3.05	0.0474	5.05	0.0064	7.05	0.0009	9.05	0.0001
1.10	0.3329	3.10	0.0450	5.10	0.0061	7.10	0.0008	9.10	0.0001
1.15	0.3166	3.15	0.0429	5.15	0.0058	7.15	0.0008	9.15	0.0001
1.20	0.3012	3.20	0.0408	5.20	0.0055	7.20	0.0007	9.20	0.0001
1.25	0.2865	3.25	0.0388	5.25	0.0052	7.25	0.0007	9.25	0.0001
1.30	0.2725	3.30	0.0369	5.30	0.0050	7.30	0.0007	9.30	0.0001
1.35	0.2592	3.35	0.0351	5.35	0.0047	7.35	0.0006	9.35	0.0001
1.40	0.2466	3.40	0.0334	5.40	0.0045	7.40	0.0006	9.40	0.0001
1.45	0.2346	3.45	0.0317	5.45	0.0043	7.45	0.0006	9.45	0.0001
1.50	0.2231	3.50	0.0302	5.50	0.0041	7.50	0.0006	9.50	0.0001
1.55	0.2122	3.55	0.0287	5.55	0.0039	7.55	0.0005	9.55	0.0001
1.60	0.2019	3.60	0.0273	5.60	0.0037	7.60	0.0005	9.60	0.0001
1.65	0.1921	3.65	0.0260	5.65	0.0035	7.65	0.0005	9.65	0.0001
1.70	0.1827	3.70	0.0247	5.70	0.0033	7.70	0.0005	9.70	0.0001
1.75	0.1738	3.75	0.0235	5.75	0.0032	7.75	0.0004	9.75	0.0001
1.80	0.1653	3.80	0.0224	5.80	0.0030	7.80	0.0004	9.80	0.0001
1.85	0.1572	3.85	0.0213	5.85	0.0029	7.85	0.0004	9.85	0.0001
1.90	0.1496	3.90	0.0202	5.90	0.0027	7.90	0.0004	9.90	0.0001
1.95	0.1423	3.95	0.0193	5.95	0.0026	7.95	0.0004	9.95	0.0000

Table 6

Student t Distribution

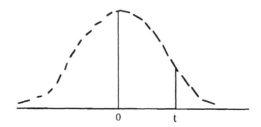

Degrees of Freedom	$t_{.75}$	$t_{.90}$	$t_{.95}$	$t_{.975}$	$t_{.99}$	$t_{.995}$	$t_{.9995}$
1	1.000	3.078	6.314	12.706	31.821	63.757	637.619
2	0.816	1.886	2.920	4.303	6.965	9.925	31.598
3	0.765	1.638	2.353	3.182	4.541	5.841	12.941
4	0.741	1.533	2.132	2.776	3.747	4.604	8.610
5	0.727	1.476	2.015	2.571	3.365	4.032	6.859
6	0.718	1.440	1.943	2.447	3.143	3.707	5.959
7	0.711	1.415	1.895	2.365	2.998	3.499	5.405
8	0.706	1.397	1.860	2.306	2.896	3.355	5.041
9	0.703	1.383	1.833	2.262	2.821	3.250	4.781
10	0.700	1.372	1.812	2.228	2.764	3.169	4.587
11	0.697	1.363	1.796	2.201	2.718	3.106	4.437
12	0.695	1.356	1.782	2.179	2.681	3.055	4.318
13	0.694	1.350	1.771	2.160	2.650	3.012	4.221
14	0.692	1.345	1.761	2.145	2.624	2.977	4.140
15	0.691	1.341	1.753	2.131	2.602	2.947	4.073
16	0.690	1.337	1.746	2.120	2.583	2.921	4.015
17	0.689	1.333	1.740	2.110	2.567	2.898	3.965
18	0.688	1.330	1.734	2.101	2.552	2.878	3.922
19	0.688	1.328	1.729	2.093	2.539	2.861	3.883
20	0.687	1.325	1.725	2.086	2.528	2.845	3.850
21	0.686	1.323	1.721	2.080	2.518	2.831	3.819
22	0.686	1.321	1.717	2.074	2.508	2.819	3.792
23	0.685	1.319	1.714	2.069	2.500	2.807	3.767
24	0.685	1.318	1.711	2.064	2.492	2.797	3.745
25	0.684	1.316	1.708	2.060	2.485	2.787	3.725
26	0.684	1.315	1.706	2.056	2.479	2.779	3.707
27	0.684	1.314	1.703	2.052	2.473	2.771	3.690
28	0.683	1.313	1.701	2.048	2.467	2.763	3.674
29	0.683	1.311	1.699	2.045	2.462	2.756	3.659
30	0.683	1.310	1.697	2.042	2.457	2.750	3.646
40	0.681	1.303	1.684	2.021	2.423	2.704	3.551
60	0.679	1.296	1.671	2.000	2.390	2.660	3.460
120	0.677	1.289	1.658	1.980	2.358	2.617	3.373
∞	0.674	1.282	1.645	1.960	2.326	2.576	3.291

Table 7

Compound Interest Multiplication Factors

Period	1%	2%	4%	6%	8%	10%
1	1.0100	1.0200	1.0400	1.0600	1.0800	1.1000
2	1.0201	1.0404	1.0816	1.1236	1.1664	1.2100
3	1.0303	1.0612	1.1249	1.1910	1.2597	1.3310
4	1.0406	1.0824	1.1699	1.2625	1.3605	1.4641
5	1.0510	1.1041	1.2167	1.3382	1.4693	1.6105
6	1.0615	1.1262	1.2653	1.4185	1.5869	1.7716
7	1.0721	1.1487	1.3159	1.5036	1.7138	1.9487
8	1.0829	1.1717	1.3686	1.5938	1.8509	2.1436
9	1.0937	1.1951	1.4233	1.6895	1.9990	2.3579
10	1.1046	1.2190	1.4802	1.7908	2.1589	2.5937
11	1.1157	1.2434	1.5395	1.8983	2.3316	2.8531
12	1.1268	1.2682	1.6010	2.0122	2.5182	3.3184
13	1.1381	1.2936	1.6651	2.1329	2.7196	3.4523
14	1.1495	1.3195	1.7317	2.2609	2.9372	3.7975
15	1.1610	1.3459	1.8009	2.3966	3.1722	4.1772
16	1.1726	1.3728	1.8730	2.5404	3.4259	4.5950
17	1.1843	1.4002	1.9479	2.6928	3.7000	5.0545
18	1.1961	1.4282	2.0258	2.8543	3.9960	5.5599
19	1.2081	1.4568	2.1068	3.0256	4.3157	6.1159
20	1.2202	1.4859	2.1911	3.2071	4.6610	6.7275
21	1.2324	1.5117	2.2788	3.3996	5.0338	7.4002
22	1.2447	1.5460	2.3699	3.6035	5.4365	8.1403
23	1.2572	1.5769	2.4647	3.8197	5.8715	8.9543
24	1.2697	1.6084	2.5633	4.0489	6.3412	9.8497
25	1.2824	1.6406	2.6658	4.2919	6.8485	10.8347
30	1.3478	1.8114	3.2434	5.7435	10.0627	17.4494
35	1.4166	1.9999	3.9461	7.6861	14.7853	28.1024
40	1.4889	2.2080	4.8010	10.2857	21.7245	45.2593
45	1.5648	2.4379	5.8412	13.7646	31.9204	72.8905
50	1.6446	2.6916	7.1067	18.4202	49.9016	117.3909

Table 8

Single Payment Present Value Factors

Period	1%	2%	4%	6%	8%	10%	12%	14%	15%	16%	18%
1	0.990	0.980	0.962	0.943	0.926	0.909	0.893	0.877	0.870	0.862	0.847
2	0.980	0.961	0.915	0.890	0.857	0.826	0.797	0.769	0.756	0.743	0.718
3	0.971	0.942	0.889	0.840	0.794	0.751	0.712	0.675	0.658	0.641	0.609
4	0.961	0.924	0.885	0.792	0.735	0.683	0.636	0.592	0.572	0.552	0.516
5	0.951	0.906	0.822	0.747	0.681	0.621	0.567	0.519	0.497	0.476	0.437
6	0.942	0.888	0.790	0.705	0.630	0.564	0.507	0.456	0.432	0.410	0.370
7	0.933	0.871	0.760	0.665	0.583	0.513	0.452	0.400	0.376	0.354	0.314
8	0.923	0.853	0.731	0.627	0.540	0.467	0.404	0.351	0.327	0.305	0.266
9	0.914	0.837	0.703	0.592	0.500	0.424	0.361	0.308	0.284	0.263	0.225
10	0.905	0.820	0.676	0.558	0.463	0.386	0.322	0.270	0.247	0.227	0.191
11	0.896	0.804	0.650	0.527	0.429	0.350	0.287	0.237	0.215	0.195	0.162
12	0.887	0.788	0.625	0.497	0.397	0.319	0.257	0.208	0.187	0.168	0.137
13	0.879	0.773	0.601	0.469	0.368	0.290	0.229	0.182	0.163	0.145	0.116
14	0.870	0.758	0.577	0.442	0.340	0.263	0.205	0.160	0.141	0.125	0.099
15	0.861	0.743	0.555	0.417	0.315	0.239	0.183	0.140	0.123	0.108	0.084
16	0.853	0.728	0.534	0.394	0.292	0.218	0.163	0.123	0.107	0.093	0.071
17	0.844	0.714	0.513	0.371	0.270	0.198	0.146	0.108	0.093	0.080	0.060
18	0.836	0.700	0.494	0.350	0.250	0.180	0.130	0.095	0.081	0.069	0.051
19	0.828	0.686	0.475	0.331	0.232	0.164	0.116	0.083	0.070	0.060	0.043
20	0.820	0.673	0.456	0.312	0.215	0.149	0.104	0.073	0.061	0.051	0.037
21	0.811	0.660	0.439	0.294	0.199	0.135	0.093	0.064	0.053	0.044	0.031
22	0.803	0.647	0.422	0.278	0.184	0.123	0.083	0.056	0.046	0.038	0.026
23	0.795	0.634	0.406	0.262	0.170	0.112	0.074	0.049	0.040	0.033	0.022
24	0.788	0.622	0.390	0.247	0.158	0.102	0.066	0.043	0.035	0.028	0.019
25	0.780	0.610	0.375	0.233	0.146	0.092	0.059	0.038	0.030	0.024	0.016
30	0.742	0.552	0.308	0.174	0.099	0.057	0.033	0.020	0.020	0.012	0.007
35	0.706	0.500	0.253	0.130	0.068	0.036	0.019	0.010	0.010	0.006	0.003
40	0.672	0.453	0.208	0.097	0.046	0.022	0.011	0.005	0.005	0.003	0.001
45	0.639	0.410	0.171	0.073	0.031	0.014	0.006	0.003	0.003	0.001	0.001
50	0.608	0.372	0.141	0.054	0.021	0.009	0.003	0.001	0.001	0.001	0.000

Table 9

Uniform Series Present Value Factors

Period	1%	2%	4%	6%	8%	10%	12%	14%	15%	16%	18%
1	0.990	0.980	0.962	0.943	0.926	0.909	0.893	0.877	0.870	0.862	0.847
2	1.970	1.942	1.886	1.833	1.783	1.736	1.690	1.647	1.626	1.605	1.566
3	2.941	2.884	2.775	2.673	2.577	2.487	2.402	2.322	2.283	2.246	2.174
4	3.902	3.808	3.630	3.465	3.312	3.170	3.037	2.914	2.855	2.798	2.690
5	4.853	4.713	4.452	4.212	3.993	3.791	3.605	3.433	3.352	3.274	3.127
6	5.795	5.601	5.242	4.917	4.623	4.355	4.111	3.889	3.784	3.685	3.498
7	6.728	6.472	6.002	5.582	5.206	4.868	4.564	4.288	4.160	4.039	3.812
8	7.652	7.325	6.733	6.210	5.747	5.335	4.968	4.639	4.487	4.344	4.078
9	8.566	8.162	7.435	6.802	6.247	5.759	5.328	4.946	4.772	4.607	4.303
10	9.471	8.983	8.111	7.360	6.710	6.145	5.650	5.216	5.019	4.833	4.494
11	10.368	9.787	8.760	7.887	7.139	6.495	5.938	5.453	5.234	5.029	4.656
12	11.255	10.575	9.385	8.384	7.536	6.814	6.194	5.660	5.421	5.197	4.793
13	12.134	11.348	9.986	8.853	7.904	7.103	6.424	5.842	5.583	5.342	4.910
14	13.004	12.106	10.563	9.295	8.244	7.367	6.628	6.002	5.724	5.468	5.008
15	13.865	12.849	11.118	9.712	8.559	7.606	6.811	6.142	5.847	5.575	5.092
16	14.718	13.578	11.652	10.106	8.851	7.824	6.974	6.265	5.954	5.668	5.162
17	15.562	14.292	12.166	10.477	9.122	8.022	7.120	6.373	6.047	5.749	5.222
18	16.398	14.992	12.659	10.828	9.372	8.201	7.250	6.467	6.128	5.818	5.273
19	17.226	15.678	13.134	11.158	9.604	8.365	7.366	6.550	6.198	5.877	5.316
20	18.046	16.351	13.590	11.470	9.818	8.514	7.469	6.623	6.259	5.929	5.353
21	18.857	17.011	14.029	11.764	10.017	8.649	7.562	6.687	6.312	5.973	5.384
22	19.660	17.658	14.451	12.042	10.201	8.772	7.645	6.743	6.359	6.011	5.410
23	20.456	18.292	14.857	12.303	10.371	8.883	7.718	6.792	6.399	6.044	5.432
24	21.243	18.914	15.247	12.550	10.529	8.985	7.784	6.835	6.434	6.073	5.451
25	22.023	19.523	15.622	12.783	10.675	9.077	7.843	6.873	6.464	6.097	5.467
30	25.808	22.396	17.292	13.765	11.258	9.427	8.055	7.003	6.566	6.177	5.517
35	29.409	24.999	18.665	14.498	11.655	9.644	8.176	7.070	6.617	6.215	5.539
40	32.835	27.355	19.793	15.046	11.925	9.779	8.244	7.105	6.642	6.233	5.548
45	36.095	29.490	20.720	15.456	12.108	9.863	8.283	7.123	6.654	6.242	5.552
50	39.196	31.424	21.482	15.762	12.233	9.915	8.304	7.133	6.661	6.246	5.554

Table 10

Capital Recovery Factors

Period	1%	2%	4%	6%	8%	10%	12%	14%	15%	16%	18%
1	1.010	1.020	1.040	1.060	1.080	1.100	1.120	1.140	1.150	1.160	1.180
2	0.508	0.515	0.530	0.545	0.561	0.576	0.592	0.607	0.615	0.623	0.639
3	0.340	0.347	0.360	0.374	0.388	0.402	0.415	0.431	0.438	0.445	0.460
4	0.256	0.263	0.275	0.289	0.302	0.315	0.329	0.343	0.350	0.357	0.372
5	0.206	0.212	0.225	0.237	0.250	0.264	0.277	0.291	0.298	0.305	0.320
6	0.173	0.179	0.191	0.203	0.216	0.230	0.243	0.257	0.264	0.271	0.286
7	0.149	0.155	0.167	0.179	0.192	0.205	0.219	0.233	0.240	0.248	0.262
8	0.131	0.137	0.149	0.161	0.174	0.187	0.201	0.216	0.223	0.230	0.245
9	0.117	0.123	0.134	0.147	0.160	0.174	0.188	0.202	0.210	0.217	0.232
10	0.106	0.111	0.123	0.136	0.149	0.163	0.177	0.192	0.199	0.207	0.223
11	0.096	0.102	0.114	0.127	0.140	0.154	0.168	0.183	0.191	0.199	0.215
12	0.089	0.095	0.107	0.119	0.133	0.147	0.161	0.177	0.184	0.192	0.209
13	0.082	0.088	0.100	0.113	0.127	0.141	0.155	0.171	0.179	0.187	0.204
14	0.077	0.083	0.095	0.108	0.121	0.136	0.151	0.167	0.175	0.183	0.200
15	0.072	0.078	0.090	0.103	0.117	0.131	0.147	0.163	0.171	0.179	0.196
16	0.068	0.074	0.086	0.099	0.113	0.128	0.143	0.160	0.168	0.176	0.194
17	0.064	0.070	0.082	0.095	0.110	0.125	0.140	0.157	0.165	0.174	0.191
18	0.061	0.067	0.079	0.092	0.107	0.122	0.138	0.155	0.163	0.172	0.190
19	0.058	0.064	0.076	0.090	0.104	0.120	0.136	0.153	0.161	0.170	0.188
20	0.055	0.061	0.074	0.087	0.102	0.117	0.134	0.151	0.160	0.169	0.187
21	0.053	0.059	0.071	0.085	0.100	0.116	0.132	0.150	0.158	0.167	0.186
22	0.051	0.057	0.069	0.083	0.098	0.114	0.131	0.148	0.157	0.166	0.185
23	0.049	0.055	0.067	0.081	0.096	0.113	0.130	0.147	0.156	0.165	0.184
24	0.047	0.053	0.066	0.080	0.095	0.111	0.128	0.146	0.155	0.165	0.183
25	0.045	0.051	0.064	0.078	0.094	0.110	0.127	0.145	0.155	0.164	0.183
30	0.039	0.045	0.058	0.073	0.089	0.106	0.124	0.143	0.152	0.162	0.181
35	0.034	0.040	0.054	0.069	0.086	0.104	0.122	0.141	0.151	0.161	0.181
40	0.030	0.037	0.051	0.066	0.084	0.102	0.121	0.141	0.151	0.160	0.180
45	0.028	0.034	0.048	0.065	0.083	0.101	0.121	0.140	0.150	0.160	0.180
50	0.026	0.032	0.047	0.063	0.082	0.101	0.120	0.140	0.150	0.160	0.180

Table 11

Durbin-Watson 95% Significance Table

N	k=1 d_L	k=1 d_U	k=2 d_L	k=2 d_U	k=3 d_L	k=3 d_U	k=4 d_L	k=4 d_U	k=5 d_L	k=5 d_U
15	1.08	1.36	0.96	1.54	0.82	1.75	0.69	1.97	0.56	2.21
16	1.10	1.37	0.98	1.54	0.86	1.73	0.74	1.93	0.62	2.15
17	1.13	1.38	1.02	1.54	0.90	1.71	0.78	1.90	0.67	2.10
18	1.16	1.39	1.05	1.53	0.93	1.69	0.82	1.87	0.71	2.06
19	1.18	1.40	1.08	1.53	0.97	1.68	0.86	1.85	0.75	2.02
20	1.20	1.41	1.10	1.54	1.00	1.68	0.90	1.83	0.79	1.99
21	1.22	1.42	1.13	1.54	1.03	1.67	0.93	1.81	0.83	1.96
22	1.24	1.43	1.15	1.54	1.05	1.66	0.96	1.80	0.86	1.94
23	1.26	1.44	1.17	1.54	1.08	1.66	0.99	1.79	0.90	1.92
24	1.27	1.45	1.19	1.55	1.10	1.66	1.01	1.78	0.93	1.90
25	1.29	1.45	1.21	1.55	1.12	1.66	1.04	1.77	0.95	1.89
26	1.30	1.46	1.22	1.55	1.14	1.65	1.06	1.76	0.98	1.88
27	1.32	1.47	1.24	1.56	1.16	1.65	1.08	1.76	1.01	1.86
28	1.33	1.48	1.26	1.56	1.18	1.65	1.10	1.75	1.03	1.85
29	1.34	1.48	1.27	1.56	1.20	1.65	1.12	1.74	1.05	1.84
30	1.35	1.40	1.28	1.57	1.21	1.65	1.14	1.74	1.07	1.83
31	1.36	1.50	1.30	1.57	1.23	1.65	1.16	1.74	1.09	1.83
32	1.37	1.50	1.31	1.57	1.24	1.65	1.18	1.73	1.11	1.82
33	1.38	1.51	1.32	1.58	1.26	1.65	1.19	1.73	1.13	1.81
34	1.39	1.51	1.33	1.58	1.27	1.65	1.21	1.73	1.15	1.81
35	1.40	1.52	1.34	1.58	1.28	1.65	1.22	1.73	1.16	1.80
36	1.41	1.52	1.35	1.59	1.29	1.65	1.24	1.72	1.18	1.80
37	1.42	1.53	1.36	1.59	1.31	1.66	1.25	1.72	1.19	1.80
38	1.43	1.54	1.37	1.59	1.32	1.66	1.26	1.72	1.21	1.79
39	1.43	1.54	1.38	1.60	1.33	1.66	1.27	1.72	1.22	1.79
40	1.44	1.54	1.39	1.60	1.34	1.66	1.29	1.72	1.23	1.79
45	1.48	1.57	1.43	1.62	1.38	1.67	1.34	1.72	1.29	1.78
50	1.50	1.59	1.46	1.63	1.42	1.67	1.38	1.72	1.34	1.77
55	1.53	1.60	1.49	1.64	1.45	1.68	1.41	1.72	1.38	1.77
60	1.60	1.62	1.51	1.65	1.48	1.69	1.44	1.73	1.41	1.77
65	1.57	1.63	1.54	1.66	1.50	1.70	1.47	1.73	1.44	1.77
70	1.58	1.64	1.55	1.67	1.52	1.70	1.49	1.74	1.46	1.77
75	1.60	1.65	1.57	1.68	1.54	1.71	1.51	1.74	1.49	1.77
80	1.61	1.66	1.59	1.69	1.56	1.72	1.53	1.74	1.51	1.77
85	1.62	1.67	1.60	1.70	1.57	1.72	1.55	1.75	1.52	1.77
90	1.63	1.68	1.61	1.70	1.59	1.73	1.57	1.75	1.54	1.78
95	1.64	1.69	1.62	1.71	1.60	1.73	1.58	1.75	1.56	1.78
100	1.65	1.69	1.63	1.72	1.61	1.74	1.59	1.76	1.57	1.78

N = Number of observations
k = Number of indpenedent variables

Source: J. Durbin and G. S. Watson, "Testing for Serial Correlation in Least Squares Regression," *Biometrika,* Vol. 38 (1951), pp. 159-177.

Index

Milton Keynes UK
Ingram Content Group UK Ltd.
UKHW020005071024
449327UK00031B/2656